MULTISCALE SIMULATION METHODS FOR NANOMATERIALS

BICENTENNIAL 1807 WILEY 2007 BICENTENNIAL

THE WILEY BICENTENNIAL–KNOWLEDGE FOR GENERATIONS

Each generation has its unique needs and aspirations. When Charles Wiley first opened his small printing shop in lower Manhattan in 1807, it was a generation of boundless potential searching for an identity. And we were there, helping to define a new American literary tradition. Over half a century later, in the midst of the Second Industrial Revolution, it was a generation focused on building the future. Once again, we were there, supplying the critical scientific, technical, and engineering knowledge that helped frame the world. Throughout the 20th Century, and into the new millennium, nations began to reach out beyond their own borders and a new international community was born. Wiley was there, expanding its operations around the world to enable a global exchange of ideas, opinions, and know-how.

For 200 years, Wiley has been an integral part of each generation's journey, enabling the flow of information and understanding necessary to meet their needs and fulfill their aspirations. Today, bold new technologies are changing the way we live and learn. Wiley will be there, providing you the must-have knowledge you need to imagine new worlds, new possibilities, and new opportunities.

Generations come and go, but you can always count on Wiley to provide you the knowledge you need, when and where you need it!

William J. Pesce
President and Chief Executive Officer

Peter Booth Wiley
Chairman of the Board

MULTISCALE SIMULATION METHODS FOR NANOMATERIALS

Edited by

RICHARD B. ROSS

SANAT MOHANTY

WILEY-INTERSCIENCE

A JOHN WILEY & SONS, INC., PUBLICATION

Published by John Wiley & Sons, Inc., Hoboken, New Jersey.
Published simultaneously in Canada.

Library of Congress Cataloging-in-Publication Data:

Multiscale simulation methods for nanomaterials / edited by Richard B. Ross and Sanat Mohanty.
p. cm.
Includes index.
ISBN 978-0-470-10528-3 (cloth)
1. Molecular electronics–Simulation methods. 2. Nanotechnology. I. Ross, Richard B., 1958– II. Mohanty, Sanat
TK874.8.M85 2007
620.1′1–dc22

2007023227

Printed in the United States of America

10 9 8 7 6 5 4 3 2 1

CONTENTS

CONTRIBUTORS

Kelly E. Anderson, Departments of Chemistry and of Chemical Engineering and Material Science, University of Minnesota, Minneapolis, Minnesota

Michael D. Barnes, Department of Chemistry, University of Massachusetts, Amherst, Massachusetts

Divesh Bhatt, Department of Chemistry and Supercomputing Institute, University of Minnesota, Minneapolis, Minnesota

Donald W. Brenner, Department of Materials Science and Engineering, North Carolina State University, Raleigh, North Carolina

Gustavo A. Carri, Department of Polymer Science and The Maurice Morton Institute of Polymer Science, The University of Akron, Akron, Ohio

Fiona Case, Case Scientific, Essex Junction, Vermont

J. Wesley Crill, Department of Materials Science and Engineering, North Carolina State University, Raleigh, North Caroilina

Taner E. Dirama, Department of Polymer Science and The Maurice Institute of Polymer Science, The University of Akron, Akron, Ohio; now at Universal Technology Corporation, Dayton, Ohio

Michael L. Drummond, Computer Science and Mathematics Division, Oak Ridge National Laboratory, Oak Ridge, Tennessee

Brett I. Dunlap, Theoretical Chemistry Section, U.S. Naval Research Laboratory, Washington, DC

Becky L. Eggimann, Departments of Chemistry and of Chemical Engineering and Material Science, University of Minnesota, Minneapolis, Minnesota

Bernd Ensing, Van't Hoff Institute for Molecular Sciences, University of Amsterdam, Amsterdam, The Netherlands

Jiali Gao, Department of Chemistry, Supercomputing Institute, and Digital Technology Center, University of Minnesota, Minneapolis, Minnesota

Robert J. Harrison, Computer Science and Mathematics Division, Oak Ridge National Laboratory, Oak Ridge, Tennessee

Ahren W. Jasper, Department of Chemistry and Supercomputing Institute, University of Minnesota, Minneapolis, Minnesota

Michael L. Klein, Center for Molecular Modeling, Department of Chemistry, University of Pennsylvania, Philadelphia, Pennsylvania

Vadim D. Knyazev, Department of Chemistry, The Catholic University of America, Washington, DC

Matthew J. McGrath, Departments of Chemistry and of Chemical Engineering and Material Science, University of Minnesota, Minneapolis, Minnesota

Vincent Meunier, Computer Science and Mathematics Division and Center for Nanophase Materials Sciences, Oak Ridge National Laboratory, Oak Ridge, Tennessee

Sanat Mohanty, Corporate Research Materials Laboratory, 3M Company, St. Paul, Minnesota

Preston B. Moore, Department of Chemistry and Biochemistry, University of the Sciences in Philadelphia, Philadelphia, Pennsylvania

Kwangho Nam, Department of Chemistry, Supercomputing Institute, and Digital Technology Center, University of Minnesota, Minneapolis, Minnesota

Steven O. Nielsen, Department of Chemistry, University of Texas at Dallas, Richardson, Texas

Marc R. Nyden, National Institute of Standards and Technology, Gaithersburg, Maryland

Clifford W. Padgett, Department of Materials Science and Engineering, North Carolina State University, Raleigh, North Carolina

Jake L. Rafferty, Departments of Chemistry and of Chemical Engineering and Material Science, University of Minnesota, Minneapolis, Minnesota

S. Rasmussen, Los Alamos National Laboratory, Los Alamos, New Mexico

Richard B. Ross, Corporate Research Materials Laboratory, 3M Company, St. Paul, Minnesota

J. David Schall, Department of Chemistry, United States Naval Academy, Annapolis, Maryland

William A. Shelton, Jr., Computer Science and Mathematics Division, Oak Ridge National Laboratory, Oak Ridge, Tennessee

Nathan E. Schultz, Department of Chemistry and Supercomputing Institute, University of Minnesota, Minneapolis, Minnesota

J. Ilja Siepmann, Departments of Chemistry and of Chemical Engineering and Material Science, University of Minnesota, Minneapolis, Minnesota

Stanislav I. Stoliarov, SRA International, Egg Harbor Township, New Jersey

Bobby G. Sumpter, Computer Science and Mathematics Division and Center for Nanophase Materials Sciences, Oak Ridge National Laboratory, Oak Ridge, Tennessee

A. Tamulis, Vilnius University Institute of Theoretical Physics and Astronomy, Vilnius, Lithuania

V. Tamulis, Vilnius University Institute of Theoretical Physics and Astronomy, Vilnius, Lithuania

Donald G. Truhlar, Department of Chemistry and Supercomputing Institute, University of Minnesota, Minneapolis, Minnesota

Darrin M. York, Department of Chemistry and Supercomputing Institute, University of Minnesota, Minneapolis, Minnesota

Ling Zhang, Departments of Chemistry and of Chemical Engineering and Material Science, University of Minnesota, Minneapolis, Minnesota

Nikolaj D. Zhuravlev, Departments of Chemistry and of Chemical Engineering and Material Science, University of Minnesota, Minneapolis, Minnesota

H. Ziock, Los Alamos National Laboratory, Los Alamos, New Mexico

Rajendra R. Zope, Department of Chemistry, George Washington University, Washington, DC, and Department of Physics, University of Texas at El Paso, El Paso, Texas

PREFACE

This book is a result of a symposium, "Large Scale Molecular Dynamics, Nanoscale, and Mesoscale Modeling and Simulation: Bridging the Gap," which was held at the Fall 2005 National American Chemical Society Meeting in Washington, DC. The symposium featured 40 presentations in eight symposium sessions over four days. A broad range of cutting-edge large-scale time and/or length simulation papers were presented on topic areas spanning biological, inorganic, organic, polymer, nanomaterial, and hybrid materials. The chapters in this book are expanded contributions of 14 of the groups presenting material at the symposium.

This collection of studies does not exhaustively cover all the various methods—such an effort is impossible, owing to the breadth of methods and the ever-evolving approaches. However, it does present a breadth of useful methods that have been used to solve mesoscale problems. In addition, it presents a wide variety of examples where mesoscale phenomena are important. Most important, it presents strategies of tackling mesoscale problems.

It has been shown that mesoscale problems are each unique—it is difficult to develop a recipe and plug in parameters to solve them. Each problem has its own critical length and time scales—often, multiple sets of scales. Each problem may allow approximations in different regimes; and each problem demands different degrees of rigorousness: Thus, each problem requires different sets of strategies. This set of examples provides a broad set of strategies and suggests ways in which a researcher may design her or his simulations to attack the problem of interest while ensuring that critical information and knowledge is passed across the various scales.

Some of these contributions relate to tools that can help passage of knowledge across these scales or other tools that can help better approximations in these processes. An example of one such strategy is presented in Chapter 5. The authors present an algorithm to fill in the detail when a coarse-grained representation of a molecular system is replaced by an atomistic representation. The methodology is demonstrated through implementation to liquid dodecane.

Another approach to passing knowledge across scales is described in Chapter 8. The authors describe the development of a coarse-grained methodology that incorporates current flow and Joule heating into a large-scale atomistic simulation. An example simulation is presented for an electrically "hot" and "cold" metal–metal contact that represents a contact in a microelectromechanical system device.

Another, easily visualized example is presented in Chapter 13. Although we understand decomposition reactions in simple gas-phase systems that can be described by elementary reactions, it is difficult to predict the thermal decomposition of nanostructures or of large molecules. Often, they break up into smaller aggregates and clusters, taking a number of different pathways. Certainly the characteristics of the individual atoms are important, but properties of the clusters also play an important role in their breakdown pathways. This study presents one way to incorporate these multiple scales in evolving our understanding of such processes.

Another example of critical mesoscale phenomena is presented in Chapter 6. What are the dynamics of proteins in a solvent (e.g., glycerol or trehalose)? The authors describe large-scale molecular dynamics simulations that probe and provide further insight into molecular mechanisms and the importance of dynamic coupling of the protein and solvent dynamics for lysozyme in pure glucose and trehalose. Understanding these dynamics on a pico- to nanosecond scale can help explicate longer-term phenomena such as denaturization.

Similarly, the effect of water and fatty acids on photosynthesis is explored in Chapter 2. The authors employ large-scale quantum mechanical and molecular mechanics methodologies to investigate self-assembly and the functioning of the photosynthetic process. They find that a critical role is played by the quantum effects associated with hydrogen bonds and van der Waals interactions resulting from increasing the number of water and fatty acid molecules. In both this case and Chapter 6, phenomena at atomistic scales and at scales including clusters of atoms (cells, aggregates) influence macroscopic properties of these materials and must be understood.

Examples of applications in nanomaterials can be seen in Chapter 3, whose authors discuss large-scale quantum electronic structure calculations coupled with a Green's function formulation for determining conductance as applied to carbon nanotubes doped with organic molecules. The studies are part of an effort to develop a framework for the design of nanotube-based electronic devices.

In an additional application to nanomaterials, multiscale and experimental studies on poly(*p*-phenylene vinylene) derivative nanostructures are discussed in Chapter 4, including the effects of high-level order and how these could also affect the creation of ordered domains in other polymer environments.

An industrial contribution applied to nanomaterials is described in Chapter 7, where multiscale methods to understand and simulate the effects of various solvents on silica nanoparticles with tethers of different chemistries are described. For these types of materials, important interactions to account for range from the molecular level between tethers and solvent molecules up to particle–particle interactions of entire tethered silica nanoparticles in a given solvent medium. Methodology developed for successive multiscaling from the molecular to the particle–particle interaction level is described and applied to several solvent and tether chemistry systems.

Continuing on a theme of method development and applications in Chapter 9, a fast variational fully analytical density functional method that has been developed for large molecules is covered in Chapter 9. The methodology has been applied to optimize the geometries of fullerines, fullerine-like cages, and nanotubes of more than 2000 atoms.

In an application to inorganic nanomaterials, the development of analytic potential energy functions and their use in for simulating aluminum nanoparticles are described in Chapter 10. Applications included simulations of liquid aluminum nanodroplets to study the size dependence of their densities, thermal expansivities, and particle shapes. Also investigated and reported are size-dependent effects on aluminum particle shape.

Multiscale modeling methodology development and application to biological-related reactions are described in Chapter 12. The authors describe a new quantum model based on a modified semiempirical AM1/d Hamiltonian that has been recently developed to model phosphoryl transfer reactions in solution, catalyzed by enzymes and ribozymes. The model has been integrated with a recently developed linear-scaling Ewald method to calculate long-range electrostatic interactions efficiently in combined quantum mechanical/molecular mechanical simulations. Applied studies to predict hydrolysis rates of dimethyl phosphate and ethylene phosphate in solution are discussed.

Several large-scale simulation applications are described in Chapter 11. Large-scale simulations employing transferable force fields are discussed for applications including retention in reversed-phase liquid chromatography, solubility of helium in n-hexadecane, entrainer effects on solubility in supercritical carbon dioxide, and segregation due to size effects for anions at aqueous vapor–liquid interfaces. In addition, first-principles Monte Carlo simulations are reported for water vapor–liquid coexistence curves.

In Chapter 14 current mathematical and molecular thermodynamical modeling techniques to predict thermodynamically stable surfactant mesoscale structures in solution are reviewed. The author then discusses the application of simulation-based approaches, dissipative particle dynamics in particular, which provide surfactant structure information and in addition provide predictions of dynamic behavior, nonequilibrium structures, and modeling of more concentrated systems, including mixtures of surfactants. Dissipative particle dynamics simulations for several surfactant systems, including ethoxylate, sodium dodecyl sulfate, and mixtures of the two, are then discussed.

Collectively, the fourteen chapters in this volume provide a snapshot of the broad range of mesoscale methodologies currently being developed and applied to a wide range of applications. The book should be very useful to current modeling experts seeking a more detailed understanding of current major mesoscale modeling approaches as well as to scientists and engineers new to the field who are seeking a rapid understanding of the current state-of-the-art approaches, their applications, and their scope: which types of material properties and structures are currently accessible via modeling and simulation and to what degree of accuracy.

RICHARD B. ROSS
SANAT MOHANTY

1

OVERVIEW OF MULTISCALE SIMULATION METHODS FOR MATERIALS

Sanat Mohanty and Richard B. Ross
Corporate Research Materials Laboratory, 3M Company, St. Paul, Minnesota

Modeling has increasingly become a tool for better proactive design, besides being a method to explain phenomena. In these functions it must be validated with experimental observations before its results can be applied with confidence. Modeling at the continuum scale is well understood, and the accuracy, precision, and robustness of results using continuum methods have been analyzed. Numerous methods, including finite-element analysis, finite-difference methods, and boundary element methods, are now used routinely in a variety of areas, including flow visualization, design of materials and structures, and in industry for automotive design, construction and structure applications, and plant design.

Modeling at the atomistic scale has also begun to be accepted as part of routine research protocol. Results from such methods are now accepted with a reasonable understanding of their accuracy and robustness. Quantum mechanical calculations are now used to understand the properties of semiconductor materials. Atomistic calculations have been used to better understand phase behavior of components and mixtures for the chemical industry. Numerous process design tools used routinely in the chemical industry include information based on atomistic calculations. Increasingly, however, design of materials comes up against phenomena that are in the mesoscale: too large for atomistic simulations and too small for continuum analysis. This is the scale where one cannot use continuum-scale assumptions, yet the system is often larger than can be modeled tractably with atomistic methodologies. How do we study systems that are made of clusters of

Multiscale Simulation Methods for Nanomaterials, Edited by Richard B. Ross and Sanat Mohanty

thousands of molecules that interact in specific ways? How do we take results from atomistic simulations and correlate them with parameters we need for continuum studies?

1. MESOSCALE MODELING

Examples of mesoscale phenomena include engineering of molecular clusters or of functionalized nanoparticles, crystals, and liquid-crystalline structures [such as clathrates[1] and micelles] that are used as templates for porous membranes,[2] materials with specific enzymatic,[3] or catalytic properties,[4] control of flow behavior,[5] nanocontainers with controlled release,[6] or materials with specific thermal, electromagnetic,[7] or mechanical properties.[8] Since a number of these materials are designed by self-assembly of atoms or molecules to form nano-sized clusters, modeling becomes a critical tool in engineering such materials. Modeling is also used to understand conditions to optimize processes to functionalize or engineer aspects of these structures, such as size of cluster, surface properties, or charges. Figure 1 shows some examples of nanoscale ordering and self-assemblies. Modeling mesoscale phenomena in materials design has thus become increasingly significant as mesoscale engineering of materials becomes critical in developing superior materials.

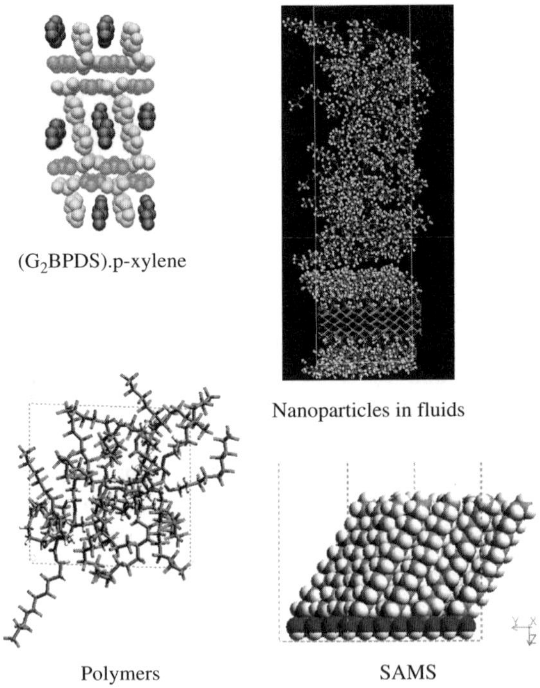

Figure 1 Mesoscale phenomena. (*See insert for color representation figure.*)

From a modeling perspective, there are a number of aspects to be addressed. Modeling helps to translate macroscopic observations based on nano- or mesoscale phenomena. Judicious use of simulations helps demonstrate why certain behavior is seen which is critical knowledge for engineering such behavior. For example, how can one predict the behavior of silicon nanoparticles whose surfaces have been modified with specific functional groups when they are in a polymer matrix;[9] or alternatively, how can one engineer the surfaces of nanoparticles so that polymer–particle composites may manifest a certain set of desired properties? Similarly, what types of structures do micelles made up of specific amphiphiles form, and how do they behave at different concentrations; or alternatively, what additives and amphiphiles could one use to engineer and design certain rheological properties in fluids, specific surface properties, or specific micellar structures?[10] How do molecules cluster in various environments,[11] and how does that affect their thermodynamic properties and mobility? Similar questions have been addressed for chromonic systems.[12] Another area with significant modeling effort is that of biomolecules.[13] Simulations have been used to understand the binding mechanisms of biomolecules on surfaces[14] and with each other, enzymatic pathways, mobility of molecules, and design of encapsulants—all with the aim of engineering drugs for specific action as well as methods of targeted delivery. Nano- and mesoscale models can be used not only to look at the equilibrium behavior of materials but also to study the mobility of particles and the rupture behavior of adhesives (essentially nonequilibrium phenomena).

Modeling mesosystems is difficult since the phenomena of interest are neither atomistic (so that solutions can be grasped by understanding the behavior of a few to hundreds of atoms or molecules) nor macroscopic (so that continuum properties of the material can be assumed without losing events occurring in the smaller-scale regime). Modeling phenomena that span different orders of length and time requires the use of multiscale models[15] (schematic in Figure 2). Yet mesoscale phenomena often define macromolecular properties (Figure 3). It is necessary to understand the mechanisms of the phenomena at these length scales

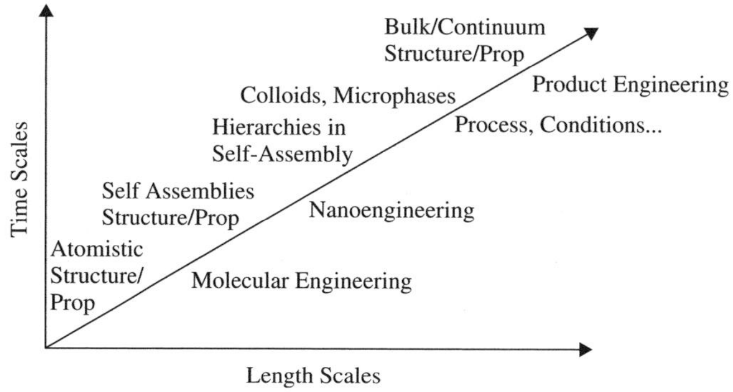

Figure 2 Modeling methods for various length and time scales.

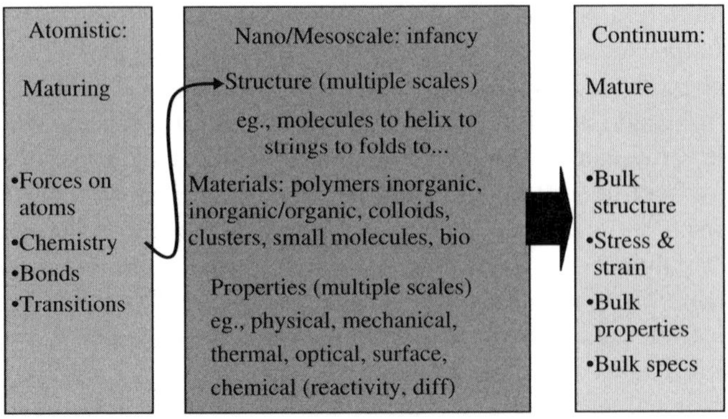

Figure 3 Modeling methods: development status and properties encompassed.

and to characterize them in order to understand or predict the macro properties that emerge.

For example, the specific folding of proteins can affect their macromolecular reaction rates. The manner in which a polymer configures itself in a stress field may define the mechanical properties of the material. Often, these systems are even more complex-showing hierarchies of phenomena. For example, the configuration of small molecules defines how they self-assemble and hence the structure of a cluster. The structure of the cluster will define its interactions with solvent that may surround it or with other clusters. These clusters may aggregate to form networks. The interacting networks in a solvent or without a solvent will have properties related to transport, absorbency, or chemical or physical reactivity, depending on phenomena at all these length scales (Figure 4).

There are additional difficulties with modeling phenomena in the mesoscale. First, it is often difficult to measure them experimentally. Often, the self-assemblies or clusters that form in solutions are transient. In addition, the exact

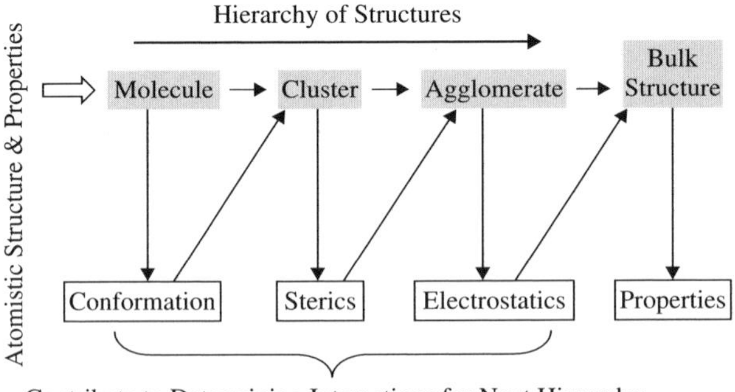

Figure 4 Examples of effects of different time and length scale influencing specific properties and next hierarchical structures.

configurations of the molecules in the clusters are not always visualizable directly, although they can sometimes be estimated from nuclear magnetic resonance, infrared, or other analysis. The effects on continuum properties of these mesoscale phenomena are also nonlinear or nondeterministic, and hence cannot be estimated easily from continuum properties. Thus, validation of models of these mesoscale events is difficult at best.

2. SIMULATION METHODS

Two types of multiscale methods—hierarchical and handshake—have provided feasible solutions to real problems. Approximating a number of polymer repeat units or solvent molecules as single beads with tunable interactions is an example of a hierarchical model.[16] Such models use information lumped from a smaller size scale or a shorter time scale to predict phenomena at the scale of simulation and/or to predict parameters or observations at larger scales. The parameters of the bead, in this case, are obtained from atomistic or quantum calculations on the molecules. In turn, the beads and chains are used to predict viscosity and material properties, which can then be used in flow models or structure/stress models to predict the macroscopic behavior of the material. Such methods are used to study equilibrium structure of polymers[17] and composites as well as nonequilibrium processes such as relaxation and/or fracture.[18] More complex versions of this algorithm have also been used in the study of nanoparticle interactions with polymers as well as a variety of solvents.[19]

Lattice models consist of spanning the system of interest by meshes or nodes.[20] Molecules that are moving or reacting are lumped into coarser particles, and the nodes or meshes keep track of the extent of mobility or reaction. Examples include curing of polymers,[17] diffusion in porous structures, and stress calculations in materials.

Handshake methods describe the problem by identifying one area where continuum assumptions hold while focusing on atomistic or mesoscale models to solve another aspect such that these two regions influence each other. Dynamic fracture is a very good example.[21] Energy from large-scale elastic fields is concentrated on the angstrom scale of the electrons that participate in atomic bonding. A simulation of this phenomenon requires an accurate description of atoms bonding at the crack tip and at the same time requires a proper description for very large volumes of strained material, the resolution varying with distance from the crack tip. Far away, it is adequate to use the equations of motion for a macroscopic-averaged continuum field. This spatial decomposition makes it possible to combine various simulation methods describing the different physical regions into a single, powerful simulation tool. Another example is the use of a handshake model to predict micelle free energies: An analytical solution suffices in the continuum assumption of the hydrocarbon core, whereas molecular simulations are required to understand the head groups and their interactions with the solvent.[22] Alternatively, different molecular models may be used in different regions. The combination of quantum mechanics (close to the docking site) and molecular mechanics (in the rest of the protein)(QM/MM) methods allows one to predict the behavior of molecules docking on proteins.[13] Particular attention must also be paid to an accurate joining of the two regimes at the boundary region.

Today, a significant fraction of related literature is of model systems which explain trends and behavior of ball–spring models but are of little use in predicting the behavior of systems with specific chemistry. Although there is a growing effort to solve real systems of industrial importance, this certainly is an area of challenge requiring development in the modeling toolbox today.

REFERENCES

1. Plaut, D. J.; Holman, T. K.; Pivovar, A. M.; Ward, M. D. Building molecular frameworks with tailored pore structures, *J. Phys. Org. Chem.*, 2000, 13: 858–869.
2. Yan, H.; Blanford, C. F.; Lytle, J. C.; Carter, C. B.; Smyrl, W. H.; Stein, A. Influence of processing conditions on structures of 3D ordered macroporous metals prepared by colloidal crystal templating, *Chem. Mater.*, 2001, 13: 4314–4321.
3. Gillies, E. R.; Fréchet, J. M. J. A new approach to acid sensitive copolymer micelles for drug delivery, *Chem. Commun.*, 2003, 14: 1640–1641.
4. Lim, M. H.; Blanford, C. F.; Stein, A. Synthesis of ordered microporous silicates with organosulfur surface groups and their applications as solid acid catalysts, *Chem. Mater.*, 1998, 10: 467–470.
5. Lu, B.; Li, X.; Scriven, L. E.; Davis, H. T.; Talmon, Y.; Zakin, J. L. Effect of chemical structure on viscoelasticity and extensional viscosity of drag reducing cationic surfactant solutions, *Langmuir*, 1998, 14: 8–16.
6. Kartik, A.; Radhakrishna, S.; Matthew, P. S.; Justin, L. P. Surfactant induced effects on turbulent swirling flows, *Rheol. Acta*, 2002, 41: 25–34.
7. Kietzke, T.; Neher, D.; Landfester, K.; Montenegro, R.; Guntner, R.; Scherf, U. Novel approaches to polymer blends based on polymer nanoparticles, *Nat. Mater.*, 2003, 2: 408–412.
8. Hasegawa, N.; Okamoto, H.; Kawasumi, M.; Usuki, A. J. Preparation and mechanical properties of polystyrene–clay hybrids, *Appl. Polym. Sci.*, 1999, 74: 3359.
9. Balazs, A.; Ginzburg, V. V.; Qiu, F.; Peng, G.; Jasnow, D. Multiscale model for binary mixtures containing nanoscopic particles, *J. Phys. Chem. B*, 2000, 104: 3411–3422.
10. Bell, R. C.; Wang, H.; Iedema, M. J.; Cowin, J. P. Sculpting nano-scale liquid interfaces, in *Mesoscale Phenomena in Fluid Systems*, Case, F., Ed., ACS Symposium Series, American Chemical Society, Washington, DC, 2003.
11. Puvvada, S.; Blankschtein, D. Thermodynamic description of micellization, phase behavior and phase separation of aqueous solutions of surfactant mixtures, *J. Phys. Chem*, 1992, 96: 5567–5579.
12. Mohanty, S. Molecular simulation of self-assembly of hydrophilic functionalized aromatics in aqueous solutions, *Mol. Simul.*, 2006, 32: 633–642. Mohanty, S.; Chou, S.-H.; Brostrom, M.; Aguilera, J. Predictive modeling of self-assembly of chromonics materials, *Mol. Simul.*, 2006, 32: 1179–1185.

13. http://kandinsky.chem.wisc.edu/~qiang/, accessed July 2003.
14. Wang, P.; Vaidehi, N.; Tirrell, D. A.; Goddard, W. A., III Virtual screening for binding of phenylalanine analogues to phenylalanyl-tRNA synthetase, *J. Am. Chem. Soc.*, 2002, 124: 14442.
15. Thomas, C. U.; Caldwell, G.; Ross, R. B.; Mohanty, S.; Freedman, M. Importance of mesoscopic structures in the development of advanced materials, in *Mesoscale Phenomena in Fluid Systems*, Case, F., Ed., ACS Symposium Series, American Chemical Society, Washington, DC, 2003.
16. Clementi, E. Global scientific and engineering simulations on scalar, vector and parallel LCAP-type supercomputers, *Philos. Trans. R. Soc. London Ser. A*, 1998, 326: 445–470.
17. Theodorou, D.; Dodd, L. R. Atomistic Monte Carlo simulation and continuum mean field theory of the structure and equation of state properties of alkane and polymer melts., *Adv. Polym. Sci.*, 1994, 116: 249.
18. Rottler, J.; Robbins, M. O. Yield conditions for deformation of amorphous polymer glasses, *Phys. Rev. E*, 2001, 64: 051801.
19. Wen, M.; McCormick, A. V.; Scriven, L. E. Kinetic gelation modeling: structural inhomogeneity during cross-linking polymerization, *Macromolecules*, 2003, 36: 4140.
20. Van Tassel, P. R.; Davis, H. T.; McCormick, A. V. New lattice model for adsorption of small molecules in zeolite micropores, *AIChE J.*, 1994, 40: 925–934.
21. Bernstein, N. Multiscale modeling of materials, in *Multiscale Simulations of Brittle Fracture and the Quantum-Mechanical Nature of Bonding in Silicon*, MRS 2000 Fall Meeting Proceedings, Pittsburgh, PA, Kubin, L. P.; Bassani, J. L.; Cho, K.; Gao, H.; Selinger, R. L. B., Eds., Materials Research Society, Warrendale, PA, 2001, p. 653.
22. Mohanty, S.; Davis, H. T.; McCormick, A. V. Complementary use of simulations and free energy models for CTAB/NaSal systems, *Langmuir*, 2001, 17: 7160–7171.

2

INFLUENCE OF WATER AND FATTY ACID MOLECULES ON QUANTUM PHOTOINDUCED ELECTRON TUNNELING IN SELF-ASSEMBLED PHOTOSYNTHETIC CENTERS OF MINIMAL PROTOCELLS

A. Tamulis and V. Tamulis
Vilnius University Institute of Theoretical Physics and Astronomy, Vilnius, Lithuania
H. Ziock and S. Rasmussen
Los Alamos National Laboratory, Los Alamos, New Mexico

The nano-protoorganisms that are proposed in ref. 1 are only a few nanometers in size. In their simplest form, these protoorganisms consist of a micelle that acts as the container, a light-driven metabolism, and a genetic system, whose functions are all very tightly coupled. The container consists of amphiphilic fatty acid (FA) molecules that self-assemble into a micelle. The hydrophobic interior of the micelle provides an alternative thermodynamic environment from the aqueous exterior and acts as a sticking point for the photosensitizer, fatty acid precursors (food), and the genetic material. Peptide nucleic acid (PNA) is chosen as the genetic material, as it is far less polar than RNA or DNA and therefore should stick to the micelle. It is also capable of undergoing the same Watson–Crick pairing and replication as RNA and DNA (Figure 1). The metabolism involves the photoexcitation of an electron in the photosensitizer which is stabilized by the donation of an electron from one of the PNA bases. The excited electron is, in turn, used to cleave a fatty acid precursor (pFA) to yield another fatty acid

Multiscale Simulation Methods for Nanomaterials, Edited by Richard B. Ross and Sanat Mohanty

molecule, thereby allowing the container to grow until it reaches an unstable size and divides. The protocell could be fed PNA monomers or use an essentially identical metabolism to convert a PNA precursor monomer into a true monomer, thereby also providing the material to allow the double-stranded PNA "gene" to replicate when it undergoes a random dehybridization to yield two complementary single-stranded templates.[1] Finally, as the different nucleobases have different electron donor and electron relay capabilities, there is also a mechanism for natural selection, with some bases and base orderings being superior to others in their ability to facilitate the metabolism.

The minimal protocell contains on the order of 10^3 atoms. Due to its small size, all its processes, including its self-assembly from component molecules, its absorption of light, and its metabolism, could at least in principle be investigated using quantum (wave) theory. The main difference from classical (Newtonian) theory is that quantum particles, including the self-organized bioorganic supramolecules and their components, behave as waves, not as discrete particles with definitive positions and momenta. For example, valence electrons of supramolecules (which mainly determine features of nanocells) possess strict quantum states and discrete quantum electronic ultraviolet–visible, vibrational, nuclear magnetic resonance, and electron para-magnetic resonance

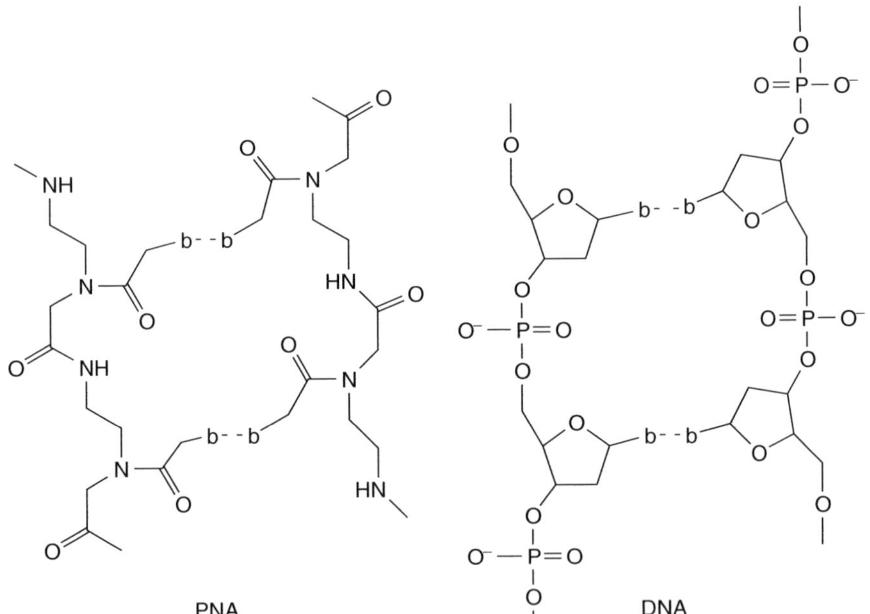

Figure 1 PNA (left side) is similar to DNA (right side) but has a nonpolar peptide-based backbone as opposed to DNA's sugar–phosphate polar backbone. Both form double-stranded chains with hydrogen bonds (dotted lines) between complementary bases (b).

spectra. The entire protocell might be considered to be a molecular electronics device that self-assembles according to quantum-based electron interaction potentials and that absorbs light and carries on its metabolism according to quantum electron excitation and tunneling equations. Therefore, in this picture, the photoinduced electron charge transfer in the protocell may be viewed as a biocomputing quantum particle–wave trace.

The main purpose of this chapter is to report the results of quantum mechanical (QM) modeling of the self-assembly and charge transfer in a minimal protocell[1] that might have implications for the first living organism on the Earth.[2] The current work uses larger supramolecular systems and more exact quantum mechanical methods than our earlier works.[3,4] This article uses a collection of quantum mechanical tools and applies them to a variety of protocell photosynthetic problems while providing a perspective of the requirements for success in the synthesis of new forms of living organisms.

1. QUANTUM MECHANICAL MOLECULAR PHYSICS APPROACH

Use of the Born–Oppenheimer approximation for decoupling the electron and nuclear motions in molecular derivatives is fully justified for most cases.[5,6] Subsequently, one only needs to solve the Schrödinger equation for the electron system, with nuclear positions being entered as parameters. Quantum chemical methods provide practical recipes for approximate solutions of many-electron molecular systems.[5,6] In particular, the Hartree–Fock (HF) approximation maps the complex many-body problem onto an effective one-electron problem in which electron–electron repulsion is treated in an average (mean field) way and the simplest antisymmetric wave function is assumed for the N-electron molecule (i.e., a single Slater determinant):

$$\psi = \langle \chi_1 \chi_2 \cdots \chi_N \rangle \tag{1}$$

Here $\chi_i(x)$ are the molecular orbitals (MOs). Following Roothaan's procedure,[7] they are expanded as linear combinations of localized atomic basis functions $\varphi_k(x)$:

$$\chi_i(x) = \sum_k \varphi_k(x) C_{ki} \tag{2}$$

In the unrestricted Hartree–Fock (UHF) approach, x refers to both coordinate and spin variables:

$$\varphi_k(x) = \phi_k(\vec{r})\xi(s) \quad \text{where } \xi(s) = \alpha \text{ or } \xi(s) = \beta \tag{3}$$

(α and β correspond to spin up and down, respectively).

The HF secular eigenvalue equation is derived variationally by minimizing the energy with respect to the choice of the MOs (i.e., the coefficients C_{ki}):

$$\mathbf{FC} = \mathbf{SCE} \tag{4}$$

where **C** is the matrix of the coefficients C_{ki}, **S** the overlap matrix coming from the nonorthogonality of the atomic basis functions $\varphi_k(x)$, **E** the eigenvector of the respective MOs energies, and **F** a Fock operator (an effective Hamiltonian for one electron), which depends on the electronic density matrix ρ_{ij}, given by

$$\rho_{ij} = \sum_{k}^{\text{occupied}} C_{ki} C_{kj}^{+} \tag{5}$$

where the sum runs over the occupied molecular orbitals. Equation (4) is nonlinear and usually is solved iteratively using the self-consistent field procedure.

The HF approximation is not very accurate because it does not include electron correlation effects.[5,6] Density function theory (DFT) makes it possible to treat these correlations by mapping the complex many-electron problem into an effective mean field problem with the same energy, which is a functional of the electron density. As a result, in DFT one solves the same one-electron problem [Eqs. (1)–(5)], but the Fock operator is replaced by the Kohn–Sham operator **h**, which is a functional of the electron density. In principle this mapping is exact, but the functional is unknown. However, extensive research has formulated accurate functionals suitable for many complex cases.[8,9] Detailed descriptions of the DFT, self-consistent field (SCF) procedure, expressions for the ground-state SCF energy, and the Fock and Kohn–Sham operators are readily available from modern quantum chemical textbooks.[5,6,8,9]

Our quantum simulations of single bioorganic molecules possessing closed electronic shells start from a trial geometry (Cartesian coordinates of the nuclei). Using restricted HF, DFT approaches, and the SCF procedure in the Gaussian03[10] or GAMESS-US[11] program packages, we obtain the lowest molecular energy, which depends on these coordinates parametrically. A subsequent standard geometry optimization procedure[5,9–11] minimizes the energy with respect to the nuclear positions. Special care is required to verify that the optimal molecular structure obtained is a global minimum in the phase space of the nuclear ($3n - 6$, n being the number of atoms) degrees of freedom.

To obtain accurate results when investigating supermolecules, two factors need to be accounted for: the quality of the density functional and the quality of the molecular orbitals (extent of the phase space of the single-electron states). For simple, covalently bonded molecules, we chose Becke's three-parameter exchange functional[12] with nonlocal Lee–Yang–Parr electron correlations[13] for the (DFT B3LYP), PBEPBE, and PBELYP models.[10,11,14–16] Currently, the B3LYP model is considered to be the most appropriate model for taking into account electron correlations in large closed-shell supermolecules where atoms are linked by covalent bonds (i.e., there are no van der Waals nor hydrogen bonds between atoms in a single molecule).[17,18] For simulations of the self-assembly of bioorganic supramolecules where separate molecules are associated by hydrogen bonds or van der Waals forces, we used the PBEPBE model[10,14] (in Gaussian03) and the PBELYP model[11,14] (in GAMESS-US). In these two models,

TABLE 1. Function Availability and Range of Applicability for Each Built-in Basis Set in Gaussian03

Basis Set	Applies to Atoms	Polarization Functions	Diffuse Functions
3-21G	H-Xe	* or **	+
6-31G	H-Kr	(3df,3pd)	++
6-311G	H-Kr	(3df,3pd)	++

the exchange functionals include some electron correlation effects at larger distances that provide relatively good descriptions of the van der Waals forces and hydrogen bonds. To obtain accurate optimal molecular geometries for neutral radical molecules, we use the 6-31G** basis set which includes 5 d and 7f polarized atomic orbitals. For self-assembly of bioorganic molecules we add diffusion orbitals of the 6-31++G** basis set (the standard tables[19] give the appropriate basis set description). The properties of the models used are given in Table 1.

2. QUANTUM MODELING OF WATER AND FATTY ACID–DEPENDENT PHOTOINDUCED ELECTRON TRANSFER IN PNA-BASED PROTOCELLS

The main parts of one of the variants of the proposed artificial protoorganism[1] are a PNA double-helix fragment which is covalently bonded to a light-absorbing 1,4-bis(*N,N*-dimethylamino)naphthalene sensitizer molecule, a pFA, an SH anion (see Figure 2, drawn using MOLDEN software[20]), and fatty acid molecules (the fatty acid molecules are not shown). The geometric and electronic structure (without the presence of water and FA molecules) was optimized by the QM semiempirical PM3 method installed in the GAMESS-US package[11] on the Vilnius Linux PC cluster. The chemical formulas of the molecules used in the simulated protocell are 1,4-bis(*N,N*-dimethylamino)naphthalene, $(CH_3)_2-N-C_8H_6-N-(CH_3)_2$; cytosine, $C_4H_5N_3O$; PNA backbone segment, $NH-C_5H_8NO-CO$; fatty acid, $HO-CO-(CH_2)_n-CH_3$; and a precursor of fatty acid, $C_6H_5-CO-(CH_2)-O-CO-(CH_2)_n-CH_3$.

The geometries of the individual molecules that constitute the PNA double helix [1,4-bis(*N,N*-dimethylamino)naphthalene]$_3$ supermolecule are then optimized separately using the PBELYP model[11,14] with the 6-311++G** basis set implemented in the GAMESS-US program package on the Vilnius Linux PC cluster and with the PBEPBE model[10,14] using the 6-311++G** basis set[19] implemented via the SGI64-G03RevC.02 Gaussian03 program package on the Poznan Supercomputing and Networking Center's SGI Origin 3000 supercomputer. Next, the geometric and electronic structure of the entire covalently bonded PNA double helix [1,4-bis(*N,N*-dimethylamino)naphthalene supermolecule] is optimized using the QM semiempirical PM3 method in the GAMESS-US program package

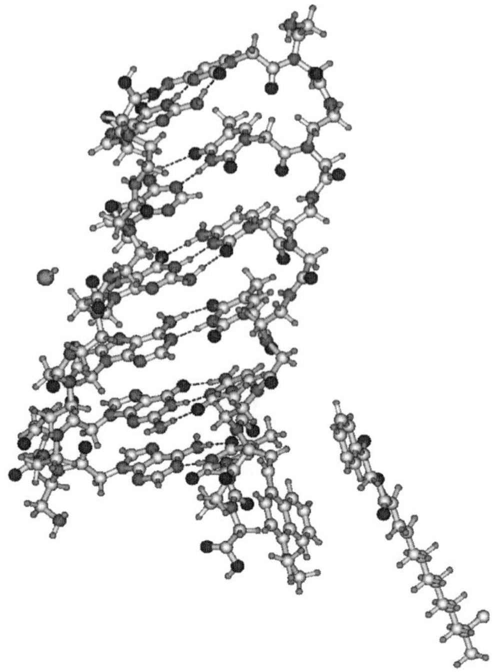

Figure 2 With the exception of the fatty acid molecules, the main parts of one of the artificial proto-organisms[1] includes a PNA double-helix molecule that is covalently bonded to the 1,4-bis(*N,N*-dimethylamino)naphthalene sensitizer molecule shown at bottom center, a pFA molecule (bottom right), and an SH anion molecule (center left). Carbon atoms are shown as white/gray spheres, hydrogens are small white/gray, oxygens are dark/gray, nitrogens are gray, and sulfur is large gray. Hydrogen bonds are depicted by dashed lines.

installed on the Vilnius laboratory $6 + 2$ nodes heterogeneous PC Linux cluster. The process was repeated using other sensitizer molecules.

Finally, we used our time-dependent (TD) DFT B3PW91 model[10] with the 6-31G basis set[19] to calculate the absorption spectra and the relative positions of the highest occupied molecular orbital (HOMO) and lowest unoccupied molecular orbital (LUMO) eigenvalues of the pFA and various sensitizer molecules: 1,4-bis(*N,N*-dimethylamino)naphthalene, 1,4-dihydroquinoxaline, and 7,8-dimethylisoalloxazine. Analysis of the results show that these sensitizers are good candidates for use in artificial living organisms because their HOMO states are energetically high enough relative to the LUMO state of the pFA molecule, and their absorption spectra are in the visible region.[4] However, as these investigations were performed without the presence of surrounding water molecules, the absorption spectra calculated are shifted strongly into the blue region. Thus, the resulting analysis is valuable only for comparison of HOMO–LUMO relationships, but not for obtaining the real wavelengths of the absorption spectra. For example, for the self-assembled supramolecule

TABLE 2. Details of the Self-Assembly Calculations

Row	Figure Number Corresponding to Calculations	Optimization Method	Basis Set	Number of FA	Number of pFA	Number of Water Molecules
1	None	B3PW91	6-31G	None	1	None
2	2	PM3	None	None	1	72
3	None	MM/GROMACS	None	4	1	7311
4	5	PM3	None	4	1	101
5	6	PBEPBE	6-31G	2	1	18
6	7	PM3	None	6	2	37

complex,[4] which consists of a cytosine–PNA fragment covalently bonded to a 1,4-bis(*N,N*-dimethylamino)naphthalene sensitizer and a pFA molecule (see Table 2, row 1), the most intense absorption line as calculated by TD DFT B3PW91 with the 6-31G basis set is for the second excited state and is equal to 355 nm, while the most intense experimentally determined absorption line for this photosynthetic center is at 450 nm.[21]

Thus, the model predicts that (at least without water and FA molecules present) if one were to construct such an artificial photosynthetic system, it would be a very poor choice for the metabolic process desired.

The QM modeling above is performed without the presence of surrounding water molecules, which resulted in a strong shift of the absorption spectra to the blue region. To correct for this shortcoming, we add some water molecules to the simulation to obtain more realistic wavelengths. The optimized geometrical structure of a 345-atom self-assembled derivative system consisting of a PNA fragment, a 1,4-bis(*N,N*-dimethylamino)naphthalene sensitizer, and a pFA molecule surrounded by water molecules is obtained using the quantum mechanical semiempirical PM3 method implemented in the GAMESS-US program package, which is installed on the Vilnius PC Linux cluster (see Table 2, row 2 and Figure 3, drawn using MOLEKEL[22]).

Most of the 72 water molecules self-organize into clusters of nano ice-like substructures. Furthermore, all the interatomic distances between the 1,4-bis(*N,N*-dimethylamino)naphthalene sensitizer and the pFA molecule become reduced (i.e., the photosynthetic system become more compressed due to the presence of the water molecules). Starting with this geometry, the individual water molecules are removed and the absorption spectrum is calculated by applying the new water cavity generation technique in the IEFPCM solvent shell model,[23] together with the electron correlation TD DFT PBEPBE model[14] using the 6-31+G* basis set, which includes diffusion atomic orbitals[19] and which is implemented using the SGI64-G03RevC.02[10] Gaussian03 program package on the LANL SGI Altix 3000 machine. This PBEPBE model more accurately describes the hydrogen bonding and van der Waals interactions. The calculation gives a value of 573.33 nm for the HOMO–LUMO (first excited state) transition and indicates that the most intense transition is associated with the sixth excited

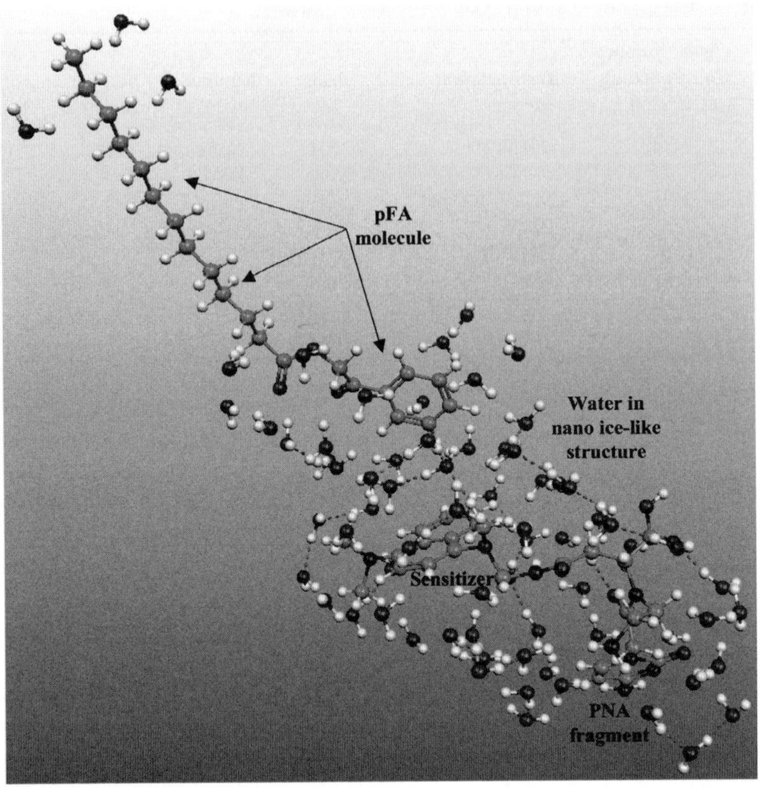

Figure 3 Image of the geometric and electronic structure of a cytosine–PNA fragment that is covalently bonded to a 1,4-bis(*N,N*-dimethylamino)naphthalene sensitizer moleecule, a pFA, and water molecules, whose positions were optimized using the PM3 method. The water molecules organized to nano icelike substructures. Carbon atoms and their associated covalent bonds are shown as gray spheres and sticks, hydrogens are in white/gray, oxygens are in dark/gray, and nitrogens are in black. Hydrogen bonds are depicted by dashed lines.

state, located at 425.91 nm (see Table 3). We note that the first excited state (573.33 nm) is shifted to the red significantly compared to the earlier 439.21-nm value when no water was included. Similarly, the wavelength of the most intense transition is shifted to 425.91 nm, compared to the earlier 355-nm value.[4] For comparison, the experimental value for the most intense transition is 450 nm.[21]

Using the calculation method mentioned above, we determine the electron charge transitions of the first six excited states by analyzing the HOMO $-\, m$ and LUMO $+\, n$ images generated for the same system (one example is shown in Figure 4), which were drawn using MOLDEN. Each excited state is found to be composed primarily of a single HOMO $-\, m$ to LUMO $+\, n$ transition, and therefore it is possible to say that an electron in a certain excited state tunnels

TABLE 3. Excitation Energies of a Cytosine–PNA Fragment[a]

Excited State	Transition HOMO − m → LUMO + n	Energy (eV)	Wavelength (nm)	Oscillator Strength (arbitrary units)
1	HOMO → LUMO	2.163	573.33	0.00001
2	HOMO − 1 → LUMO	2.320	534.32	0.00001
3	HOMO − 2 → LUMO	2.606	475.85	0.00001
4	HOMO − 3 → LUMO	2.651	467.70	0.00001
5	HOMO → LUMO + 1	2.755	450.09	0.00001
6	HOMO → LUMO + 2	2.911	425.91	0.00010

[a]Conditions: 1,4-bis(*N,N*-dimethylamino)naphthalene sensitizer and a pFA molecule using the water cavity shell IEFPCM solvent model,[23] as calculated using the PBEPBE model with the 6-31+G* with basis set.

Figure 4 The electron charge transfer transition associated with the first excited state is from (left half of image) the HOMO of the 1,4-bis(*N,N*-dimethylamino)naphthalene sensitizer molecule to (right half of image) the LUMO of the pFA molecule as calculated using the PBEPBE model with the 6-31+G* basis set. The system is calculated using the water cavity generation technique in the IEFPCM solvent shell model.[23] Black regions correspond to negative values of the wave function volume and gray to positive values.

from a certain HOMO − m to a certain LUMO + n: ($m = 0$ to 3, $n = 0$ to 2). The results are summarized in Table 3.

One should note that using the water cavity generation technique in the IEFPCM solvent shell model, the electron charge transfer transition associated with the first, second, and third excited states is from the HOMO − m located on the sensitizer-1,4-bis(*N,N*-dimethylamino)naphthalene molecule to the LUMO located on the pFA molecule (see Figure 4). We find that when using the water cavity IEFPCM solvent shell model together with the PBEPBE model with the 6-31+G* basis set, the electron charge transfer transition for the fourth excited state is from the HOMO − 3 located on the peptide group near the cytosine to the LUMO located on the pFA molecule. We also find that in this model, the electron charge transfer transition for the fifth excited state is from the HOMO located on the 1,4-bis(*N,N*-dimethylamino)naphthalene sensitizer molecule to the LUMO + 1 located on the cytosine–PNA fragment molecule.

The most intensive electron charge transfer transition (see Table 3), which is associated with the sixth excited state (wavelength of 425.91 nm), is from the HOMO located on the 1,4-bis(*N,N*-dimethylamino)naphthalene sensitizer molecule to the LUMO + 2 located on the same sensitizer molecule. Unfortunately, this shows a problem with the water cavity IEFPCM solvent shell model, as the experimental results have the most intense transition at 450 nm,[21] with the electron tunneling being to the pFA molecule. These discrepancies are probably explained by two factors: (1) the water cavity IEFPCM solvent shell model does not properly simulate the real water shell around the photosynthetic center, and (2) the initial geometry of the photosynthetic system provided by Gaussian03 was not fully accurate, which was amplified further by the shortcomings of the water cavity IEFPCM solvent shell model, resulting in problems with both the energy levels and the transition strengths. Detailed experimental investigations using ultrashort monochromatic laser pulses to perform femtosecond scanning of electron tunneling directions for the different excited states may be able to provide an understanding of these issues.

Next we perform a molecular mechanics optimization of the geometry of a protocell section containing a cytosine-bearing PNA fragment with a covalently attached 1,4-bis(*N,N*-dimethylamino)naphthalene sensitizer, a pFA molecule, and four FA molecules in a cubic 80-Å box that contains 7311 water molecules (see Table 2, row 3). The initial step is performed using the GROMACS software.[24] Later, most of the water molecules are removed and the geometry of the remaining 498 atoms of this bioorganic complex, which included 101 water molecules, is reoptimized using the QM semiempirical PM3 method implemented using GAMESS-US[11] on the Vilnius PC Linux cluster and then with the electron-correlated PBEPBE model[14] using the 3-21G basis set[19] implemented via the SGI64-G03RevC.02 Gaussian03 program package on the LANL SGI Altix 3000 machine (see Table 2, row 4). The reduced water complex is shown in Figure 5. It is only because of the hydrogen bonds between the lipid and water molecules (these bonds are marked by dashed lines in Figure 5) that the FA micelle exists.

Since PNA interacts with other FA and water molecules through their correlated electrons, intermolecular distances and surface area calculations are critical to understanding the time-dependent electron tunneling processes associated with the various excited states of the protocell. The distances between the separated sensitizer, pFA precursor, and water molecules are comparable to van der

Figure 5 Image of the geometric and electronic structure of a system consisting of a cytosine–PNA fragment (center) covalently bonded to a 1,4-bis(*N,N*-dimethylamino)naphthalene sensitizer molecule (above the PNA), a pFA (upper right), four of the six-carbon FA molecules (bottom center) that would comprise the micelle, and 101 water molecules that are optimized using the PM3 and PBEPBE models with the 3-21G basis set. The water molecules organize into nano icelike substructures. Carbon atoms and their associated covalent bonds are shown as gray spheres and sticks, hydrogens are in white/gray, oxygens are in dark/gray, and nitrogens are in black. Hydrogen bonds are depicted by dashed lines.

Waals and hydrogen-bonding radii, and we may therefore regard the protocell as a single electron-conjugated supramolecule that we can deal with using an electron-correlated model.

Having used the water molecules to establish the geometry of the protocell supramolecule, we remove all but 18 of the water molecules of the photosynthetic complex drawn in Figure 5 to yield a reduced system for which electron transfers can be more easily calculated. The water molecules we keep are those that (by hydrogen bonds) are both self-connected and connected to the sensitizer and/or the pFA molecule in the region of any likely electron transfer. Unfortunately, when not surrounded by water, FA molecules have a tendency to separate during the electron correlation geometry reoptimization relative to their initially constructed regular structure. This is due to a weak Coulomb repulsion between neighboring FA molecules that is normally negated by hydrogen-bonds involving the surrounding water molecules. The re-reoptimized geometry of this further-reduced water self-assembled complex is calculated using the electron correlation PBEPBE model with the 6-31G basis set implemented via the SGI64-G03RevC.02 Gaussian03 program package on the LANL SGI Altix 3000 machine (see Table 2, row 5 and Figure 6).

Due to the absence of water molecules in the lower half of the photosynthetic complex, two FA molecules dissociate from the complex during the geometry reoptimization, resulting in a final system size of 209 atoms. Furthermore, the carboxyl ends of two other FA molecules are now facing the cytosine–PNA–sensitizer assembly, thereby reducing all the interatomic distances compared with the similar complex drawn in Figure 5. The absorption

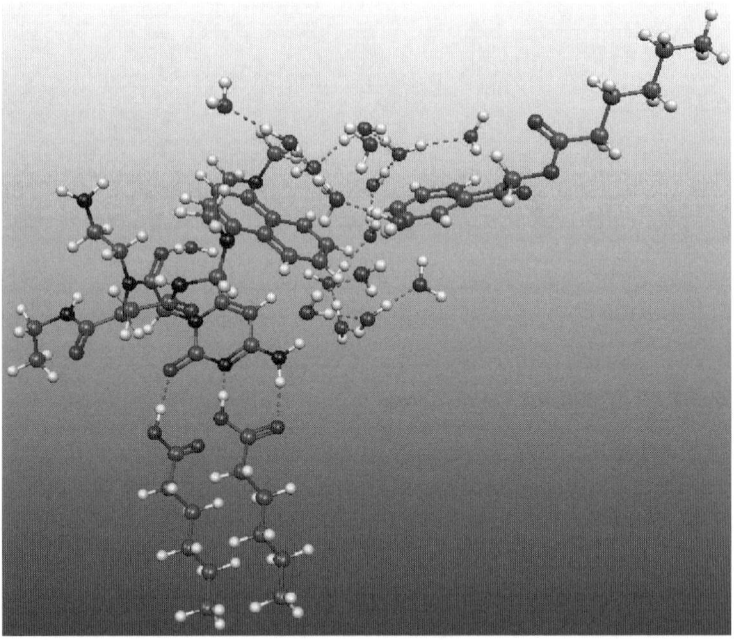

Figure 6 Reoptimized geometry of the reduced protocell photosynthetic center performed using the PBEPBE model with the 6-31G basis set. The water molecules organize into nano icelike substructures. Carbon atoms and their associated covalent bonds are shown as gray spheres and sticks, hydrogens are in white/gray, oxygens are in dark/gray, and nitrogens are in black. Hydrogen bonds are depicted by dashed lines.

spectrum of this reduced water 209-atom photosynthetic system is calculated by running the electron correlation TD DFT PBEPBE model[15,16] with the 6-31G basis set[19] on the Vilnius PC Linux cluster. The calculation gives a value for the HOMO–LUMO (first excited state) transition of 716.8 nm (see Table 4). This is shifted significantly to the red compared with the 573.33-nm value calculated for the 129-atom complex (see Table 3) using the water cavity IEFPCM solvent shell model and the 439.21-nm value calculated without water molecules.

The hydrogen bonds and van der Waals interactions that result from the addition of water and FA molecules clearly play a critical role in the functioning and performance of the photosynthetic system. These quantum interactions, which do not show up in the water cavity model, further compress the overall system, resulting in a smaller gap between the HOMO and LUMO states. The shift of the absorption spectrum shift to the red for the artificial protocell photosynthetic center might be considered as the measure of the complexity of this system.

We next perform calculations using the TD DFT PBEPBE model[15,16] with the 6-31G basis set on the Vilnius PC cluster of the difference of electron charge density (excited state-ground state) for the conjugated lipid–cytosine–1,4-bis(*N,N*-dimethylamino)naphthalene supermolecule and pFA molecule and visualized the electron charge tunneling associated with certain excited-state transitions. We find that the electron tunneling transitions of the first to fifth and seventh to ninth excited states should induce metabolic photodissociation of the pFA molecule because the transferred electron cloud is located on the head (the waste piece) of the pFA molecule. An example is shown in Figure 7.

TABLE 4. Excitation Transition Energies of a Cytosine–PNA Fragment[a]

Excited State	Transition HOMO − m → LUMO + n	Weight of Individual Transition	Energy (eV)	Wavelength (nm)	Oscillator Strength (arbitrary units)
1	HOMO → LUMO	1.000000	1.730	716.8	0.00000006
2	HOMO − 1 → LUMO	0.999744	1.879	660.0	0.00041141
3	HOMO − 2 → LUMO	0.999989	2.160	574.1	0.00065522
4	HOMO − 3 → LUMO	1.000000	2.250	551.0	0.00000194
5	HOMO − 4 → LUMO	0.999999	2.301	538.8	0.00000107
6	HOMO → LUMO + 1	0.999988	2.316	535.4	0.00006185
7	HOMO − 5 → LUMO	0.999927	2.409	514.7	0.00030617
8	HOMO − 6 → LUMO	0.999954	2.514	493.3	0.00001499
9	HOMO − 7 → LUMO	0.999971	2.670	464.3	0.00001616
10	HOMO − 2 → LUMO + 1	0.929090	2.758	449.5	0.05468100
	HOMO − 1 → LUMO + 1	0.057840			
11	HOMO − 10 → LUMO + 1	0.031776	2.789	444.5	0.60693889
	HOMO − 2 → LUMO + 1	0.064548			
	HOMO − 1 → LUMO + 1	0.838353			
	HOMO − 1 → LUMO + 10	0.011296			

[a]Condition: a 1,4-bis(*N,N*-dimethylamino)naphthalene sensitizer and a pFA, two FA molecules, and 18 water molecules, as calculated using the TD PBEPBE model with the 6-31G basis set. The weight of an individual excitation is given if it is larger than 0.01.

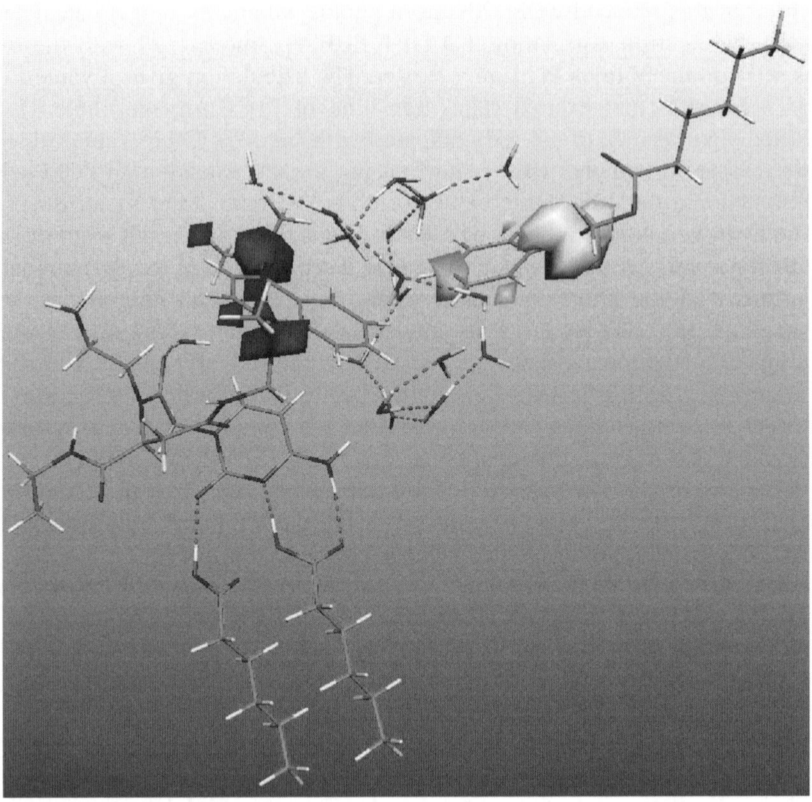

Figure 7 Visualization of the electron charge transfer associated with the second excited state. The tunneling is from the sensitizer 1,4-bis(*N,N*-dimethylamino)naphthalene molecule to the pFA molecule. The electron cloud hole is indicated in blue, and the transferred electron cloud location is shown in gray. (*See insert for color representation of figure.*)

The redistribution of the electron associated with the sixth, tenth, and eleventh excited states should not induce metabolic photodissociation of the pFA molecule because the electron tunneling is from the FA–cytosine–1,4-bis(*N,N*-dimethylamino)naphthalene supramolecule to the same sensitizer molecule. This is an internal electron charge redistribution on the same supramolecule. The most intense electron transition, which has a 444.5-nm wavelength (see Table 4), belongs to the internal electron charge redistribution on the same sensitizer molecule, which is again in conflict with experimental data.[21] We also note that for some of the configurations, the electron tunneling distances appear to be unreasonably large. Summarizing all these negative facts, we can state that this photosynthetic system model with its very limited number of water molecules and with only two FA molecules and one pFA molecule is too small to obtain reasonable spectra and electron tunneling results.

Finally, we perform a geometry optimization of another photosynthetic system, this one consisting of a cytosine–PNA fragment with a covalently bonded 1,4-bis(*N,N*-dimethylamino)naphthalene sensitizer, two pFA, six FA molecules, and 37 water molecules (total of 357 atoms) using the quantum mechanical semiempirical PM3 method installed in the GAMESS-US package on the Vilnius Linux PC cluster (see Table 2, row 6 and Figure 8). This system is again allowed to self-assemble. The absorption spectrum (see Table 5) of this system

Figure 8 Image of the geometric and electronic structure of a system consisting of a cytosine–PNA fragment covalently bonded to a 1,4-bis(*N,N*-dimethylamino)naphthalene sensitizer molecule (in the center), two pFA molecules (bottom and top right), six FA molecules, and water molecules that are optimized using the PM3 method. The water molecules organize into nano icelike substructures. Carbon atoms and their associated covalent bonds are shown as green spheres and sticks, hydrogens are shown in white/gray, oxygens are red, and nitrogens are a blue. Hydrogen bonds are depicted by dashed lines. (*See insert for color representation of figure*.)

TABLE 5. Excitation Transition Energies of a Cytosine–PNA Fragment[a]

Excited State	Transition HOMO − m → LUMO + n	Weight of Individual Transition	Energy (eV)	Wavelength (nm)	Oscillator Strength (arbitrary units)
1	HOMO → LUMO	0.998147	2.203	562.7	0.005800
2	HOMO → LUMO + 1	0.998154	2.208	561.6	0.030042
3	HOMO → LUMO + 2	0.999998	2.368	523.7	0.000522
4	HOMO → LUMO + 3	0.999786	2.370	523.2	0.029103
5	HOMO → LUMO + 4	0.999789	2.373	522.5	0.039266
6	HOMO → LUMO + 5	1.000000	2.472	501.5	0.000133
7	HOMO − 1 → LUMO HOMO − 1 → LUMO + 1	0.584669 0.415306	2.656	466.9	0.019864
8	HOMO → LUMO + 6	1.000000	2.692	460.6	0.000659
9	HOMO − 1 → LUMO + 2 HOMO − 1 → LUMO + 3	0.959621 0.024393	2.818	440.0	0.236093
10	HOMO − 1 → LUMO + 2 HOMO − 1 → LUMO + 3	0.028294 0.962863	2.820	439.7	0.077127
11	HOMO − 1 → LUMO + 4	0.981380	2.823	439.2	0.181830

[a]Conditions: a 1,4-bis(*N,N*-dimethylamino)naphthalene sensitizer and two pFA, six FA molecules, and 37 water molecules, as calculated using the TD PBEPBE model with the 6-31G basis set. The weight of an individual excitation is given if it is larger than 0.01.

is calculated by running the TD DFT PBEPBE model[15,16] with the 6-31G basis set on the Vilnius Linux PC cluster.

The calculation indicates that the most intense electron charge transition from the 1,4-bis(*N,N*-dimethylamino)naphthalene sensitizer molecule to one of the pFA molecules is for the ninth excited state (HOMO − 1 → LUMO + 2) transition and has an energy and wavelength of 2.818 eV and 440.0 nm, respectively (see Table 5 and Figure 9). This result is close to the experimental value of 450 nm.[21] The improved agreement with experiment is probably due to the increased complexity of the model, which now includes even more water and FA molecules and hence is closer to the real situation. We also find a relatively intense electron charge transition for the tenth excited state to the second pFA molecules (HOMO − 1 → LUMO + 3), equal to 439.7 nm (2.820 eV).

3. CONCLUSIONS

We are able to identify how the quantum mechanical calculations of charge transfer from our photosensitizer to our precursor fatty acid molecule are quantitatively improved as we include more detailed environmental information in terms of the presence of water and fatty acid molecules. The critical role played by the quantum effects associated with hydrogen bonds and van der Waals interactions is the main finding of our computational investigations. This results directly from the addition of water and FA molecules to the environment around the self-assembling charge transfer pathway. These quantum interactions compress the overall system, resulting in a smaller gap between the

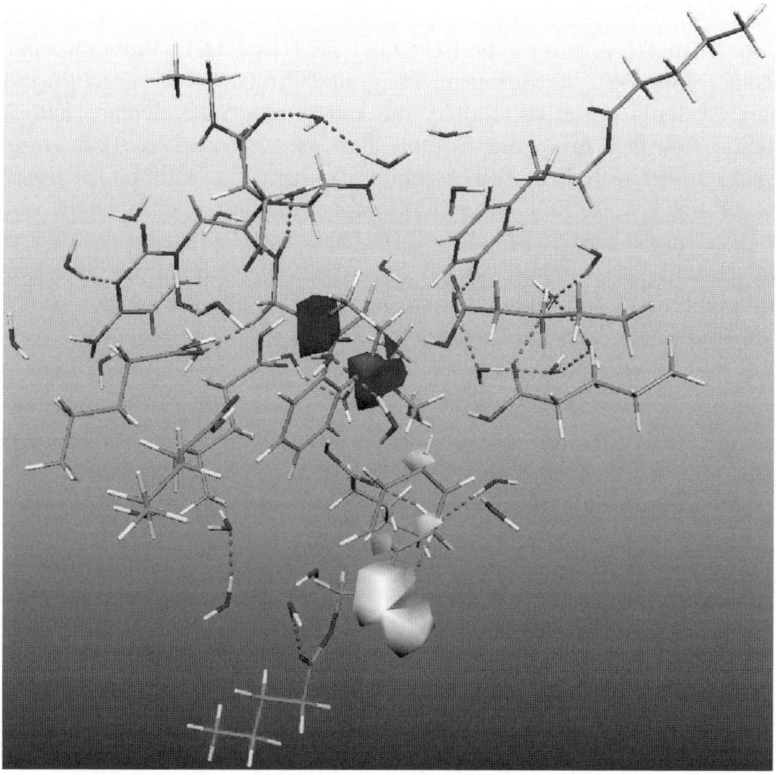

Figure 9 Visualization of the electron charge tunneling associated with the ninth excited state. The transition is from the sensitizer 1,4-bis(*N,N*-dimethylamino)naphthalene molecule (in the center) to the first pFA molecule (bottom). The electron cloud hole is indicated in blue, and the transferred electron cloud location is visualized in white/gray. (*See insert for color representation of figure*.)

HOMO and LUMO states. By investigating a number of different models of the photosynthetic system, each subsequent model including more water and FA molecules, we are able to quantify this effect. The last and most realistic model used gives an excitation photon wavelength of 440.0 nm, which compares well with the experimental value of 450.0 nm. This quantum self-assembled model of the photosynthetic system included a PNA fragment that is covalently linked to a 1,4-bis(*N,N*-dimethylamino) naphthalene sensitizer molecule, two pFA molecules, and six of the FA molecules constituting the micellar container. The system also includes 37 water molecules. It is possible to understand the remaining small (10 nm) wavelength difference as stemming from the greater number of water and FA molecules found in the real system. The slightly shorter wavelength given by this model is also consistent with our findings that the inclusion of more water molecules in similar models results in longer wavelengths for the absorption spectrum.

Acknowledgments

The work is funded in part by the EU COST D27, by PACE (Programmable Artificial Cell Evolution), the European Integrated Project in the EU FP6-IST-FET Complex Systems Initiative, and by the Lithuanian State Science and Studies Foundation. Further, this work is funded in part by the Los Alamos National Laboratory LDRD-DR Protocell Assembly project. The authors are grateful for the use of the LANL SGI Altix 3000 supercomputer and to Dr. R. L. Martin for installing the SGI64-G03RevC.02 Gaussian03 program package on that machine. Finally, the authors would like to thank Professor A. Graja and the Poznan Supercomputing and Networking Center for the use of their SGI Origin 3000 supercomputer and the Gaussian03 package.

REFERENCES

1. Rasmussen, S.; Chen, L.; Nilsson, M.; Abe, S. *Artif. Life*, 2003, 9: 267–316.
2. Nelson, K.; Levy, M.; Miller, S., *Proc. Natl. Acad Sci. USA*, 2000, 97: 3868–3871.
3. Tamulis, A.; Tamuliene, J.; Tamulis, V. Quantum mechanical design of photoactive molecular machines and logical devices, in *Handbook of Photochemistry and Photobiology*, Vol. 3,*Supramolecular Photochemistry*, Nalwa, H.S., Ed., American Scientific Publishers, 2003, pp. 495–553.
4. Tamuliene, J., Tamulis, A. Quantum mechanical investigations of self-assembled system consisting of peptide nucleic acid, sensitizer, and lipid precursor molecules, *Lithuanian J. Phys.*, 2005, 45(3): 167–174.
5. Cook, D. B. *Handbook of Computational Quantum Chemistry*, Oxford University Press, New York, 1998.
6. Szabo, A.; Ostlund, N. S. *Modern Quantum Chemistry: Introduction to Advanced Electronic Structure Theory*, McGraw-Hill, New York, 1989.
7. Roothaan, C. C. J.; *Rev. Mod. Phys.*, 1960, 32: 179.
8. Parr, R. G.; Yang, W. *Density Functional Theory of Atoms and Molecules*, Oxford University Press, New York, 1989.
9. Dreizler, R. M.; Gross, E. K. U. *Density Functional Theory*, Springer-Verlag, Berlin, 1990.
10. Frisch, M. J.; Trucks, G. W.; Schlegel, H. B.; Scuseria, G. E.; Robb, M. A.; Cheeseman, J. R.; Montgomery, J. A., Jr.; et al. *Gaussian03, Revision C.01,* Gaussian, Inc., Wallingford CT, 2004.
11. Schmidt, M. W.; Baldridge, K. K.; Boatz, J. A.; Elbert, S. T.; Gordon, M. S.; Jensen, J. H.; Koseki, S.; et al. General atomic and molecular electronic structure system, *J. Comput. Chem.*, 1993, 14: 1347–1363; http://www.msg.ameslab.gov/GAMESS/GAMESS.html.
12. Becke, A. D. *Phys. Rev, A.* 1988, 38: 3098.
13. Lee, C., Yang, W.; Parr, R. G. *Phys. Rev. B*, 1988, 37: 785.
14. Perdew, J. P.; Burke, K.; Ernzerhof, M. *Phys. Rev. Lett.*, 1996, 77: 3865–3868.
15. Neese, F. ORCA: An Ab Initio, Density Functional and Semiempirical Program Package, Version 2.4, Max-Planck-Institut für Bioanorganische Chemie, Mülheim an der Ruhr, Germany, 2004.

16. Neese, F. *J. Chem. Phys.* 2003, 119: 9428.
17. Springborg, M., Ed. *Density-Functional Methods in Chemistry and Materials Science*, Wiley, Chichester, England, 1997.
18. Jensen, F. *Introduction to Computational Chemistry*, Wiley, Chichester, England, 1999.
19. *Extensible Computational Chemistry Environment Basis Set Database*, Environmental and Molecular Sciences Laboratory at Pacific Northwest Laboratory, Richland, WA.
20. http://www.cmbi.ru.nl/molden/molden.html.
21. Rasmussen, S., et al., Assembly of a minimal protocell, preprint February 2006, to appear in *Protocells: Bridging Nonliving and Living Matter*, Rasmussen, S., et al., Eds., MIT Press, Cambridge, MA.
22. http://www.cscs.ch/molekel/.
23. Cances, E.; Mennucci, B. *J. Chem. Phys.*, 2001, 114: 4744.
24. http://www.gromacs.org/contributions/index.php.

3

OPTIMIZING THE ELECTRONIC PROPERTIES OF CARBON NANOTUBES USING AMPHOTERIC DOPING

BOBBY G. SUMPTER AND VINCENT MEUNIER
Computer Science and Mathematics Division and Center for Nanophase Materials Sciences, Oak Ridge National Laboratory, Oak Ridge, Tennessee

In recent years, molecular electronics has become one of the premier fields in nanoscience.[1] A common theme within this field is the detailed exploration of the electronic structure, response, and transport, along with the development of new electronic devices and applications that depend on properties of matter at the molecular scale. The fundamental importance of molecules in electronic device applications stems not only from their electronic properties but also from the ability to bind to one another and to self-organize into larger structures. This, combined with the large diversity in chemical variability, where an extremely large number of molecules can be synthesized, allowing for an essentially continuous variation in properties as well as entirely new functionality, presumably opens the door to enormous flexibility in possible new types of nanoscale devices. The exploration into these possibilities was initially fueled by the theorectical analysis of Aviram and Ratner,[2] who proposed the idea of using a single molecule containing a donor and acceptor group separated by an insulating spacer as an

Multiscale Simulation Methods for Nanomaterials, Edited by Richard B. Ross and Sanat Mohanty

electric rectifier. From that time, initial progress in the molecular electronics field came mainly from theoretical and modeling efforts until experimental techniques allowing manipulation and measurements of single molecules became possible. In a pioneering experiment, Reed and co-workers[3] were first to clearly demonstrate an experimental technique based on self-assembled monolayers of benzene-1,4-dithiol and a mechanical break junction: that a single molecule can show differential conductance similar to a diode, thereby providing experimental proof of the potential of molecules to act as electronic devices. The field has now seen a large increase in experimental research, but this work is often complicated by problematical procedures and difficulties in interpretation of the results.[4] At the same time, theoretical developments and computational modeling have progressed alongside experiment, and with current computing capabilities, we are now in a position to offer results that are based entirely on first-principles calculations and take into account that the environment is of an open system.[4]

Although considerable attention has been given to the transport properties of a molecule of benzene-1,4-dithiol, fueled by the seminal work of Reed et al.,[3] less work has been focused on open molecular devices composed of doped, nanostructured carbon materials. In particular, the unique structural and electronic properties of carbon nanotubes offer tremendous potential for applications in the field of molecular electronics. As such, single-walled carbon nanotubes (SWCNTs) have been examined optimistically for their use as molecular wires and field-effect transistors. In addition, low-temperature electronic transport measurements have revealed that SWCNTs are ideal quasi-one-dimensional mesoscopic systems. They can behave both as quantum waveguides and as quantum dots, and a number of interesting effects, such as Kondo and shell filling, have been observed. The remarkable electrical properties of SWCNTs originate from the unique electronic structure of the two-dimensional material, graphene (sp^2-bonded carbon atoms arranged in a honeycomblike structure), from which they can be modeled. In general, the conducting properties of a material are determined by the nature of the electronic states near the Fermi energy. For graphene, the band structure (the energy of the electronic states as a function of the wave vector), lies somewhere between the extremes of a metal and a semiconductor. In most directions, electrons at the Fermi energy are backscattered by atoms in the lattice, giving rise to a bandgap similar to that of a semiconductor. In contrast, in one direction the scattered electrons undergo destructive interference with other electrons, leading to metalliclike behavior. When graphene is rolled along a chiral vector to form a nanotube, the resulting periodic boundary condition on the wave function leads to quantization of the wave vector perpendicular to the tube axis. Thus, depending on how the tube is formed from a graphene sheet, the transport properties can appear like a semiconductor or a metal. The theoretical condition for metallic behavior or an (n,m) single-walled carbon nanotube is $(n-m)/3 = \text{integer}$, a relationship that has been verified experimentally.

Although electronic transport of pristine metallic nanotubes has been shown to be ballistic, practical realizations of new nanotube-based electronic devices hinge on a number of outstanding problems, such as the capability of achieving large-scale, air-stable, and controlled doping. Recent experimental evidence[5] suggests that amphoteric doping of single-walled carbon nanotubes might be achieved by encapsulating organic molecules possessing different electron affinities and ionization energies. Particularly exciting is the apparent stability and fine tunability at low carrier density of n-doped nanotubes, something that to date has been difficult to achieve by other processing methodologies. To verify and understand these experimental conclusions, which were inferred from x-ray diffraction, Raman, and optical absorption spectroscopy of film samples, we have performed large-scale electronic structure calculations coupled with quantum conductance evaluation within the Green's function formulation.[6] From a theoretical and computational perspective, there is still an orders-of-magnitude discrepancy between experimental and theoretically determined transport properties for single molecules, and new computational approaches that allow quantitative predictions for molecular-based systems, such as the one presented herein, are a fundamentally important and crucial step toward advancement of the molecular electronics field.

To investigate and optimize the electronic transport processes in carbon nanotubes doped with organic molecules, we have performed large-scale quantum electronic structure calculations coupled with a Green's function formulation for determining the quantum conductance.[6] Other computational investigations on amphoteric doping of carbon nanotubes have to date focused mainly on computing the band structure using periodic plane-wave pseudopotential methods.[7] Unfortunately, this relatively straightforward approach does not allow neither the examination of quantum conductance for these systems nor the possibility of investigating nonperiodically filled nanotubes. Our approach is based on a scheme where quantum chemistry calculations on finite systems are recast to infinite, nonperiodic systems, therefore mimicking actual working devices.[6] Results from our calculations strongly suggest that the electronic structure of a carbon nanotube can easily be manipulated by encapsulating appropriate organic molecules, leading to charge transfer processes that induce efficient n- or p-type doping of the carbon nanotube. Even though a molecule may cause n- or p-doping, compared to a pristine tube we have found it to have a minor effect on the transport properties of the nanotube. In addition, the efficient process of charge transfer between the organic molecules and the nanotube is found to substantially reduce the susceptibility of the π-electrons of the nanotube to modification by oxygen while maintaining stable doping at room temperature.

1. METHODS AND COMPUTATIONAL APPROACH

All-electron density function theory (DFT) calculations of hydrogen-terminated (10,10)-SWCNTs with encapsulated organic molecules were performed using the NWChem[8] package with the local density approximation (LDA). A number of different atom-centered, contracted Gaussian basis sets (the Pople split-valence basis sets: 3-21G,[9] 6-31G*,[10] 6-311G*[10]) were used along with charge density fitting of the Coulomb potential during calculation of the self-consistent solution. For this purpose we used the auxiliary basis set DGAUSS A2.[11] The charge density fitting technique enables the evaluation of four- and two-center electron integrals to be conducted using at most three- or two-center electron integrals, thereby reducing the formal scaling of the computational cost from fourth order to third order. Implementation of the Schwarz inequality further reduces the computational cost by allowing evaluation and storage of small integrals to be avoided (Schwarz screening for the Coulomb integrals was 10^{-10}). The total energy without charge fitting was computed once calculations with charge fitting converged, although there are only very minor differences when using such a small Schwarz screening. Fermi energies and charge density isosurface plots which give the overall charge distribution of the system were obtained from these calculations. The total charge transfer density was estimated as the difference between the charge density of the total system (nanotube + molecule) and the (10,10)-SWCNT and individual molecules:

$$\rho_{\mathrm{CT}}(r) = \rho_{\mathrm{tot}}(r) - \rho_{\mathrm{nanotube}}(r) - \rho_{\mathrm{mol}}(r) \tag{1}$$

Although this approach can give only a rough estimate of the charge density resulting from the charge transfer processes, in particular since the direct influence of nano-confinement on the molecule's electronic structure (electronic confinement) is neglected, it does provide a reasonable procedure for getting a qualitative idea of the magnitude and spatial distribution of the charge density due to transfer.

It is worth noting that calculations using a triple-ζ-quality basis set, 6-311G*, included as many as 8058 basis functions (there are 440 atoms in the nanotube, chosen to allow encapsulation of up to four molecules, and 38, 20, and 14 atoms for the encapsulated molecules TDAE, F_4-TCNQ, and TTF of Table 1) and represented very large and challenging computations. Calculations that included four encapsulated molecules approach 10,000 basis functions. The encapsulated molecules were based on those examined in the experiments of Takenobu et al.[5]

TABLE 1. Summary of Some Results Obtained from All-Electron DFT/LDA Calculations

	Fermi Level(eV)		Fermi Level Shift(eV)			
System	3-21G	6-31G*	3-21G	6-31G*	Doping	CT/atom[a]
(10,10) nanotube	−4.13	−3.84				
Nanotube + 1 TDAE	−4.00	−3.77	0.13	0.070	*n*	0.009
Nanotube + 1 F_4-TCNQ	−4.24	−3.99	−0.11	−0.14	*p*	−0.02
Nanotube + 3 F_4-TCNQ	−4.26	−4.03	−0.13	−0.19	*p*	−0.03
Nanotube + 1 TTF	−4.06	−3.96	0.073	0.12	*n*	0.02
Nanotube + 3 TTF	−4.03	−3.76	0.10	0.08	*n*	0.02

[a] The charge transfer (CT) per dopant atom was computed from a Mulliken population analysis.

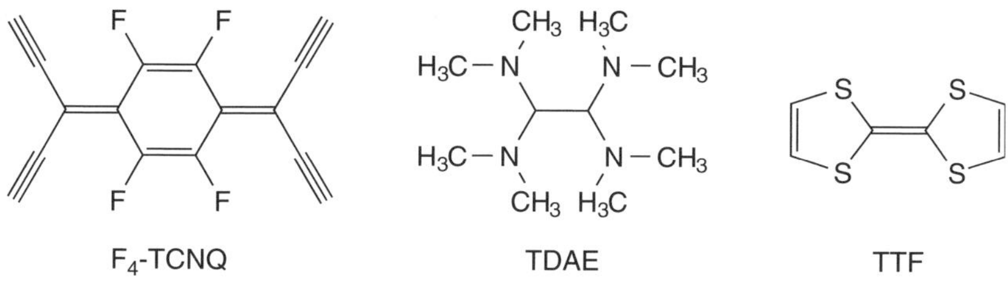

Figure 1 Organic molecules studied in this work (the hydrogen atoms are not shown on TTF). From left to right: F_4-TCNQ (tetrafluorotetracyano-*p*-quinodimethane), TDAE [tetrakis(dimethylamino)ethylene], and TTF (tetrahiafulvalene).

and are shown in Figure 1. This selection appropriately spans the amphoteric *n*- and *p*-doping systems.

The initial starting geometries for the different SWCNT-based systems were obtained by first optimizing the geometry of the individual molecules using DFT/LDA with the 6-31G* basis set, followed by optimizing their orientation inside a fixed (10,10)-SWCNT geometry using molecular mechanics and the MM3 potential. The MM3 parameterization[12] has proven to give very accurate results for structural optimizations of many organic molecules.[13–15] Of particular importance is the capability of the MM3 model to account for intermolecular interactions of the π-electron densities through the dependence of the stretching and torsion terms on iterative self-consistent field evaluations for the relevant π-conjugated bonds. The overall reliability of this model for structural calculations has been demonstrated continually for numerous aromatic compounds (benzene, biphenyl, annulene) and conjugated systems (*trans*-stilbene and even multiple oligomers of PPV).[16,17] This approach is expected to generate a reasonably accurate packing of the molecules since the main force governing their orientation, is through charge and dispersion interactions. To ensure the stability of the optimized molecular positions and orientations we carried out classical molecular dynamics simulations at room temperature and pressure.

It is also possible to perform full geometry optimization of the molecules inside the tube using the ONIOM[18–20] or a quantum mechanical/molecular mechanical (QM/MM) approach; however, for the present case one needs to use many-body electronic structure methods such as MP2[21,22] since relatively poor geometries are quite common when using semiempirical quantum methods (i.e., AM1, PM3, etc.), and current exchange-correlation functionals used in DFT do not accurately take into account nonlocal forces such as dispersion. One possibility is to use a two-level ONIOM method, where the high-level model is MP2/6-31G* and the low-level model is (DFT/LDA)/6-31G*. We have implemented this model with some success, and the results for a single molecule of F_4-TCNQ were in reasonable accord with those obtained from the MM3 computations. Considering the extensive computational demand of those types of calculations, we opted to use the structures determined from the MM3 optimizations, in particular because we are interested in nanotubes with many molecules encapsulated, as this corresponds to the original experiments.

To examine the charge transport properties that correspond to actual eletronic nanodevices, one must consider an open system (i.e., one that is neither isolated, as for a molecule, nor periodic, as for bulk crystalline systems). At nanometer distances it is possible for electrons to move ballistically through a device without scattering. For ballistic transport, the relation between current (I) and voltage (V) is $I = GV$, where G is the quantum conductance. The famous Landauer formula provides a fundamental correspondence between the quantum conductance and the transmission function T (i.e., probability per conducting channel)[23]:

$$G = \frac{2e^2}{h} T \tag{2}$$

Since a metallic nanotube has two extended electron bands crossing at the Fermi level, it behaves intrinsically as an ideal two-channel ballistic conductor. The theoretical conductance is thus a constant $G = 2G_0$, with $G_0 = 2e^2/h = 12.9\ k\Omega^{-1}$. When moving away from the Fermi energy, more bands are able to contribute to the current, an effect that gives a corresponding increase in G. In reality, since a nanotube is never perfect, the propagating electrons will be scattered by lattice defects, phonons, or at contacts. This leads to an unavoidable reduction of the transmission probability and in turn of the conductance. The reduction of conductance will also occur when molecules (chemisorbed or physisorbed molecules, i.e., dopants) interact with the tube. It is clear in that case that the conductance cannot simply be evaluated from counting the bands for a given electron energy but requires an explicit calculation of the transmission function.

Practically, the transmission function can be evaluated efficiently using a Green's function and transfer matrix approach for computing transport in extended systems,[24,25] generalized for multiterminal transport.[26] This method is applicable to any general Hamiltonian that can be described within a localized-orbital basis. In the present work, to simulate actual working devices, we use a scheme where the above-mentioned quantum chemistry calculations on finite systems are recast to infinite nonperiodic systems. It is worth noting that within our approach, only coherent transport is treated.

In a system consisting of n terminals meeting at a branching region (referred to hereafter as the *conductor region*) the current flowing from the ith to the jth terminal for a set of applied voltages $\tilde{v} = V_1, \ldots, V_n$ is given by the Landauer–Büttiker formula,

$$I_{ij}(\tilde{v}) = \frac{2e^2}{h} \int dE\, T_{ij}(\tilde{v}, E) \times \left[f_0(E - \mu_i - V_i) - f_0(E - \mu_j - V_j) \right] \tag{3}$$

where f_0 is the Fermi–Dirac distribution and μ_i is the chemical potential of terminal i. The two-terminal transmission is given by

$$T_{ij} = \mathrm{Tr}(\Gamma_i G_C^r \Gamma_j G_C^a) \tag{4}$$

and must be reevaluated for each different set of applied bias. The coupling operators Γ_i between the conductor and the ith lead is expressed as a function of the advanced (a) and retarded (r) self-energy terms of the semi-infinite leads $\Sigma_i^{a,r}$:

$$\Gamma_i = i(\Sigma_i^r - \Sigma_i^a) \tag{5}$$

while the conductor Green's function G_C is obtained from

$$G_C(\varepsilon) = \left[S_C(\varepsilon) - F_C - \sum_{i=1}^{n} \Sigma_i \right]^{-1} \tag{6}$$

where $\varepsilon = \mathrm{E} \pm \mathrm{i}\eta$ is a complex energy where a small imaginary part is added (or subtracted) for the retarded (advanced) Green's function. S and F are the overlap and Fock matrices of the conductor region (see Figure 2). In the present work, the self-energies are computed efficiently using the surface Green's function matching theory with a robust iterative transfer matrix procedure.[25] The main task thus consists in truncating the Fock and overlap matrices, which are computed from all-electron density functional calculations using atom-centered localized basis functions and connecting the resulting new matrices with the corresponding

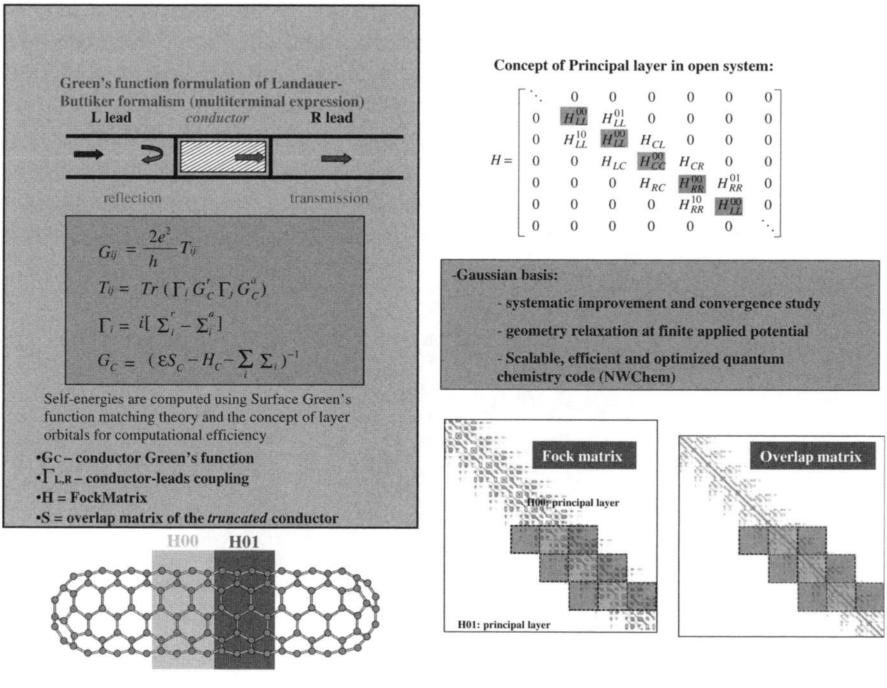

Figure 2 Division of the Fock and overlap matrices for computing the conductor Green's function.

truncated matrices obtained for a pristine system (these make up the leads). Considerable care needs to be taken to include enough basis-set-dependent buffer atoms in the initial run such that the electronic properties in the part of interest is not influenced by the finite size of the actual system.

2. RESULTS AND DISCUSSION

The position of three different organic molecules with varying ionization potentials inside a (10,10) carbon nanotube were considered during geometry optimization. Figure 3 shows one optimized system with four molecules of tetrafluorotetracyano-*p*-quinodimethane (F_4-TCNQ) inside a (10,10)-SWCNT. The energy landscape for these systems has multiple local minima for the orientation and stacking of the molecules inside the nanotube. Experimental evidence based on x-ray diffraction gives an estimated average distance of 10 Å between neighboring TNCQ molecules at a chemical concentration of C_{140}/TCNQ.[5] However, since there were no peaks observed in the x-ray diffraction data that would indicate a one-dimensional stacking of the molecules, it is likely to be randomly oriented. Indeed, we do observe local energy minima corresponding to random stacking and orientation of the molecules. These structures are stable at room temperature, as verified from molecular dynamics simulations (room-temperature stability is important to avoid de-doping). To take this into account we have considered several different orientations and positions of the molecules inside the nanotube and elucidated the effects on the electronic structure. Our main goal is to determine if charge transfer from an encapsulated molecule to the nanotube can induce efficient and stable n- or p-type doping. For this purpose we consider only the case of a single encapsulated molecule with different orientations with respect to the tube. This allows us to recast the system properties easily into an open system for the computation of the transmittance and conductance. For this, a separate calculation for perfect leads is first performed

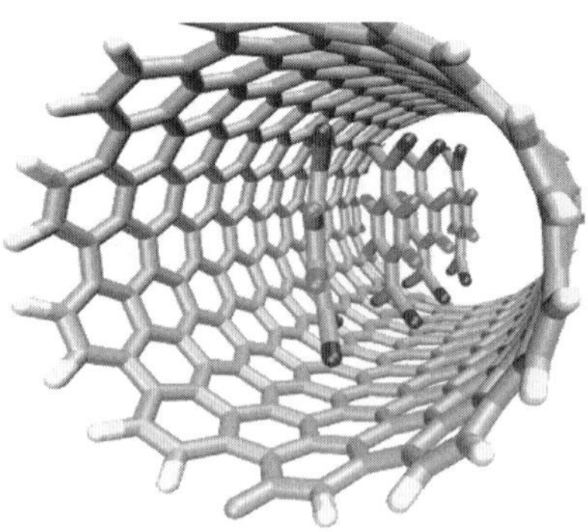

Figure 3 One possible geometry for four molecules of F_4-TCNQ inside a hydrogen-terminated (10,10) SWCNT: the average distance between the molecules is ~5 Å.

within the same basis set. The effects of additional molecules can also be examined within this same formalism.

Table 1 summarizes how the Fermi energy shifts with respect to a pristine (10,10)-SWCNT as a result of encapsulating organic molecules with different ionization energies. Clearly, molecules possessing the lower ionization energies (TDAE, TTF) tend to increase the Fermi energy [compared to that of a pristine (10,10) tube], whereas those with the highest ionization energy tend to decrease it. These changes correspond to the well-known effects related to n- and p-type doping, respectively. We verified that the trends observed are consistent with increasing basis set size (up to a triple-ζ-quality basis set) and with different orientations of the molecule inside the tube.

Direct examination of the molecular orbitals, population analysis (Mulliken and natural bond orbitals), and total charge density as defined by Eq. (1) provide further evidence of doping effects. A charge density isosurface conveniently demonstrates the process of charge transfer between the tube and the encapsulated molecules. Figure 4 shows an estimated charge density isosurface due only to charge transfer for three different molecules (blue and red represent electron depletion and accumulation, respectively). For molecules possessing the lowest ionization energies, TDAE and TTF, there is substantial transfer of electrons to the nanotube (as seen from the estimated electron density together with the results based on a population analysis that are given in Table 1). On the other hand, for the molecule possessing the highest ionization energy, F_4-TCNQ, there is significant transfer of negative charge from the nanotube to the molecule. There is also

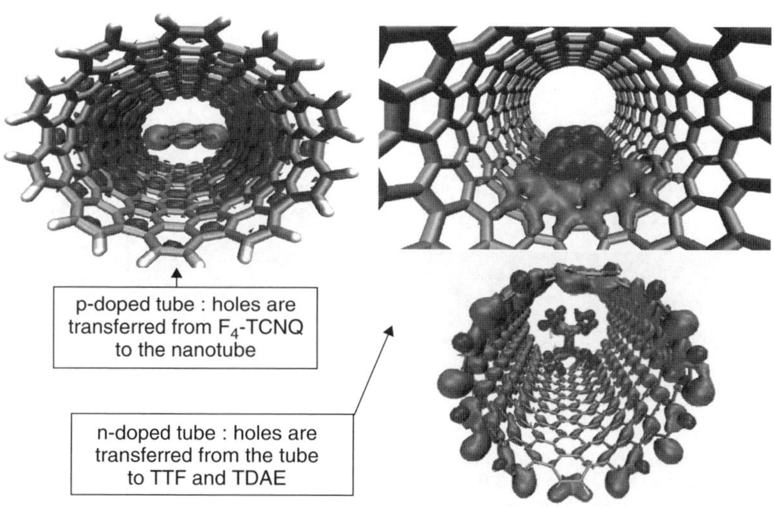

Figure 4 Charge density isosurface resulting from charge transfer as computed from DFT(LDA)/6-31G* for one molecule of F_4-TCNQ (top left panel), one molecule of TTF (top right panel), and one molecule of TDAE (bottom right panel). (*See insert for color representation figure*.)

a notable difference in the distribution of charge on the nanotube as a function of molecule type and position. For excentric TTF there is a fairly large localized charge transfer, while the charge transfer for TDAE and F_4-TCNQ, which are more oriented toward the center of the nanotube, is far more uniform. According to a population analysis, the main contribution to charge transfer for TTF occurs from the four sulfur atoms, each losing an average charge of $\sim 0.3|e|$, while the C=C bridge between the two five-membered rings gains some charge from the nanotube (also notable in Figure 4). Fundamentally, the charge distribution on the nanotube resulting from the presence of the molecule(s) is of interest since a random stacking of multiple molecules inside the nanotube can lead to molecular positions and orientations that are displaced from the center of the tube (notable in Figure 3). The results in Table 1 for three F_4-TCNQ and TTF molecules, representing a concentration of $\sim C_{140}/F_4$-TCNQ and C_{140}/TTF, show a substantial decrease in Fermi energy for F_4-TCNQ and an increase for TTF. The Fermi energy for these systems is not simply linearly related to that for a single molecule since each molecule has a different orientation and therefore has different efficiencies for charge transfer [i.e., different electronic couplings for charge transfer, the charge transfer integrals in Eq. (7) (see below)]. In addition, for a multimolecule system, charge transfer can also occur between parts of the molecules themselves. Examination of the charge density isosurface for these multimolecule systems (Figure 5) indicates a reasonably uniform charge distribution on the nanotube due to charge transfer. We can therefore expect a uniform charge distribution induced by such doping concentrations, as indicated in the experiments of Takenobu et al.[5] Overall, Figures 4 and 5 clearly show a dramatic amount of charge transfer, which agrees with the Fermi energy shifts given in Table 1. The n- and p-doping character is a function of the type of molecule that is encapsulated and correlates well with the ionization energy.

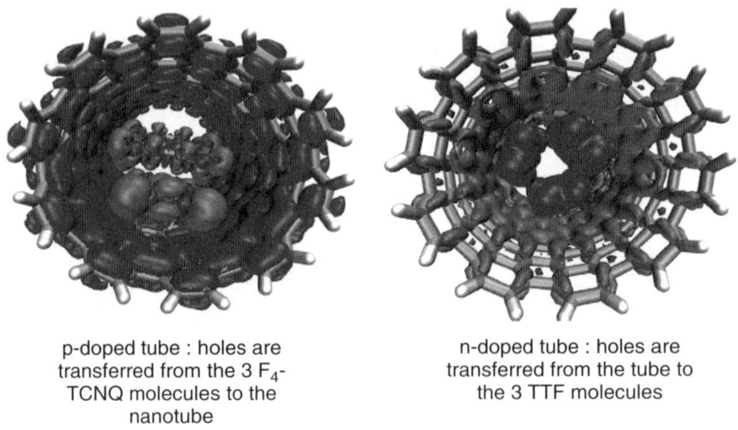

Figure 5 Charge density isosurface resulting from charge transfer as computed from DFT(LDA)/6-31G* for three molecules of F_4-TCNQ (left panel) and TTF (right panel). (*See insert for color representation figure*.)

The dynamics of the doping, in particular that due to charge transfer, can be characterized by using semiclassical electron transfer theory.[27–29] In this case the charge transfer rate K_{ij} can be described as

$$K_{ij} = \frac{4\pi^2}{h}\left(\frac{1}{(4\pi k\lambda_{ij}T)}\right)^{-1/2} t^2 \exp\left[\frac{-(\Delta E_{ij} + \lambda ij)^2}{4\lambda_{ij}kT}\right] \tag{7}$$

where T, k, h, λ, t, and ΔE are the temperature, Boltzmann's and Planck's constants, the reorganization energy (electron–phonon; can be estimated from the vibrational normal modes), the transfer integral, and the difference between the energies of the two sites i and j, respectively. Often, it is assumed that the site energies are equal, and this reduces Eq. (7) to a simpler and commonly used form. However, as discussed briefly below, this can be a rather poor approximation for many systems.

The transfer integrals reflect the strength of the interaction between molecular systems and play a central role in both the band and hopping regimes. Estimation and calculation of their value is clearly an important metric in the examination of charge transfer processes.[30–33] It is possible to compute the transfer integrals by using the matrix elements of the Kohn–Sham Hamiltonian and the spatial overlap between the highest occupied and lowest unoccupied molecular orbitals (HOMO or LUMO) on the various molecules. In our case, one is the tube and the other is the organic molecule (asymmetric system).[6] In most quantum chemistry codes, one can determine an orthonormal basis set that maintains the maximum possible character of the initial localized molecule orbitals by simply performing a symmetric transformation (i.e., like that used in a Lowden or natural orbital analysis).[34,35] Valeev et al.[32] have recently used this approach in a detailed study of charge transport parameters for systems composed of ethylene and pentacene. The expression obtained for the energy splitting is

$$\Delta E = \sqrt{(\varepsilon_1 - \varepsilon_2)^2 + 4t^2} \tag{8}$$

where the ε_i's are the site energies of the charge, t the transfer integral, and ΔE the energetic splitting between the HOMO or LUMO orbitals, depending on whether there is hole or electron transfer. When the site energies are equal, the charge transfer integral is reduced to splitting of the orbital energies as defined by Koopman's theorem. This gives a simple estimate of the charge transfer integrals corresponding to half the splitting of the HOMO and LUMO levels, where the splitting of the HOMO may be identified as two times the hole transfer integral, and the splitting of the LUMO, as twice the electron transfer integral. This type of one-electron-level approximation can unfortunately either underestimate or in some cases overestimate (electric polarization effects)[32] the transfer integrals, but the degree of this error generally decreases as the size of the molecular systems increases, and the general trends can often appear correct. However, without taking into account different site energies and spatial overlap of the HOMO orbitals of the various systems, the one-electron approach can be a dangerous approximation.

In our case, the estimated transfer integrals for the three encapsulated molecules and their different orientations and positions indicate a relatively fast charge transfer rate, generally increasing for molecular positions closer to the nanotube sidewall. There is also a strong dependence on the angular orientation with respect to the tube axis, where a certain orientation maximizes the coupling between the molecule and the nanotube.[36] This dependence suggests interesting possibilities for controlling the conductance properties by manipulation of the orientation of the encapsulated molecule. Figure 6 is the charge density isosurface for the HOMO of two different orientations of F_4-TCNQ, which illustrates that one orientation has stronger coupling between the tube and the molecule. Indeed, this interaction is clearly noted in the quantum conductivities of the two orientations.[37]

The stability of the amphoteric doping effects discussed above in air and at room temperature is very important for practical electronic device applications. Electronic structure calculations have suggested that oxygen from air physisorbs to pristine nanotubes, resulting in *p*-type doping with a binding energy determined to be in the rather broad range 0.038 to 0.25 eV.[38–46] However, the nature of the oxygen–nanotube interaction, physisorption versus chemisorption, is still a matter of some debate in the literature.[39,40] From experimental results on ultraclean nanotubes it seems clear that no chemisorption of oxygen occurs, and that physisorbed oxygen causes *p*-type doping.[47,48] We have not addressed this issue in the present case, as we are interested primarily in the doping effects of encapsulated organic molecules. On the other hand, the stability of the *n*-doped nanotubes in air at room temperature is of considerable practical importance, and we have performed some preliminary calculations[49] to determine if oxygen physisorbtion on the outside of the *n*-doped (10,10) SWCNT significantly alters the electronic transport properties. Figure 7 shows the results obtained from a spin-polarized DFT/LDA calculation for a single molecule of oxygen absorbed on a nanotube

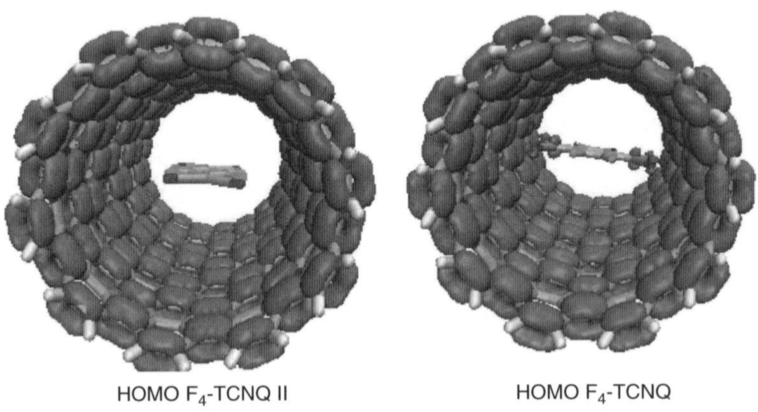

Figure 6 Charge density isosurface for the HOMO of the F_4-TCNQ molecule oriented along the tube axis (left panel) and perpendicular to the axis (right panel).

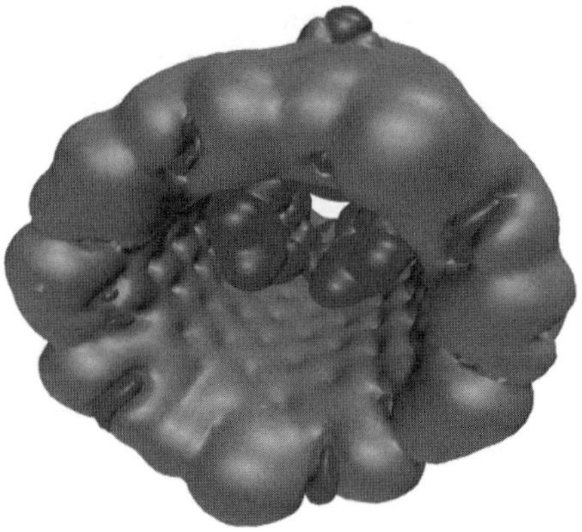

Figure 7 Charge density difference isosurface (due to charge transfer only) for one oxygen molecule absorbed on a (10,10) SWCNT with three encapsulated TTF molecules.

doped with three molecules of TTF (compare to Figure 5). Since molecular dynamics simulations at room temperature indicate that the TTF molecules remain inside the tube, this indicates that one should be able to obtain a room-temperature air-stable n-doped nanotube system.

One of the most appealing features of single-walled nanotubes for electronic device applications stems from their unique ballistic electronic transport properties. We have shown above that stable and efficient doping of carbon nanotubes can be tailored using adequate doping molecules. This indicates that complex electronic devices can be designed using a well-defined set of adsorbed agents. However, for the modified systems to be useful in a nanotube-based technology, the doping should ideally not hinder electronic transport along the system. This is not necessarily given since encapsulated molecules can be seen as scattering centers for ballistic transport channels. It follows that when seen as a mere structural defect, the doping molecule might well alter the transport property in such a way as to reduce unacceptably the transport of electrons over long distances.

To determine the ramifications of the quantum transport properties of doped-SWCNT *within the conditions of an actual working device*, the density of states, band structure, and conductance were examined. The task consists of explicitly connecting, in real space, the doped system to perfect leads made up of pristine material (i.e., a pure carbon nanotube). Since the formalism we use is based on localized orbitals, it is straightforward to remap the results discussed above into an open operational system. Using the Fock and overlap matrices from these calculations, we used the Green's function formalism to compute the transmittance and conductance that would occur for an actual nanotube-based electronic device.

As discussed in Section 1, the main task consists of *truncating* the Fock and overlap matrices, which are computed during the quantum chemistry calculation, and connecting the resulting new matrices with the corresponding truncated matrices obtained for a pristine system. Great care needs to be taken to include adequate basis-set-dependent *buffer* atoms in the initial run in such a way that the electronic properties in the *conductor* part of interest are not influenced by the finite size of the actual system. The general procedure is checked straightforwardly by comparing the band structure of the leads with the corresponding structure obtained using a periodic code. Figure 8 shows the band structure computed for the pristine nanotube based on the finite quantum chemistry calculation. The band structure is in good agreement with that obtained from a periodic plane-wave calculation. Alternatively, the density of states computed can be compared with experimental data. The examples in this chapter have been obtained using a modest 3-21G basis set, which in the largest case correspond to a total of 3000 basis functions for the transport calculation. Careful checks show that a principal layer extending over at least two periodic cells should be used to reproduce experimental data. This is due to the fact that the spatial extent of the basis functions used extends over at least one unit cell. From Figure 9 we also note that the density of states of a perfect periodic (10,10) nanotube compares very well with experimental[50,51] and tight-binding data.[52,53] For instance, the well-known metallic plateau around the Fermi energy closely matches the tight-binding plateau for a hopping parameter of 2.55 eV, in very close agreement with reported values.

Next, we have computed the transmission properties of electrons injected from a perfect semi-infinite tube (lead) through a doped nanotube section and into another semi-infinite nanotube. Density of states and quantum conductance through a nanotube n-doped with a TDAE molecule are shown on Figure 9 (top and bottom panels, respectively) and are compared to those of a pristine tube (left panel). Corresponding information for a TTF n-doped nanotube is shown in Figure 10. Interestingly, we see that in each case, the conductance spectra,

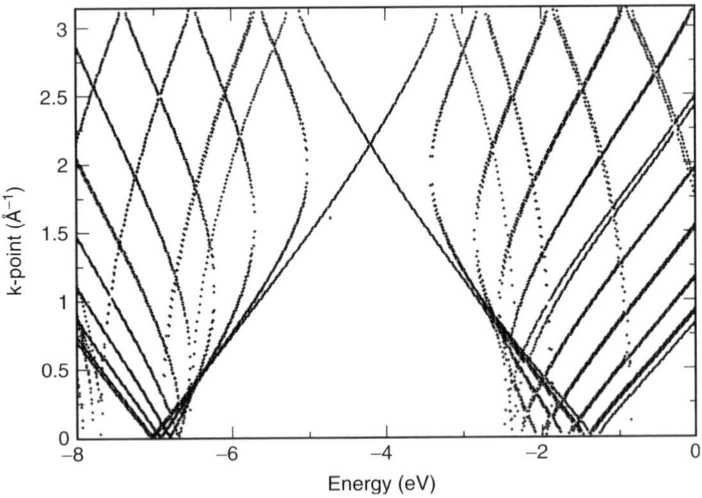

Figure 8 Band structure for a pristine (10,10) carbon nanotube computed from finite all-electron DFT calculations using atom-centered localized basis functions.

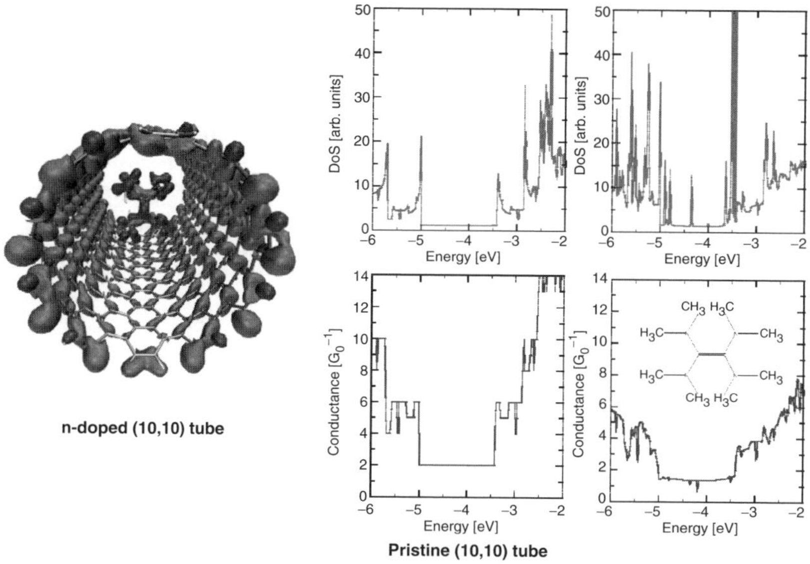

Figure 9 Charge transfer density isosurface (left panel) for TDAE inside a SWCNT as in Figure 4 along with the electronic properties (density of states, top, and quantum conductance, bottom) for a pristine (10,10) SWCNT (left panels) and a TDAE-doped case (right panels).

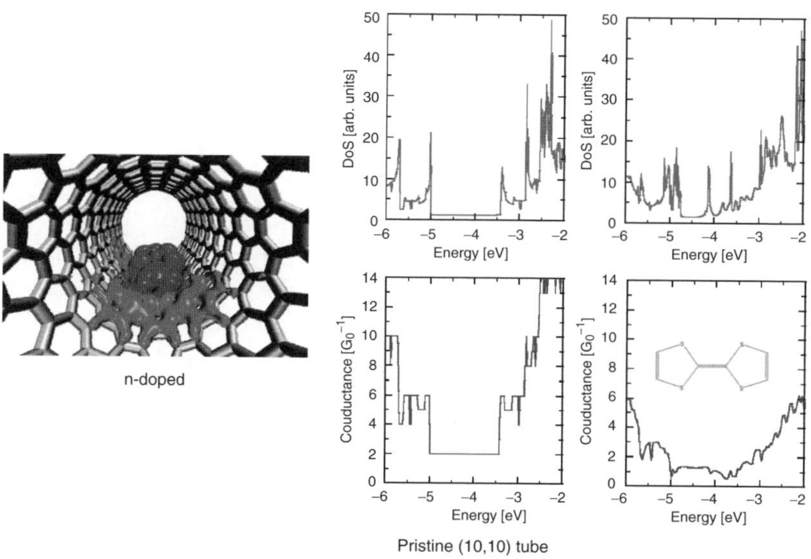

Figure 10 Charge transfer density isosurface (left panel) for TTF inside a SWCNT as in Figure 4 along with the electronic properties (density of states, top, and quantum conductance, bottom) for a pristine (10,10) SWCNT (left panels) and a TTF-doped case (right panels).

still display the same overall shape as that of a pristine tube, with a noticeable slight reduction of the plateau at Fermi energy. This reduction arises from the two lead Bloch states coming from the nanotube, which are partially reflected by the physisorbed molecule. However, the reduction is small enough that we may infer that we keep a good conductor even when electrons have to travel through a nanotube doped by the presence of a physisorbed molecule. The robustness of the eigenchannels originating from a defect-free tube and passing though the modified region constitutes a clear indication of the possibility of realizing complex carbon nanotube–based nanoelectronics by using organic molecules as dopants.

3. CONCLUSIONS

The present study complements the experimental evidence provided by Takenobu et al.[5] by showing how the electronic structure of a carbon nanotube can be manipulated by encapsulating organic molecules. Charge transfer processes induced by encapsulated organic molecules lead to efficient n- or p-type doping of a single-walled carbon nanotube. It is promising that even though a molecule can induce n- or p-type doping, it has a rather small effect on the transport properties of the doped tube compared to a pristine SWCNT. In addition to these desired properties, the process of charge transfer between the organic molecules and the SWCNT appears to reduce the susceptibility of the π orbitals of the nanotube to oxidation. Here we note that from our calculations to date, oxygen physisorbtion does not appear to neutralize the desired effects induced from TTF donor molecules (i.e., it remains an n-doped system). Coupled with room-temperature stability of the encapsulated molecules as examined by molecular dynamics simulations, this correlates well with the experimental observations of Takenobu et al.,[5] suggesting long-time air stability of n-doped nanotubes.

Acknowledgments

This research was sponsored by the Laboratory Directed Research and Development Program of Oak Ridge National Laboratory (ORNL), the Center for Nanophase Materials Science, and the Division of Materials Sciences and Engineering, U.S. Department of Energy under contract DEAC05-00OR22725 with UT–Battelle, at Oak Ridge National Laboratory. The extensive computations were performed using the resources of the National Center for Computational Sciences at ORNL.

REFERENCES

1. Aviram, A.; Ratner, M. *Molecular Electronics: Science and Technology (Annals of the New York Academy of Sciences)*, Vol. 852, New York Academy of Sciences, New York, 1998.

2. Aviram, A.; Ratner, M. A. *Chem. Phys. Lett.*, 1974, 29: 277.
3. Reed, M. A.; Zhou, C.; Muller, C. J.; Burgin, T. P.; Tour, J. M. *Science*, 1997, 278: 252.
4. Cuniberti, G.; Gagas, G.; Richter, K., Eds. *Introducing Molecular Electronics*, Springer-Verlag, New York, 2005.
5. Takenobu, T.; Takano, T.; Shiraishi, M.; Murakami, Y.; Ata, M.; Kataura, H.; Achiba, Y.; Iwasa, Y. *Nat. Mater.* 2003, 2: 683–688.
6. Meunier, V.; Sumpter, B. G. *J. Chem. Phys.*, 2005 123: 024705–024701.
7. Lu, J.; Nagase, S.; Yu, D.; Ye, H.; Han, R.; Gao, Z.; Zhang, S.; Peng, L. *Phys. Rev. Lett.*, 2004, 93: 116804–116801.
8. Kendall, R. A.; Apra, E.; Bernholdt, D. E.; Bylaska, E. J.; Dupuis, M.; Fann, G. I.; Harrison, R. J.; et al. *Comput. Phys. Commun.*, 2000, 128: 260–283.
9. Binkley, J. S.; Pople, J. A.; Hehre, W. J. *J. Am. Chem. Soc.*, 1980, 102: 939–947.
10. Hehre, W. J.; Ditchfie, R; Pople, J. A. *J. Chem. Phys.*, 1972, 56: 2257.
11. Godbout, N.; Salahub, D. R.; Andzelm, J.; Wimmer, E. *Can. J. Chem.*, 1992, 70: 560-571.
12. Allinger, N. L.; Yuh, Y. H.; Lii, J.-H. *J. Am. Chem. Soc.*, 1989, 111: 8551.
13. Tai, J. L.; Allinger, N. L. *J. Comput. Chem.*, 1998, 19: 475.
14. Lii, J.-H.; Allinger, N. L. *J. Am. Chem. Soc.*, 1989, 111: 8576.
15. Fratini, A. V.; Baker, K. N.; Resch, T.; Knachel, H. C.; Adams, W. W.; Socci, E. P.; Farmer, B. L. *Polymer*, 1993, 43: 1571.
16. Nevins, N.; Lii, J.-H.; Allinger, N. L. *J. Comput. Chem.*, 1996, 17: 695.
17. Claes, L.; Deleuze, M. S.; Francois, J.-P. *J. Mol. Struct. (Theochem)*, 2001, 549: 63.
18. Maseras, F.; Morokuma, K. *J. Comput. Chem.*, 1995, 16: 1170.
19. Humbel, S.; Sieber, S.; Morokuma, K. *J. Chem. Phys.*, 1995, 105: 1959.
20. Svensson, M.; Humbel, S.; Froese, R. D. J.; Matsubar, T.; Sieber, S.; Morokuma, K. *J. Phys. Chem. B.*, 1996, 100: 19357.
21. Moller, C.; Plesset, M. S. *Phys. Rev., A* 1934, 10: 1927.
22. Pople, J. A.; Binkley, J. S.; Seeger, R. *J. Quantum Chem. Symp.*, 1976, 10: 1.
23. Landauer, R. *Philos. Mag.*, 1970, 21: 863.
24. Nardelli, M. B.; Fattebert, J. L.; Bernholc, J. *Phys. Rev. B*, 2001, 64.
25. Nardelli, M. B. *Phys. Rev. B*, 1999, 60: 7828–7833.
26. Meunier, V.; Nardelli, M. B.; Bernholc, J.; Zacharia, T.; Charlier, J. C. *Appl. Phys. Lett.*, 2002, 81: 5234–5236.
27. Marcus, R. A. *Rev. Mod. Phys.*, 1993, 65: 599.
28. Marcus, R. A.; Sutin, N. *Biochim. Biophys. Acta*, 1985, 811: 265.
29. Malagoli, M.; Bredas, J. L. *Chem. Phys. Lett.*, 2000, 327: 13.
30. Bredas, J. L.; Calbert, J. P.; Da Silva Filho, D. A.; Cornil, J. *Proc. Natl. Acad. Sci. USA*, 2002, 99: 5804.
31. Bredas, J. L.; Belijonne, D.; Coropceanu, V.; Cornil, J. *Chem. Rev.*, 2004, 104: 4971.
32. Valeev, E. F.; Coropceanu, V.; Da Silva Filho, D. A.; Bredas, J. L. *J. Am. Chem. Soc.*, 2006. Submitted.
33. Senthilkumar, K.; Grozema, F. C.; Bickelhaupt, F. M.; Siebbeles, L. D. A. *J. Chem. Phys.*, 2003, 119: 9809.

34. Lowdin, P. O. *J. Chem. Phys.,* 1950, 18: 365.
35. Reed, A. E.; Curtiss, L. A.; Weinhold, F. *Chem. Rev.*, 1988, 88: 899.
36. Meunier, V.; Sumpter, B. G. *Nanotech*, 2007, 18: 424032.
37. Meunier, V.; Kalinin, S. V.; Sumpter, B. G. *Phys. Rev. Lett.*, 2007, 98: 056401.
38. Giannozzi, P.; Car, R.; Scoles, G. *J. Chem. Phys.*, 2003, 118: 1003.
39. Ricca, A.; Drocco, J. A., *Chem. Phys. Lett.*, 2002, 362: 217.
40. Ricca, A.; Bauschlicher, C. W. J. *Phys. Rev. B*, 2003, 68: 035433.
41. Zhao, J.; Xie, R.-H. *J. Nanosci. Nanotech.*, 2003, 3: 459.
42. Zhao, J.; Buldum, A.; Han, J.; Lu, J. P. *Nanotechnology*, 2002, 13: 195.
43. Jhi, S.-H.; Louie, S. G.; Cohen, M. L. *Phys. Rev. Lett.*, 2000, 85: 1710.
44. Peng, S.; Cho, K. *Nanotechnology,* 2000, 11: 85.
45. Sorescu, D. C.; Jordan, K. D.; Avouris, P. *J. Phys. Chem. B*, 2001, 105: 11227.
46. Zhu, X. Y.; Lee, S. M.; Lee, Y. H.; Grauenheim, T. *Phys. Rev. Lett.*, 2000, 85: 2757.
47. Collins, P. G.; Bradley, K.; Ishigami, M.; Zetti, A. *Science* 2000, 287: 1801.
48. Ulbricht, H.; Moos, G.; Hertel, T. *Phys. Rev. B*, 2002, 66: 075404.
49. Meunier, V.; Sumpter, B. G. 2005.
50. Odom, T. W.; Huang, J. L.; Kim, P.; Lieber, C. M. *Nature,* 1998, 391: 62–64.
51. Wildoer, J. W. G.; Venema, L. C.; Rinzler, A. G.; Smalley, R. E.; Dekker, C. *Nature*, 1998, 391: 59–62.
52. Charlier, J. C.; Lambin, P. *Phys. Rev. B*, 1998, 57: R15037–R15039.
53. White, C. T.; Mintmire, J. W. *Nature*, 1998, 394: 29–30.

4

USING ORDER AND NANOCONFINEMENT TO TAILOR SEMICONDUCTING POLYMERS: A COMBINED EXPERIMENTAL AND MULTISCALE COMPUTATIONAL STUDY

Michael L. Drummond and Bobby G. Sumpter
Computer Science and Mathematics Division, Oak Ridge National Laboratory, Oak Ridge, Tennessee

Michael D. Barnes
Department of Chemistry, University of Massachusetts, Amherst, Massachusetts

William A. Shelton, Jr. and Robert J. Harrison
Computer Science and Mathematics Division, Oak Ridge National Laboratory, Oak Ridge, Tennessee

Poly(*p*-phenylene vinylene) (PPV) and its derivatives (see Figure 1 for some common examples) represent one of the most studied families of the semiconducting organic polymers first discovered by Nobel laureates Shirakawa, MacDiarmid, and Heeger 30 years ago.[1,2] As with all such polymers, the PPV framework provides a means to combine the useful characteristics of traditional inorganic semiconductors with the cost-effectiveness and improved flexibility of organic polymers, in both a literal sense and in the sense of their more versatile and tunable syntheses. Indeed, the list of applications, both potential and already realized, reflects the impact these chemical systems have had and are continuing

Multiscale Simulation Methods for Nanomaterials, Edited by Richard B. Ross and Sanat Mohanty

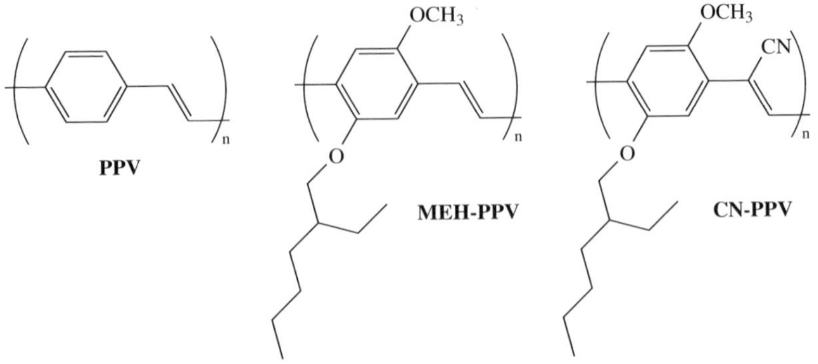

Figure 1 Three commonly used poly(*p*-phenylene vinylene) monomer units.

to have, and also speaks to the tremendous amount of ongoing collaborative research. Some of these applications include light-emitting diodes (LEDs),[3–6] photovoltaics,[7,8] photodiodes,[9] transistors,[10] solid-state lasers,[11] chemical sensors,[12] molecular wires,[13] storage media,[14] and quantum computing.[15] Essentially all of these uses are dependent, to varying extents, on the ability of these polymers both to transport a charge[16] and to absorb and/or emit light in the visible spectrum.

Unfortunately, a great deal of research has revealed that adjusting the properties of a polymer to optimize one of these beneficial characteristics often results in a diminution of the other. That is, a PPV-based film can either fluoresce brightly or conduct readily, but cannot, in general, do both.[17] A great deal of research has indicated that the primary cause of this fundamental dichotomy is polymer morphology: the three-dimensional structure of the individual polymer molecules.[18] The polymer arrangement influences both *intrachain communication*, through the bonds of the conjugated backbone, and *interchain communication*, through the space in between neighboring polymer strands. In principle, the latter can be further separated into interactions between two different polymer molecules and between two conjugated segments of the same molecule separated by a small ($\approx$3 to 5 Å) vacuum; in practice, this distinction has proven elusive. Species where π-electron density is delocalized through space across multiple strands are termed *aggregates* if the attraction holds in both the ground and excited states, whereas *excimers* are species that are dissociative in the former state but attractive in the latter.[19] (However, it should be noted that use of the word *aggregate* in the literature often reverts to the layperson's definition, thereby including both species.) The presence of aggregates causes the mobility/luminescence balance mentioned above, as they enhance carrier mobility by complementing a polymer's standard intrachain conductance with an interchain transmission mechanism. These aggregates, however, possess long-lived, low- or nonemissive excited states,[18] and thereby quench fluorescence and serve to diminish the potential of applications such as LEDs.

The nature of aggregated species has been the subject of intense research, particularly among the Schwartz group at UCLA.[17–21] Briefly, this research has determined that the most commonly utilized derivative of PPV, poly(2-methoxy-5-(2′-ethylhexyloxy)-1,4-phenylene vinylene) (MEH-PPV; Figure 1), assumes vastly different morphologies in solution, depending on the solvent. A "good" solvent for MEH-PPV, defined in their research as chlorobenzene (but see below for further discussion), causes the polymer to assume an elongated morphology generally termed a *defect coil*.[22] This structure has few opportunities for self-aggregation, but experiences enhanced through-space communication between separate molecules. In contrast, a "bad" solvent such as tetrahydrofuran (THF) causes MEH-PPV to assume a coiled and much more compact form. This coiling causes increased aggregation between segments on the same molecule, but this smaller molecular volume leads to a decrease in aggregation between different polymer molecules. The latter type of aggregation dominates in dilute solution, and therefore MEH-PPV possesses a greater degree of aggregation in dilute chlorobenzene solution than in dilute THF solution, as evidenced by the signature red shift of aggregated species in both absorption and emission spectra. In contrast, in single-molecule spectroscopy studies,[23] self-aggregation is by definition the only mechanism operative, and thus the bad solvent, toluene, shows a stronger red shift than that of the good solvent, chloroform. A third environment, and the one most relevant to industrial applications, is that of thin films cast from solution. As such films are made, the solution-phase morphology is largely retained, and thus chlorobenzene-cast films possess a significant low-energy tail, especially compared to the smaller tail found in THF-cast films.

Given the importance of interchain interactions, the pertinent question is how to manipulate their prevalence to produce the desired result, which varies depending on the application. There is a large body of literature, summarized in this chapter, describing myriad approaches to control the degree of solution-phase aggregation and thereby optimize the performance of created devices. Complementary to these efforts are studies utilizing recent and powerful advances in single-molecule spectroscopy, which elucidate the fundamental photophysics of PPV-based polymers entirely separate from the influence of intermolecular aggregation. These findings are significant and detailed[24–29] but for the purposes of this discussion, the most important result is the confirmation of the heuristic model of the "energy funnel."[30] In this picture, PPV is described as a long chain of chromophores of varying lengths; conjugation is not continuous through the entire molecule, but rather is disrupted either through defects introduced in synthesis[31] or through significant intermonomer torsion ($\approx 40^\circ$).[32] The variety of conjugation lengths among the chromophores provides an effective antenna, spanning much of the visible spectrum. As light is absorbed, the exciton created is transmitted along the conjugated chain via a through-bond Dexter energy transfer mechanism[33] as well as via a through-space Förster energy transfer mechanism,[34] with the latter generally regarded as faster than the former for MEH-PPV.[35] Energy is transferred to low-lying segments possessing longer conjugation lengths until a minimum is found and either emission occurs or energy is dissipated nonradiatively. In this description, the degree of aggregation is crucial in a number of ways: (1) highly ordered regions are intrinsically low in energy and thus are likely emission sites, (2) the increased order at such sites tends to favor longer conjugation lengths low in energy, and (3) effective Förster energy transfer is contingent upon well-organized through-space interactions. These effects are especially important in the crowded films used in most device applications.

Given this model of the energy funnel, and given that the extent of its functioning depends on the degree of aggregation, it is necessary to develop approaches whereby polymer–polymer interactions can be controlled and modified to suit the application at hand. Such approaches, which have been thoroughly explored, range from immediately obvious tactics such as adjusting temperature, solvent, and polymer concentration to more exotic controlling mechanisms such as dendrification or pore confinement. In the next section, a number of different approaches are reviewed. No judgment is made concerning the efficacy of any approach; only through attempted implementation will industry develop the method or combination of methods that will prove to be easy, scalable, reproducible, and cost-effective. Following this review, both experimental and computational work performed in our laboratory involving nanoconfined and oriented semiconducting polymers is detailed. Our data, along with indications emerging in the literature,[36–41] seem to show that a direct correlation between aggregation and poor luminescence efficiency is overly simplistic; the situation appears to be complicated by the degree of order in the aggregated domains. Thus, there is hope that the trade-off between effective charge mobility and strong emissive properties can be ameliorated, thereby leading to greatly improved devices fabricated from semiconducting organic polymers.

1. CONTROLLING THE DEGREE OF AGGREGATION

The following review is not intended to be comprehensive, neither in the array of techniques addressed nor in the listing of researchers working on each methodology. Further references can, of course, be found in the representative citations given below. In addition, the influence of factors such as solvent, temperature, and polymer concentration is commonly investigated in almost all of the studies cited below, the relative effects of which depend on the main variable being investigated. Finally, it should be stressed that many of these papers are also interested in basic research into the fundamental photophysics of PPV-based polymers and are often significant in that respect as well.

Polydispersity of Polymer Molecular Weights

Like most polymers, PPV and its derivatives do not possess well-defined molecular weights, due to the stochastic nature of polymer synthesis and chain termination events. Thus, a given polymer sample contains chains of varying molecular weights, and samples also tend to vary depending on the chemical supplier or synthesis methodology used (see below). It was found[42] via measurement of fluorescence intensity that, presumably because of their structure and degree of aggregation, some individual polymer molecules in the ensemble afforded quantum yields far surpassing that of the bulk. Thus, if these strands could be separated, or if synthetic conditions could be modified to produce such strands preferentially, there would be obvious ramifications for the performance of devices, particularly in organic LED applications.

Polymer Synthesis

The synthesis of MEH-PPV occurs mainly through the *Gilch route*, which is increasingly being supplanted by the *Horner route*.[43] The former possesses well-documented conjugation defects,[31] whereas the latter produces much more regular polymers. Despite this difference, fluorescence quantum yields of films formed from each type of polymer are similar, as are traveling-wave laser behaviors. The largest practical difference determined between the two polymers is the ease with which high-quality optical films can be formed from the cleaner Horner polymers, which makes them more suitable for potential industrial applications in that respect. However, the level of aggregation, which directly influences quantum yield, does not appear to vary significantly with respect to the method of synthesis.

Deliberate Control of the Prevalence of Conjugational Defects

Given the changes in behavior mentioned above that depend on the prevalence of conjugation breaks, it follows that defects can be incorporated intentionally to modify polymer characteristics and degree of aggregation. In one such study[44] it was determined, via analysis of fluorescence spectra, that reduced quantum yields are caused by both an intrachain energy transfer mechanism to segments with long conjugation lengths and by an interchain energy transfer mechanism to weakly emissive aggregates. Furthermore, the relative importance of these two mechanisms depends directly on the amount of deliberate conjugation defects introduced in synthesis, with a high degree of conjugation leading to the dominance of aggregate-based fluorescence quenching. This effect is exacerbated when poor solvents for MEH-PPV are used. Similar results were found using single-molecule fluorescence,[45] although under such conditions it was determined that intrachain energy transfer is inefficient at both high and low degrees of conjugation.

Another control of defect concentration was afforded through postsynthesis heating.[46] This approach not only disrupts aggregation but also serves to selectively remove *cis* linkages in the MEH-PPV backbone, as indicated by ^{1}H-NMR. Decreased absorbance, increased photoluminescence, and overall greatly improved electroluminescent behaviors were found.

Modification of the Polymer Backbone

A number of studies have investigated the effects of incorporating different chemical moieties into the conjugated polymer backbone. In one such study[47] a segmented block copolymer was formed between a PPV-like trimer unit and a nonconjugated hexanedioxy unit. Through measurement of emission spectra it was concluded that a flexible alkyl chain prevents the formation of excimers in solution but not in thin films. In another study[48] the phenylene–vinylene backbone was interrupted with phenylene–ethynylene and anthracene–ethynylene conjugated units. Although a study of the effects of aggregation was complicated by different side-group substitutions brought about by synthetic circumstances (see below), it is clear that these backbone substitutions provide a means to shift both absorption and emission spectra. Similar changes have been found in the incorporation of, for example, pyrazine[49] or oxadiazole[50] units into the PPV backbone.

Modification of Polymer Substituents

One of the potentially most powerful and tunable ways to influence polymer properties as well as aggregation characteristics is afforded via substitution of the PPV backbone, both at various positions on the phenyl ring and at both vinyllic positions. Such changes are readily accomplished by changing the identity of the monomer units prior to polymerization. One of the more successful changes involves the incorporation of a cyano group (CN-PPV; see Figure 1) into the vinyl moiety.[49,50] This strong electron-withdrawing group serves to increase the electron affinity of the conjugated backbone,[51] thereby improving electron injection characteristics. Our own research, detailed later, takes significant advantage of this characteristic of CN, as well as the effect this group has on interchain electrostatic attraction.

Backbone substitution with alkoxy groups, such as in MEH-PPV, are common to the point of near ubiquity, as they ameliorate the poor solubility of unmodified PPV. Careful investigations[48] of photoconductivity, absorption, and emission characteristics have revealed that the length, branching, and relative locations (i.e., whether chains are on the same side of the polymer from monomer to monomer) of these alkoxy groups are all influential, often by modifying the propensity to form aggregates. A similar investigation[52] involving the *ortho*, *meta*, or *para* substitution of a phenoxy side group attributes changes in photo- and electroluminescence to disruption of conjugated backbone planarity; the influence on the degree of aggregation was not considered but is also likely to play a role.

The final substituent modification noted involves the addition of a crown-ether group to PPV.[53] In addition to modifying photoluminescence properties ipso facto, these entities provide a means to further influence photophysical behavior via subsequent addition of metal salts, which generally quench fluorescence. It was conjectured that these PPV-substituted crown ethers, as well as their metallated counterparts, promote aggregate formation and therefore improved charge mobility.

Symmetry of Side-Group Substituents

A less obvious substituent effect concerns changing not the identity of the substituent, but rather the placement and substitution patterns of these groups. For example, a PPV derivative called MDMO-PPV possesses both a long and a short alkoxy group (similar to MEH-PPV), which are usually randomly oriented with respect to each other from monomer to monomer (i.e., either up or down in Figure 1). However, if regioselective MDMO-PPV is synthesized,[54] greatly improved charge transport is noted, attributed mostly to improved interchain aggregation. Additionally, symmetric substitution (i.e., identical up and down groups in Figure 1) is also noted[55] to promote interchain aggregation, whereas asymmetric patterns generally lead to self-aggregation, as noted in both photoluminescence spectra and atomic force microscopy measurements.

Modification of Polymer End Groups

Another possible polymer modification involves putting end groups on the conjugated backbone. One such study[56] added the dye coumarin to the ends of MEH-PPV and studied fluorescence and fabricated device performance. Although emissive properties were modified significantly, aggregation characteristics were apparently unchanged compared to MEH-PPV. In contrast, another study[57] capped one end of PPV with a hydrophilic poly(ethylene glycol) group and the other end with a hydrophobic alkyl group, thereby promoting self-assembly. In these systems, the degree of aggregation, and therefore charge mobility and fluorescent properties, could be controlled indirectly by varying the length of the hydrophilic group.

Use of Dendrimers

Two different approaches utilizing dendrimers can be envisioned to control the degree of aggregation. In the first,[58] dendritic substituents are added to a PPV backbone, with the goal of eliminating intermolecular aggregation. This goal was achieved, as demonstrated by photoluminescence studies of solution, film, and single-molecule environments, for in-plane interactions, but neither inter- nor intramolecular $\pi-\pi$ interactions were curtailed successfully. In the second approach,[59] the core was not PPV but rather poly(propyleneimine); PPV served as the dendritic branches. Absorption and fluorescence spectra indicate that aggregation is significant and depends strongly on the dendrimer generation number.

Use of Liquid-Crystal Environments

Similar to the dendrimer and self-assembly techniques just described, the organization afforded by liquid-crystalline solvents has a drastic effect on the degree of aggregation. For the PPV-related polymer poly(*p*-phenylene ethynylene),[33] a liquid-crystal solvent promotes intrachain exciton migration and hinders both aggregate formation and interchain charge migration. A more tunable aggregation effect is noted for a water-soluble PPV derivative, poly(sodium phenylenevinylenedicarboxylate),[60] when it is placed in a liquid-crystal network.

Use of Polymer Blends

On a macroscopic scale, PPV-based polymers can be blended with other polymers, such as the inert polystyrene,[61] a different luminescent polymer (e.g., polyfluorene[62]), or with other PPV-based polymers.[63] In the first case, essentially a very dilute solution of MEH-PPV results, with very little aggregation, as evidenced through photo- and electroluminescence studies. Similar results are also noted with the second approach, although with improved electrical properties, due to the conductive nature of the non-PPV polymer. In the third methodology, a MEH-PPV film cast from THF solution, known[18] to have few aggregates, is sandwiched in an LED device between MEH-PPV films cast from chlorobenzene solution, where aggregates are prevalent. The resulting device shows indications of combining the efficient luminescence of the middle layer with the improved charge transport of the outer layers.

Use of Pore Confinement

The use of, for example, silica-based pores[35,64] represents, in a sense, a case in the study of confinement far more extreme than in the more flexible liquid crystal and copolymer matrices described above. Control over not only the degree of aggregation, but also the direction of emission, can be afforded through the use of pores of one size[35] or even of different sizes.[64] In the latter situation, the polymer, as evidenced by photoluminescence and polarized photoluminescence measurements, was determined to be aggregated and coiled in large pores, aggregated and uncoiled in medium pores, and unaggregated in small pores.

Mixtures of Good and Poor Polymer Solvents

Extensive work[44,65–67] has been done on systems where “poor” solvents are added incrementally to a well-solvated PPV chain, to observe the resulting increase in coiling and/or aggregation. Often, these studies are combined with variable temperatures and concentrations, and in this brief review we do not attempt the formidable task of decoupling the various effects. However, we do wish to highlight that the definition of good and poor solvents is not clear a priori. Indeed, as demonstrated,[66] the position of the maximum absorption peak, which can be taken as a proxy for effective conjugation length, does not correlate with solvent polarity, except for the series of linear alcohols. Such a simple correlation is presumably muddled due to differing solubilities of the two polymer components: the aromatic backbone and the usually aliphatic sidechains. Thus, the propensity for a polymer to aggregate may depend on immiscibility between these two moieties of PPV-based systems,[37] rather than an overall degree of solubility.

Applied Electric Field

The use of PPV-based polymers in LED devices is among the most promising of possible applications, and therefore it is no surprise that the effect of the electric field applied to such devices has been investigated.[68] More relevant to the present discussion is the fact that electroluminescence of MEH-PPV is dominated by contributions from aggregated species at low electric field strength, but not at high electric fields. Higher field emission is blue-shifted compared to the aggregate emission, indicating a lack of aggregation, and thus the applied field affords a means to select for or against interchain species.

Applied Pressure

MEH-PPV blended with a variety of other polymers was subjected to pressures of up to 65 kbar.[69] Aggregates were indicated to be more prevalent at high pressure, as inferred from reversibly decreased emission intensities. However, intramolecular energy transfer was also postulated to be of increased importance at elevated pressures. Moreover, matrix effects of the "solvent" polymer were found to be significant as well, hindering simple deconvolution of the effect of pressure on luminescence and charge transport properties.

Solvent Evaporation Rate

An intriguing study[70] recently found that slow solvent evaporation ("slow" was unfortunately undefined) of MEH-PPV in a variety of solvents produced films with significant red photoluminescence, indicative of significant aggregation, but without the decreased emission intensity usually regarded as a signature of interchain congregation. The researchers speculate that techniques typically employed to generate films, such as spin casting, produce films from a solution that has not been properly equilibrated. It is, by this logic, this lack of equilibration that contributes to poor luminescence efficiency rather than aggregation in and of itself. This finding is in line with the references cited in the introduction, as well as our own results, and is discussed in greater detail below.

Variables in Spin Casting in Film Production

Spin casting is the most widely used technique for producing the thin films most needed for device applications from polymer solutions.[71] Thus, it is no surprise that conditions such as spin speed[71,72] and atmosphere used during the process[73] are attractive ways to control film morphology and the degree of aggregation. In addition, through investigations of photo- and electroluminescent behavior, these conditions may be optimized to produce films that, similar to those described above produced using the slow-solvent evaporation technique, are clearly aggregated but do not have the diminished fluorescence typically found with such species.[71]

Relevance

Beyond serving as a catalog of experimental procedures for modifying PPV-based devices, the last two means of influencing aggregation discussed above, in particular, offer evidence of situations and conditions where the typical red shift associated with aggregation is seen without a concomitant decrease in luminescence efficiency. Similar results have been found, for example, with aggregates of 2,5-dioctyloxy-PPV[74] and with alkylated polyfluorenes.[75] It seems that a common theme in these situations is the presence of order, either observed explicitly by techniques such as x-ray diffraction, neutron and x-ray scattering, or electron microscopy; inferred via the use of differential scanning calorimetry; or assumed, as with lengthy time scales, affording thermodynamic rather than kinetic minima. Thus, as order appears to be a key ingredient for achieving the sought-after balance between charge mobility and desirable luminescent properties, it is worthwhile at this point to review the various indications of order in PPV and its derivatives.

2. EVIDENCE OF ORDER IN PPV AND ITS DERIVATIVES

One of the earliest experimental indications of order in PPV-based films was found via x-ray and electron diffraction of stretched films of unsubstituted PPV and 2,5-dimethoxy-PPV.[76] Order was increased in the latter due to an interdigitation of methoxy groups in parallel coplanar neighboring chains. Electron diffraction later[77] showed indications of order in both unstretched and stretched PPV, as well as derivatives with 2,5-substitution of methoxy, ethoxy, and methyl groups. Order for the unstretched films decreased as unsubstituted > methoxy > methyl > ethoxy; stretching led to the switching of the relative positions of unsubstituted and methoxy-substituted PPV. MEH-PPV was also shown,[78] using x-ray diffraction, to possess domains of nanocrystallinity, the structures of which were similar to that described above for 2,5-dimethoxy-PPV. THF-cast films were shown to have both more and larger domains of order than those cast from chlorobenzene or *p*-xylene. Significantly, these THF-cast films proved to be the best for polymer-based laser applications.

The presence of order in MEH-PPV, as created by fine-tuning solvent, concentration, and spin-speed conditions, was also invoked to describe the efficient photoluminescent behavior of aggregated films mentioned in Section 1.[72] More specifically, speculation was offered about the relative orientations of the long, branched alkoxy groups on parallel, coplanar polymer strands. It was reasoned that in aromatic solvents miscible with the conjugated backbone, neighboring polymers would preferentially direct their long sidechains toward each other, whereas nonaromatic solvents would tend to promote a twisted geometry, to shield the aromatic backbone from the poor solvent. Spin speed was noted to be a determinative factor, especially in the former situation, as centrifugal force can be expected to disrupt any weakly bound arrangement. Concentration effects were also noted, with films cast in a concentration range termed "concentrations for loose aggregation" proving especially effective in producing efficient LED devices.

The two spectral signatures identified in this study—emission at 640 nm with high quantum efficiency and at 700 nm with low quantum efficiency—were also found in films prepared with the slow solvent evaporation technique.[70] In fact, although the latter research did not note the similarity with the work just described, comparison of, for example, Figure 4 of ref. 72 with Figure 2 of ref. 70 shows that similar species are probably responsible for the efficient luminescence noted in both studies. Although the slow solvent evaporation researchers claim that only their technique, presumably ill-suited for rapid device production, affords the well-equilibrated environment necessary for strongly luminescent aggregates, this environment can apparently be approximated, if not duplicated exactly, through appropriate control of conditions in spin casting.[72]

While the aggregated species postulated by these two studies appear to be similar, if not identical, a host of techniques, including polarized light micrographs, electron and x-ray diffraction, differential scanning calorimetry, and electron micrographs, were used to characterize a different ordered domain, termed *mesomorphous*, in MEH-PPV films.[79] This ordered domain was produced as films were annealed to between 130 and 180°C; order was removed above 290°C. The structure proposed for this nanodomain is shown in Figure 2. Briefly, it consists of "boardlike entities" 16 Å thick with about 4-Å spacing between the conjugated backbones. This structure in turn forms "beadlike" domains about 100 to 200 Å in size, conjectured to consist of single MEH-PPV molecules. These beads, in turn, form chains of length about 2000 Å. Although emission was observed at 640 nm, which was earlier[72] shown to be a signature of the efficiently luminescent aggregate, the work described here, in contrast, found the usual decreased luminescence efficiency associated with aggregated species. Additional considerations suggest

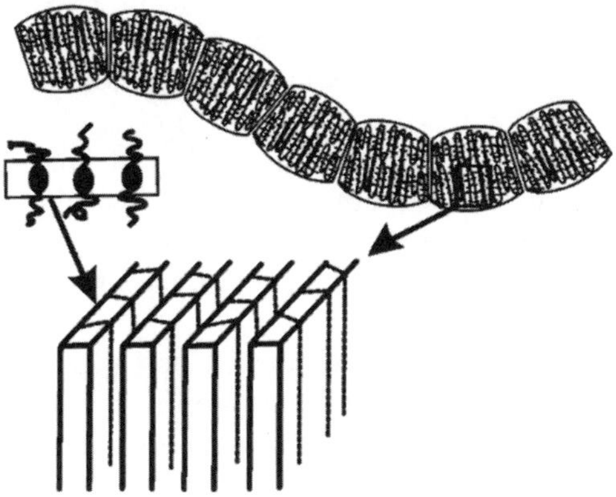

Figure 2 Mesomorphically ordered domain found in MEH-PPV after annealing a thin film. (Reprinted with permission from ref. 79. Copyright © 2004, American Chemical Society.)

that this 640-nm emission is probably a combination of aggregated emission and vibronically coupled single-chromophore emission.

Further studies by the same researchers investigated the relative roles of interbackbone attraction (i.e., $\pi-\pi$ stacking) and amphiphilicity (i.e., the differing solubilities of the aromatic core and the generally aliphatic side chains) in forming ordered domains. Cyano substitution[37] was used to investigate the influence of the former effect, due to its ability to increase electrostatic attraction between stacked chains (see below). For the latter effect,[80] long unbranched octyloxy chains were use to increase the disparity in solubility between the two parts of the polymer. In both cases, order was found following annealing. The latter system possesses a structure nearly identical, for our purposes, to that shown in the bottom half of Figure 2 for MEH-PPV. The former system also forms a lamellar mesophase structure, although the interaction between the CN group on one chain and a phenylene ring on a neighboring chain leads to a lateral shift from chain to chain. However, no beads or string of beads were reported.

Differences in photoluminescence between these two extreme cases were found to be substantial. In fact, a clear distinction was noted between *aggregate emission* in the CN-substituted system and *excimer emission* in the octyloxy system, using the explicit definitions noted in the introduction. In addition, and particularly relevant for this discussion, whereas the excimer emission intensity (and thus photoluminescence efficiency) was found to be independent of the degree of order indicated by x-ray diffraction patterns, aggregate emission intensity clearly increased as the degree of order increased.

Finally, two additional ordered domains are noted here, although the lack of published luminescence spectra prohibits the establishment of correlations, if any, between species mentioned previously. First,[40] disklike domains of nanoscale order were noted for MEH-PPV in dilute toluene solution using small-angle neutron scattering. It is postulated that such species could serve as sites of nucleation for further ordering as concentration is increased. In a second study,[39] order is observed in films of MEH-PPV without heating, similar to the earlier behavior discussed for annealed films.[79] Both wide- and small-angle x-ray scattering are used to propose the structural model shown in Figure 3. No mechanism is offered to explain the correspondence between the models of Figure 3 (before heating) and Figure 2 (after heating). It is noted, however, that the angle of orientation with respect to the surface normal (β in Figure 3) becomes random after heat is applied. Furthermore, evidence shows that this diminishment of order decreases the ability of a film to carry a charge.

To summarize, the literature shows the existence of a number of ordered domains, with differences in size, ordering motifs, preparation conditions, origins, environments, relevance, and luminescent properties. Clearly, additional research is required to clarify the nature of aggregated species in PPV-based polymers. In the remainder of this work we discuss our efforts to address this issue through both experimental techniques and in complementary computational investigations.

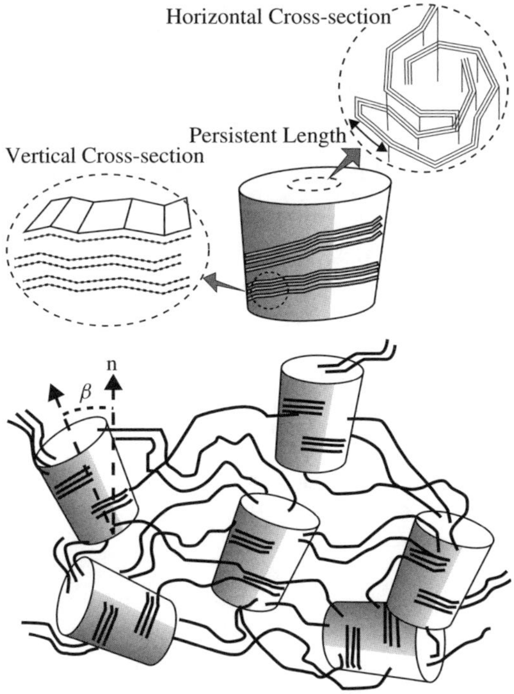

Figure 3 Mesomorphically ordered domain found in MEH-PPV before annealing of the thin film. (Reprinted with permission from ref. 39. Copyright © 2005, American Chemical Society.)

3. EXPERIMENTAL EVIDENCE OF ORIENTED MEH-PPV AND CN-PPV NANOSTRUCTURES

Over the past few years, we have employed inkjet printing methods to produce polymer-containing microdroplets that form isolated, oriented, and highly ordered nanostructures on a glass substrate.[81–85] The solvent molecules in these < 5-μm droplets evaporate rapidly, which causes collapse into a highly organized structure. In fact, the solvent leaves the polymer molecule on time scales much faster than those typical of intrachain organization,[86] and thus the initial solution-phase conformation (controlled by the choice of solvent) has a tremendous impact on the resulting unsolvated nanostructure. Therefore, poor solvents that promote a compacted polymer in solution, such as toluene, result in nanostructures that show a great degree of compression, whereas solvents that cause the polymer to be extended in solution, such as dichloromethane, lead to much less ordered nanostructures.[84]

That the nanostructure possesses significant order is indicated indirectly by a variety of techniques. The presence of orientation is clearly revealed through the use of high-resolution fluorescent imaging, the results of which are shown in

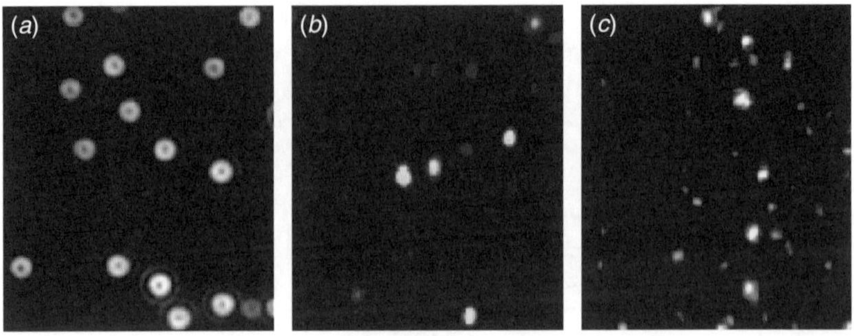

Figure 4 High-resolution fluorescence images of (a) single molecules of CN-PPV cast from dilute toluene solution using inkjet methods, (b) single molecules of CN-PPV cast from dilute dichloromethane solution using inkjet methods, and (c) single molecules of MEH-PPV spin cast from dilute THF solution. The presence of the bull's-eye-like pattern in (*a*) indicates an aligned single dipole emitter, whereas the lack of this pattern in (*b*) and (*c*) shows the lack of this emitter. (Reprinted in part with permission from refs. 83, 84, and 82, respectively. Copyright © 2004, 2004, and 2003, respectively, American Chemical Society.)

Figure 4. These images were chosen for their contrast; the conclusions drawn do not vary with polymer identity (i.e., CN-PPV or MEH-PPV). In Figure 4*a*, the pattern of concentric rings with no fluorescence in the center is indicative[87] of single dipoles *aligned perpendicular* to the glass substrate (or, more precisely, tilted from the surface normal, on average by $\approx 3.5^\circ \pm 10\%$[81]). Figure 4*b* indicates the "solvent memory" of these nanoconfined structures; dichloromethane, an effective solvent for CN-PPV, leads to isolated molecules on the substrate that do *not* possess the aligned single dipole. Similarly, the molecules in a typical thin film also lack this emission pattern, as shown in Figure 4*c*.

Further information about this *z*-orientation (i.e., perpendicular to the glass surface) is obtained through the use of tapping-mode atomic force microscopy.[84] A histogram of measured particle heights indicates a strong peak at 10 nm for nano-oriented CN-PPV, indicating a high degree of uniformity in this system. Although the size of the tip radius limits precise measurement of the expanse of the nanoparticle in the in-plane directions, estimations based on particle height and known molecular weight of the polymer indicate that they are 1.5 to 3 times longer in the *z*-direction than in the *x* and *y* directions. Electric force microscopy[84] also confirms that these nanoparticles are negatively charged, on the order of two to eight electron units. This measurement supports earlier conjecture[81] that *z*-orientation results from electrostatic repulsion between dangling SiO^- on the glass surface and charge imparted to the polymer through the microdroplet formation process. To ameliorate these repulsions, the molecules orient themselves so as to possess as little contact with the surface as possible: hence, *z*-orientation. When the glass substrate is replaced with a conductive and grounded substitute, charge interactions become favorable and thus the orientation changes to in-plane.

The characteristics just described are not concentration dependent; that is, single-dipole fluorescence, orientation, and height uniformity are found even when the starting concentration is diluted to 10 fmol.[81] The surface coverage, however, does depend on concentration, thereby indicating that the emitters possess just one nanoconfined polymer molecule. Photon antibunching measurements[83] further indicate that emission occurs not just from one molecule but also from only one *site* per molecule. This emission, for CN-PPV, is redshifted from thin-film emission by 20 nm and possesses a full-width at half-maximum of 15 nm, 25 nm less than that observed for nonoriented CN-PPV thin-film emission.[84] The z-oriented single molecules also show impressive longevity and emissive intensity. For example, nano-oriented CN-PPV shows a steady, strong emission at over 300,000 counts per second (cps) for times approaching an hour.[84] In comparison, previously reported single molecules produced by spin casting showed peak emission at a rate of less than 40,000 cps, which decayed to roughly 10,000 cps in less than a minute.[23] The reason for the improved characteristics of oriented CN-PPV can be attributed to both the reduced degree of interaction with the substrate and the intrinsically improved ordering imparted by rapid solvent evaporation in the inkjet process. The influence of the former effect on single-molecule fluorescence has been reported both for oligomers of phenylene–vinylene[88] and for MEH-PPV[89]; the latter effect is discussed in depth below in the context of our computational studies.

Each of these experimental techniques, and especially their judicious combination, points strongly toward a great deal of order within single-molecule nanoparticles. However, the specific structure of the ordered entity has not yet been discussed. In addition, the morphology imparted by the initial solution, the roles of microdroplet environment and nanoscale confinement on the final structure, the roles of competing intramolecular forces, the effect of conjugation breaks and substituents on nanoscale structure, the influence of polymer–substrate contact, and a host of other questions have been only partially explored via experimental techniques. Therefore, to augment and elucidate the findings described above, extensive computational research with a variety of multiscale techniques applicable to different size and time scales has been undertaken. In the following section we relate some of these findings as well as work in progress; additional details can be found elsewhere.[85]

4. COMPUTATIONAL STUDIES ON ORIENTED MEH-PPV AND CN-PPV NANOSTRUCTURES

We begin our discussion of computational modeling by considering an environment common to such studies but uncommon in polymer studies: the gas phase. In this phase the molecular mechanics force field MM3[90] predicts that both MEH-PPV and CN-PPV generally prefer to assume a stacked cofacial geometry. The specific energy difference between these orientations and more extended geometries varies depending on both the arrangement of the monomer units with respect to each other and the amount and placement of breaks in conjugation (preserved in these computational studies, as they are an unavoidable legacy of the synthesis[31]). Computationally, folding occurs at these defect sites, in agreement with previous simulation work.[22] In general, the energy gap is measured in tens of kcal/mol. In contrast to these MM3 calculations, the semiempirical technique AM1,[91] nominally a step up in computational complexity, indicates that gas-phase polymers prefer an extended conformation by roughly 10 kcal/mol, contrary to experimental expectations and molecular mechanics simulations. However, out of the gas phase, the combination of the two techniques [i.e., AM1 for the polymer and MM3 for explicitly added solvent molecules in a QM/MM-type approach (see below)] indicates results more in line with experiment: polymers folded in toluene and THF but extended in dichloromethane.

Armed with a feeling for the relative accuracy of these two techniques, the attempt to simulate the creation of oriented nanoparticles computationally was pieced together as follows. Initially, constructed polymers, varying in both number and placement of defect sites, were subjected to molecular dynamics on nanosecond time scales at elevated temperatures, to remove any bias imparted by the choice of starting geometry. Geometries from various trajectories were picked at random and optimized in various solvent continuum fields[92] using MM3. Solvent molecules were then explicitly added to each structure and subsequently reoptimized to account for any deficiencies in the solvent field approximation. Finally, solvent molecules were removed through a simulated annealing process. This computational process does indeed produce the geometries described previously using experimental techniques: extended structures bent at defect sites, but generally without aggregation, in good solvents such as dichloromethane versus highly ordered structures such as the representative structure shown in Figure 5 for a 35-monomer four-defect MEH-PPV molecule in the bad solvent THF. This structure clearly possesses a high degree of order, resulting from efficient chain packing due to significant $\pi-\pi$ interactions.

It is proposed that the highly packed structures typified in Figure 5 are accurate representatives of the experimental nanostructures described in Section 3. The key feature responsible for much of the luminescent behavior of the nanostructure is the fact, as verified by a number of computational approaches, that the interchain distance is shortest between the innermost chains but increases as one moves outward from the core. Thus, despite the obvious homogeneity of the structure in Figure 5, each parallel strand is in a slightly different environment, and thus they are expected to absorb at different wavelengths. Differences in intermonomer torsion angles in the chains accentuate these differences. Regardless, the efficient interchain communication promotes funneling of photoexcitations to the most tightly packed core region, which is lowest in energy both because it experiences the most $\pi-\pi$ overlap[36] and because this most efficient packing reduces torsional motion that leads to poorer conjugation and thus higher energy. Thus, as energy is funneled to the core, emission occurs in only one direction, that is, collinear with the principal axis of the conjugated backbone, as indicated previously.[93] So when these molecules are electrostatically oriented perpendicular to the substrate

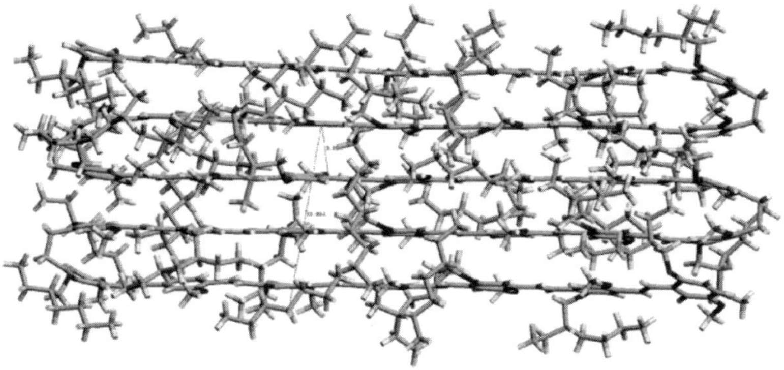

Figure 5 Final structure of a 35-unit MEH-PPV polymer with four defects, as optimized with the procedure described in the text. (Reprinted with permission from ref. 85. Copyright © 2005, American Chemical Society.)

(as described above), emission occurs only in the z-direction, yielding the telltale fluorescence image shown in Figure 4*a*. In effect, a combination of the funneling mechanism and the favorable self-solvation of the structure in Figure 5 acts as a nanoscale antenna–emitter system.

Having established this framework archetype, it is worthwhile to investigate the effects on the electronic structure of factors, such as backbone substitution and interchain shifting, toward achieving an eventual goal of polymer engineering. Quantum mechanics above the entry-level semiempirical calculations described above have proven insightful in this regard, with a few caveats. Density functional theory (DFT) is the most obvious choice to begin high-level investigations into this system, as it has a proven track record of accuracy and can be applied routinely to systems consisting of hundred(s) of atoms. However, the literature[94] indicates that DFT fails to properly describe the extent of delocalization in PPV-based systems. This failure is independent of system, as shown by unpublished calculations on both PPV and CN-PPV performed in our lab and elsewhere[95]; however, there are indications that it might not be independent of functional.[96] Nevertheless, without adequate benchmarking, it seems most prudent to advance to the MP2 level.

With this technique, aligned trimers of PPV are shifted by 0.98 Å along the chain axis and by 0.70 Å along the transverse axis, with 3.50 Å of interchain separation. MEH-PPV and CN-PPV do not appear to be shifted significantly in either direction, with interchain separations of 3.53 and 3.40 Å, respectively. The latter finding is not in agreement with experimental findings discussed earlier for the structure depicted in Figure 2. However, it should be noted that these calculations were made with a small basis set, due to computational limitations, and also probably depend on oligomer length. Thus, more accurate calculations on larger systems are in progress. Additionally, it should be pointed out that both the nanoconfined environment and the order imparted by the multiple π-stacked

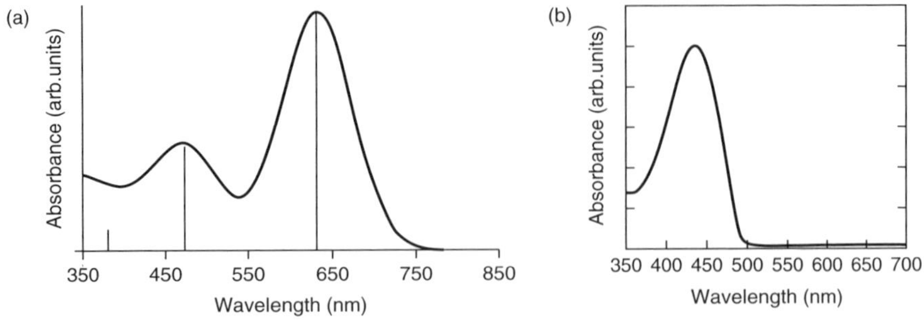

Figure 6 (a) Absorption spectrum of a pentamer of PPV calculated with TD-DFT, and (b) the experimental spectrum of the same system. (Figure 6b is reprinted with permission from ref. 98. Copyright © 2005, American Physical Society.)

chains will have a nonnegligible influence on the degree of in-plane shifting. This displacement between parallel chains, if it occurs, is likely to have a significant effect. Extensive very high level computational work[97] on the prototypical $\pi-\pi$ interaction of benzene has revealed that a parallel-displaced structure is approximately 1 kcal/mol more stable than an aligned face-to-face structure. Our research at the MP2 level with a large basis on a single monomer $\pi-\pi$ stack indicates similarly that displacement leads to a stabilization of 1.5 kcal/mol.

Complementary to the aforementioned studies, which were all performed in the electronic ground state, calculations have been performed to investigate the excited-state behavior of PPV strands. To date, the highest-level methodology that can be applied to systems of this size is time-dependent DFT (TD-DFT), which possesses a balance of speed and accuracy similar to its ground-state parent technique. Unfortunately, TD-DFT also fails to yield accurate results. Shown in Figure 6 is a comparison of the calculated absorption spectrum of a pentamer of PPV with the experimental result.[98] Although one could argue that if the spectrum calculated were shifted to higher energies, a reasonable result could be obtained, calculations on longer oligomers show that the large peak is shifted to even lower energies than for the pentamer shown in Figure 6*a*. Furthermore, these longer oligomers also have additional, higher-energy peaks; experimentally, absorption spectra of longer oligomers do shift toward lower energies, but the spectra only indicate the presence of a single peak. Thus, TD-DFT is regarded as a complete failure. Fortunately, the CIS-ZINDO/s scheme (sometimes alternatively identified as INDO/SCI[99]) can yield qualitative insight.[36]

5. OUTLOOK ON THE USE OF MULTISCALE MODELING ON POLYMERS

Further elucidation of the structure and photophysical behavior of PPV-type polymers via computational techniques will continue to require multiscale modeling of the type described above, characterized by combinations of approaches that operate on varied—sometimes vastly so—length, accuracy, and time scales. For example, the $\pi-\pi$ forces responsible for the ordering of two parallel cofacially stacked PPV chains act on a smaller length scale than do the phenyl–phenyl torsions acting along an extended intrachain segment, which in turn lead to effects more local than those induced both by the presence of occasional defect sites and through interchain sidechain interactions. Figures 2 and 3 give indications that the latter length scale often differs by orders of magnitude from the first two.

Different thresholds of accuracy are also needed to adequately describe behavior at these disparate length scales. The challenges of properly describing $\pi-\pi$ stacking, for example, are well known, proving insurmountable even for some ab initio techniques.[97] Addressing torsion angles between phenyl rings, or the balance of hydrophilic and hydrophobic interactions of polymer backbones and side chains, however, generally requires far less rigorous approaches. Given these differences, it should not be expected that any single modeling technique will yield both useful and timely results; hence, the combination of techniques of multiscale modeling should bear significant fruit.

The final scale in multiscale modeling—time—is perhaps the most difficult to judge a priori for PPV and presumably for other systems as well. Specifically, while photoabsorption and emission naturally take place on femtosecond to nanosecond time scales, it is less clear how much time is required to allow for the PPV ordering that is the subject of most of this work. If anything, indications are that for such order to develop spontaneously in solution, a macroscopically "slow" time scale is needed,[70] clearly well beyond the capability of any of the computational techniques heretofore discussed. Therefore, it seems most prudent to apply multiscale modeling operating across different length and accuracy scales and then to perform simulations for as long as is computationally feasible.

We briefly offer a few computational techniques that have proven, and may continue to prove, useful in multiscale modeling of polymeric systems. One of the most restrictive possibilities is to use periodicity, where mirror images are repeated and are described in conjunction with time-saving technical advances such as plane-wave basis sets. Although it seems counterintuitive to describe an inherently aperiodic system, such as a polymer, with a periodic approach, it should be noted that periodicity need only be applied in one direction. Thus, this approach may prove useful, for example, in describing parallel stacks of polymer chains or in investigating the concentration dependence of solvated polymers.[85]

A less restrictive multiscale modeling technique involves the definition of a core region of the polymer, which requires a rigorous computational treatment; the remainder is described with a less exacting computational technique. These techniques are generally called quantum mechanical/molecular mechanical (QM/MM), reflecting the most common approach where the core is described quantum mechanically and the remainder is investigated with molecular mechanics. However, it should be noted that any hierarchy can, in principle, be applied. Moreover, there may be more than one demarcation, as, for example, in the ONIOM approach.[100] Commonly, in such techniques, the core of the system is where electronic effects are crucial and thus must be described with time-intensive calculations; for the low-level area(s) of the calculation, generally steric effects dominate and are thus described more simply. In addition to technical hurdles such as the challenge of properly describing atoms at the boundary,[101] the main drawback of QM/MM approaches is that calculation of the core of the polymer is still often a prohibitively expensive computational bottleneck. Nevertheless, we have successfully used an AM1/MM3 hybrid approach, noted above, to accurately describe a solvated polymer, and similar successes will no doubt follow.

The final multiscale technique that will be noted herein is coarse graining. In a sense, coarse graining can be viewed as an extreme application of the QM/MM approach. In essence, the core region of QM/MM is somehow described, be it through accurate calculations, experimental evidence, or even chemical intuition, and is then frozen and coalesced into a single entity, termed a *superatom* or, colloquially, a *bead*. Interactions are then calculated between these rigid beads, often employing extremely rapid methods (e.g., using a pairwise potential calculation). Calculations of this sort have given fast yet apparently accurate results for both synthetic[102] and natural[103] polymers. The most striking limitation is that of the rigid bead, which could be extremely limiting for polymers with long flexible side chains, such as MEH-PPV. However, phenomena such as the ramifications of defect sites, the presence of a copolymer (requiring the definition of two types of beads and the successful description of their interactions), and the effects of adding small, rigid substituents (such as CN or perhaps even methoxide), could probably be investigated rapidly with coarse-graining techniques.

The choice of a specific multiscale technique to be employed—be it periodic calculations, QM/MM, coarse graining, or a method not mentioned here or yet to be developed—will depend not only on advances in methodology but also, probably, the system itself. For example, if a system proves, in preliminary simulations, to be fairly rigid or regular, periodic investigations or coarse graining might provide tremendous boosts in computational expediency. In contrast, if the interaction of flexible side chains is shown to be highly important in determining polymer structure, a QM/MM approach might be more appropriate. Regardless of the specific strategy adopted, two simple yet often overlooked aspects of multiscale modeling should be stressed. First, it is important to check, through comparison to experiment and/or higher-level calculation, that the approximations adopted are indeed appropriate. It goes without saying that this benchmarking should be performed in the initial phases, before the utilization of such approximations; however, checks should be incorporated throughout the modeling process.

For example, even if it seems reasonable that an isolated pentamer of MEH-PPV can be separated via QM/MM into given core and noncore regions, it should nonetheless be checked that the same delineation holds as, for example, pentamers are stacked upon one another or as defect sites are introduced. Second, it is important to keep in focus that the primary goal of multiscale modeling, especially in the case of polymer systems eventually destined for commercial purposes, is to supplement or, ideally, replace the costly experimental screening process. Thus, the object to be pursued is not merely the successful modeling of a fairly well described system, such as MEH-PPV, but eventually, the screening, design, and characterization of novel polymeric systems. The requisite first step, of course, is a successful demonstration of the former step, but pursuit of the latter should always be the paramount goal. Fortunately, given the extreme malleability, in more respects than one, of PPV (see Section 1), both of these goals can be pursued realistically and fruitfully.

6. CONCLUSIONS

The family of polymers based on the PPV backbone is clearly a system of tremendous interest, due primarily to a variety of useful eventual applications. The simple chemical structure shown in Figure 1 belies the incredible depth of complexity that results when one considers the interaction of a polymer molecule with both other polymer strands and with itself. The research discussed in the first two sections shows the critical importance of understanding and, to whatever extent possible, controlling the inter- and intrachain interactions responsible for determining polymer conformation and morphology.

In this vast field of research, there have been occasional indications of anomalous polymer behavior. *Anomalous* in this context is not intended as a pejorative; indeed, the unusual systems discussed above should be sought out, as they differ from the norm insofar as they exhibit behavior contradictory to that of conventional PPV wisdom: They conduct charge well, yet luminesce efficiently. The common theme highlighted in this chapter for those unusual systems is that there is always an associated degree of *order*. The order may be demonstrated explicitly through standard structural characterization techniques, as in the thermodynamically induced order of MEH-PPV,[79] or it may be reasonably assumed, such as in films produced using slow solvent evaporative techniques.[70] The system produced in our laboratory is one of the best characterized of these ordered anomalies. Even if such nanocrystalline domains are eventually and conclusively shown to be undesirable, as has been suggested,[38,104] they still warrant investigation, if only to determine how to prevent their formation. Our experiments and calculations, as well as some of the work discussed in Section 2 lead us to believe that the presence of order will instead be highly sought.

It is perhaps appropriate at this point to speculate as to the cause(s) of ordering in polymers, which are generally taken to be amorphous. One strong possibility seems to be the presence of forces that favor conglomeration, particularly nonbonded attractive interactions between π-electron densities on neighboring chains. Agglutination of largely hydrophobic solubilizing sidechains is also expected to play a role; whether this role augments or hinders π-stacking is still being investigated.[37,38,80] Perhaps it is not enough to form *some* favorable $\pi-\pi$ interactions; rather, perhaps *extensive* well-ordered behavior is needed to yield both effective conductance and fluorescence. An extreme position of such a view suggests that a certain type of π-stacking is required, the so-called "brick wall" motif, to yield this beneficial balance in properties.[36]

A final possibility is that the ordered domains, whatever their nature, are, perhaps counterintuitively, fleeting in nature. Along these lines, single-molecule studies have shown[28] that an isolated MEH-PPV molecule can switch between narrow and broad emission on a time scale of seconds at temperatures as low as 15 K. Thus, it is conceivable that ordered species must either be trapped, as via the electrostatic control afforded in our z-oriented nanostructures, or conditions must be perfect to favor their formation, as with the narrow window of favorable spin speed, solvent, and concentration discussed above.[71]

Whatever the cause of efficient conductance and luminescence is ultimately determined to be, it is clear that further research is needed to fully understand PPV and its derivatives. In the first place, it should be established conclusively, not via conjecture, what causes PPV chains to order. Second, evidence, probably obtained using continued multiscale computational techniques, needs to be gathered to elucidate how and why ordered domains lead to the beneficial behavior described here. Finally, numerous optimization studies, similar to those given in Section 1, must be conducted to find out how to make polymer-based materials with these superlative qualities, so that commercialization can occur and consumers can enjoy the realized promise of PPV-based products.

REFERENCES

1. Shirakawa, H.; Louis, E. J.; MacDiarmid, A. G.; Chiang, C. K.; Heeger, A. J. *Chem. Commun.*, 1977, 578.
2. Chiang, C. K.; Fincher, C. R., Jr.; Park, Y. W.; Heeger, A. J.; Shirakawa, H.; Louis, E. J. *Phys. Rev. Lett.*, 1977, 39: 1098.
3. Gustafsson, G.; Cao, Y.; Treacy, G. M.; Flavetter, F.; Colinari, N.; Heeger, A. J. *Nature*, 1992, 357: 477.
4. Burroughes, J. H.; Bradley, D. D. C.; Brown, A. R.; Marks, R. N.; Mackay, K.; Friend, R. H.; Burn, P. L.; Holmes, A. B. *Nature*, 1990, 347: 539.
5. Greenham, N. C.; Moratti, S. C.; Bradley, D. D. C.; Friend, R. H.; Holmes, A. B. *Nature*, 1993, 365: 628.
6. Friend, R. H.; Gymer, R. W.; Holmes, A. B.; Burroughes, J. H.; Marks, R. N.; Taliani, C.; Bradley, D. D. C.; et al. *Nature*, 1999, 397: 121.
7. Brabec, C. J.; Sariciftci, N. S.; Hummelen, J. C. *Adv. Funct. Mater.*, 2001, 11: 15.
8. Sariciftci, N. S.; Smilowitz, L.; Heeger, A. J.; Wudl, F. *Science*, 1992, 258: 1474.
9. Granstrom, M.; Petritsch, K.; Arias, A. C.; Lux, A.; Andersson, M. R.; Friend, R. H. *Nature*, 1998, 395: 257.
10. Huitema, H. E. A.; Gelinck, G. H.; van der Putten, J. B. P. H.; Kuijk, K. E.; Hart, C. M.; Cantatore, E.; Herwig, P. T.; van Breemen, A. J. J. M.; de Leeuw, D. M. *Nature*, 2001, 414: 599.
11. Kraft, A.; Grimsdale, A. C.; Holmes, A. B. *Angew. Chem. Int. Ed. Engl.*, 1998, 37: 402.
12. McQuade, D. T.; Pullen, A. E.; Swager, T. M. *Chem. Rev.*, 2000, 100: 2537.
13. Tour, J. M. *Acc. Chem. Res.*, 2000, 33: 791.
14. Möller, S.; Perlov, C.; Jackson, W.; Taussig, C.; Forrest, S. *Nature*, 2003, 426: 166.
15. Knill, E.; Laflamme, R.; Milburn, G. L. *Nature*, 2001, 409: 46.

16. Brédas, J. L.; Street, G. B. *Acc. Chem. Res.*, 1985, 18: 309.
17. Nguyen, T.-Q.; Kwong, R. C.; Thompson, M. E.; Schwartz, B. J. *Appl. Phys. Lett.*, 2000, 76: 2454.
18. Schwartz, B. J. *Annu. Rev. Phys. Chem.*, 2003, 54: 141.
19. Nguyen, T.-Q.; Schwartz, B. J.; Schaller, R. D.; Johnson, J. C.; Lee, L. F.; Haber, L. H.; Saykally, R. J. *J. Phys. Chem. B*, 2001, 105: 5153.
20. Nguyen, T.-Q.; Doan, V.; Schwartz, B. J. *J. Chem. Phys.*, 1999, 110: 4068.
21. Nguyen, T.-Q.; Yee, R. Y.; Schwartz, B. J. *J. Photochem. Photobiol. A*, 2001, 144: 21.
22. Hu, D. H.; Yu, J.; Wong, K.; Bagchi, B.; Rossky, P. J.; Barbara, P. F. *Nature*, 2000, 405: 1030.
23. Huser, T.; Yan, M.; Rothberg, L. J. *Proc. Natl. Acad. Sci. USA*, 2000, 97: 11187.
24. Vanden Bout, D. A.; Yip, W.-T.; Hu, D.; Fu, D.-K.; Swager, T. M.; Barbara, P. F. *Science*, 1997, 277: 1074.
25. Hollars, C. W.; Lane, S. M.; Huser, T. *Chem. Phys. Lett.*, 2003, 370: 393.
26. Schindler, F.; Lupton, J. M.; Feldmann, J.; Scherf, U. *Proc. Natl. Acad. Sci. USA*, 2004, 101: 14695.
27. Lee, Y. J.; Kim, D. Y.; Grey, J. K.; Barbara, P. F. *ChemPhysChem*, 2005, 6: 2404.
28. Schindler, F.; Lupton, J. M. *ChemPhysChem*, 2005, 6: 926.
29. Barbara, P. F.; Gesquiere, A. J.; Park, S.-J.; Lee, Y. J. *Acc. Chem. Res.*, 2005, 38: 602.
30. Yu, J.; Hu, D.; Barbara, P. F. *Science*, 2000, 289: 1327.
31. Becker, H.; Spreitzer, H.; Ibrom, K.; Kreuder, W. *Macromolecules*, 1999, 32: 4925.
32. Brédas, J. L.; Street, G.; Themmans, B.; Andre, J. *J. Chem. Phys.*, 1985, 83: 1323.
33. Nesterov, E. E.; Zhu, Z.; Swager, T. M. *J. Am. Chem. Soc.*, 2005, 127: 10083.
34. Claudio, G. C.; Bittner, E. R. *J. Phys. Chem. A*, 2003, 107: 7092.
35. Schwartz, B. J.; Nguyen, T.-Q.; Wu, J.; Tolbert, S. H. *Synth. Met.*, 2001, 116: 35.
36. Cornil, J.; Beljonne, D.; dos Santos, D. A.; Calbert, J. P.; Brédas, J. L. *Thin Solid Films*, 2000, 363: 72.
37. Chen, S. H.; Su, C. H.; Su, A. C.; Chen, S. A. *J. Phys. Chem. B*, 2004, 108: 8855.
38. Peng, K.-Y.; Chen, S.-A.; Fann, W.-S.; Chen, S.-H.; Su, A.-C. *J. Phys. Chem. B*, 2005, 109: 9368.
39. Jeng, U.; Hsu, C.-H.; Sheu, H.-S.; Lee, H.-Y.; Inigo, A. R.; Chiu, H. C.; Fann, W. S.; et al. *Macromolecules*, 2005, 38: 6566.
40. Ou-Yang, W.-C.; Chang, C.-S.; Chen, H.-L.; Tsao, C.-S.; Peng, K.-Y.; Chen, S.-A.; Han, C. C. *Phys. Rev. E*, 2005, 72: 031802.
41. Quan, S.; Teng, F.; Xu, Z.; Qian, L.; Hou, Y.; Wang, Y.; Xu, X. *Eur. Polym. J.*, 2006, 42: 228.
42. Mirzov, O.; Scheblykin, I. G. *Chem. Phys.*, 2005, 318: 217.
43. Holzer, W.; Penzkofer, A.; Tillmann, H.; Hörhold, H.-H. *Synth. Met.*, 2004, 155: 170.
44. Padmanaban, G.; Ramakrishnan, S. *J. Phys. Chem. B*, 2004, 108: 14933.
45. Hu, D.; Yu, J.; Padmanaban, G.; Ramakrishnan, S.; Barbara, P. F. *Nano Lett.*, 2002, 2: 1121.

46. Chou, H.-L.; Lin, K.-F.; Fan, Y.-L.; Wang, D.-C. *J. Polym. Sci. B*, 2005, 43: 1705.
47. Kyllo, E. M.; Gustafson, T. L.; Wang, D. K.; Sun, R. G.; Epstein, A. J. *Synth. Met.*, 2001, 116: 189.
48. Egbe, D. A. M.; Cornelia, B.; Nowotny, J.; Günther, W.; Klemm, E. *Macromolecules* 2003, 36: 5459.
49. Peng, Z.; Galvin, M. E. *Chem. Mater.*, 1998, 10: 1785.
50. Niazimbetova, Z. I.; Christian, H. Y.; Bhandari, Y. J.; Beyer, F. L.; Galvin, M. E. *J. Phys. Chem. B*, 2004, 108: 8673.
51. Samuel, I. D. W.; Rumbles, G.; Collison, C. J.; Moratti, S. C.; Holmes, A. B. *Chem. Phys.*, 1998, 227: 75.
52. Kim, J. Y.; Park, S. H.; Lee, K.; Yum, I. S.; Jin, S. H. *Appl. Phys. Lett.*, 2002, 81: 1732.
53. Ramachandran, G.; Simon, G.; Cheng, Y.; Smith, T. A.; Dai, L. *J. Fluoresc.*, 2003, 13: 427.
54. Mozer, A. J.; Denk, P.; Scharber, M. C.; Neugebauer, H.; Sariciftci, N. S.; Wagner, P.; Lutsen, L.; Vanderzande, D. *J. Phys. Chem. B*, 2004, 108: 5235.
55. Kemerink, M.; van Duren, J. K. J.; van Breemen, A. J. J. M.; Wildeman, J.; Wienk, M. M.; Blom, P. W. M.; Schoo, H. F. M.; Janssen, R. A. J. *Macromolecules*, 2005, 38: 7784.
56. Huang, Y.; Lu, Z.-Y.; Peng, Q.; Jiang, Q.; Xie, R.-G.; Han, S.-H.; Dong, L.-G.; Peng, J.-B.; Cao, Y.; Xie, M.-G. *Mater. Chem. Phys.*, 2005, 93: 95.
57. Hulvat, J. F.; Sofos, M.; Tajima, K.; Stupp, S. I. *J. Am. Chem. Soc.*, 2005, 127: 366.
58. Huser, T.; Yan, M. *J. Photochem. Photobiol. A*, 2001, 144: 43.
59. Meskers, S. C. J.; Bender, M.; Hübner, J.; Romanovskii, Y. V.; Oestreich, M.; Schenning, A. P. H. J.; Meijer, E. W.; Bässler, H. *J. Phys. Chem. A*, 2001, 105: 10220.
60. Gin, D.; Yonezawa, K. *Synth. Met.*, 2001, 121: 1291.
61. He, G.; Li, Y.; Liu, J.; Yang, Y. *Appl. Phys. Lett.*, 2002, 80: 4247.
62. Liu, J.; Shi, Y.; Yang, Y. *Appl. Phys. Lett.*, 2001, 79: 578.
63. Nguyen, T.-Q.; Kwong, R. C.; Thompson, M. E.; Schwartz, B. J. *Synth. Met.*, 2001, 119: 523.
64. Cadby, A. J.; Tolbert, S. H. *J. Phys. Chem. B*, 2005, 109: 17879.
65. Zhang, H.; Lu, X.; Li, Y.; Ai, X.; Zhang, X.; Yang, G. *J. Photochem. Photobiol. A*, 2002, 147: 15.
66. Traiphol, R.; Sanguansat, P.; Srikhirin, T.; Kerdcharoen, T.; Osotchan, T. *Macromolecules*, 2006, 39: 1165.
67. Menon, A.; Galvin, M.; Walz, K. A.; Rothberg, L. *Synth. Met.*, 2004, 141: 197.
68. Sinha, S.; Monkman, A. P. *J. Appl. Phys.*, 2003, 93: 5691.
69. Yang, G.; Li, Y.; White, J. O.; Drickamer, H. G. *J. Phys. Chem. B*, 1999, 103: 7853.
70. Arnautov, S. A.; Nechvolodova, E. M.; Bakulin, A. A.; Elizarov, S. G.; Khodarev, A. N.; Martyanov, D. S.; Paraschuk, D. Y. *Synth. Met.*, 2004, 147: 287.
71. Shi, Y.; Liu, J.; Yang, Y. *J. Appl. Phys.*, 2000, 87: 4254.
72. Liu, J.; Shi, Y.; Ma, L.; Yang, Y. *J. Appl. Phys.*, 2000, 88: 605.

73. Lipson, S. M.; O'Brien, D. F.; Drury, A.; Byrne, H. J.; Blau, W. J. *Synth. Met.*, 2001, 119: 569.
74. Chang, R.; Hsu, J. H.; Fann, W. S.; Yu, J.; Lin, S. H.; Lee, Y. Z.; Chen, S. A. *Chem. Phys. Lett.*, 2000, 317: 153.
75. Teetsov, J.; Vanden Bout, D. A. *J. Phys. Chem. B*, 2000, 104, 40: 9378.
76. Martens, J. H. F.; Bradley, D. D. C.; Burn, P. L.; Friend, R. H.; Holmes, A. B.; Marseglia, E. A. *Synth. Met.*, 1991, 41: 301.
77. Martens, J. H. F.; Marseglia, E. A.; Bradley, D. D. C.; Friend, R. H.; Burn, P. L.; Holmes, A. B. *Synth. Met.*, 1993, 55: 449.
78. Yang, C. Y.; Hide, F.; Díaz-García, M. A.; Heeger, A. J.; Cao, Y. *Polymer*, 1998, 39: 2299.
79. Chen, S. H.; Su, A. C.; Chou, H. L.; Peng, K. Y.; Chen, S. A. *Macromolecules*, 2004, 37: 167.
80. Chen, S. H.; Su, A. C.; Han, S. R.; Chen, S. A.; Lee, Y. Z. *Macromolecules*, 2004, 37: 181.
81. Mehta, A.; Kumar, P.; Dadmun, M. D.; Zheng, J.; Dickson, R. M.; Thundat, T.; Sumpter, B. G.; Barnes, M. D. *Nano Lett.*, 2003, 3: 603.
82. Kumar, P.; Mehta, A.; Dadmun, M. D.; Zheng, J.; Peyser, L.; Bartko, A. P.; Dickson, R. M.; Thundat, T.; Sumpter, B. G.; Noid, D. W.; Barnes, M. D. *J. Phys. Chem. B*, 2003, 107: 6252.
83. Kumar, P.; Lee, T.-H.; Mehta, A.; Sumpter, B. G.; Dickson, R. M.; Barnes, M. D. *J. Am. Chem. Soc.*, 2004, 126: 3376.
84. Kumar, P.; Mehta, A.; Mahurin, S. M.; Dai, S.; Dadmun, M. D.; Sumpter, B. G.; Barnes, M. D. *Macromolecules*, 2004, 37: 6132.
85. Sumpter, B. G.; Kumar, P.; Mehta, A.; Barnes, M. D.; Shelton, W. A.; Harrison, R. J. *J. Phys. Chem. B*, 2005, 109: 7671.
86. Chu, B.; Ying, Q.; Grosberg, A. Y. *Macromolecules*, 1995, 28: 180.
87. Bartko, A. P.; Dickson, R. M. *J. Phys. Chem. B*, 1999, 103: 11237.
88. Summers, M. A.; Bazan, G. C.; Buratto, S. K. *J. Am. Chem. Soc.*, 2005, 127: 16202.
89. Pullerits, T.; Mirzov, O.; Scheblykin, I. G. *J. Phys. Chem. B*, 2005, 109: 19099.
90. Lii, J.-H.; Allinger, N. L. *J. Comput. Chem.*, 1998, 19: 1001.
91. Dewar, M. J. S.; Zoebisch, E. G.; Healy, E. F.; Stewart, J. J. P. *J. Am. Chem. Soc.*, 1985, 107: 3902.
92. Cramer, C. J.; Truhlar, D. G. *Chem. Rev.*, 1999, 99: 2161.
93. Bredas, J. L.; Beljonne, D.; Cornill, J.; Calbert, J. P.; Shuai, Z.; Silbey, R. *Synth. Met.*, 2001, 125: 107.
94. Grozema, F. C.; Candeias, L. P.; Swart, M.; van Duijnen, P. T.; Wildeman, J.; Hadziioanou, G.; Siebbeles, L. D. A.; Warman, J. M. *J. Chem. Phys.*, 2002, 117: 11366.
95. Moro, G.; Scalmani, G.; Cosentino, U.; Pitea, D. *Synth. Met.*, 2000, 108: 165.
96. Geskin, V. M.; Dkhissi, A.; Brédas, J. L. *Int. J. Quantum Chem.*, 2003, 91: 350.
97. Sinnokrot, M. O.; Sherrill, C. D. *J. Phys. Chem. A*, 2004, 108: 10200.
98. Yang, X.; Dykstra, T. E.; Scholes, G. D. *Phys. Rev. B*, 2005, 71: 045203.
99. Ridley, J.; Zerner, M. C. *Theor. Chim. Acta*, 1973, 32: 111.

100. Dapprich, S.; Komaromi, I.; Byun, K. S.; Morokuma, K.; Frisch, M. J. *J. Mol. Struct. (Theochem)*, 1999, 461: 1.

101. Maseras, F.; Morokuma, K. *J. Comput. Chem.*, 1995, 16: 1170.

102. Müller-Plathe, F. *ChemPhysChem*, 2002, 3: 754.

103. Queyroy, S.; Neyertz, S.; Brown, D.; Müller-Plathe, F. *Macromolecules*, 2004, 37: 7338.

104. Chen, S. H.; Su, A. C.; Chang, C. S.; Chen, H. L.; Ho, D. L.; Tsao, C. S.; Peng, K. Y.; Chen, S. A. *Langmuir*, 2004, 20: 8909.

5

COARSE GRAINED-TO-ATOMISTIC MAPPING ALGORITHM: A TOOL FOR MULTISCALE SIMULATIONS

STEVEN O. NIELSEN

Department of Chemistry, University of Texas at Dallas, Richardson, Texas

BERND ENSING

Van't Hoff Institute for Molecular Sciences, University of Amsterdam, Amsterdam, The Netherlands

PRESTON B. MOORE

Department of Chemistry and Biochemistry, University of the Sciences in Philadelphia, Philadelphia, Pennsylvania

MICHAEL L. KLEIN

Center for Molecular Modeling, Department of Chemistry, University of Pennsylvania, Philadelphia, Pennsylvania

The many approaches of modeling chemical and biological systems at the mesoscopic level, which include coarse-grained (CG) dynamics, dissipative particle dynamics (DPD), and Brownian dynamics, are being linked increasingly to the underlying molecular description. Historically, these models used a description of molecular interactions that was rather simple and arbitrary, in part for computational efficiency but also because of limited knowledge concerning the molecular forces. In the past few years, various groups have developed methods to obtain the potential energy terms in these models from either experimental data or from fully atomistic simulation (FA) data. We refer the interested reader to two review articles[1,2] on this subject and give some brief examples. Ortiz et al.[3] have

Multiscale Simulation Methods for Nanomaterials, Edited by Richard B. Ross and Sanat Mohanty

developed a DPD model for aqueous diblock copolymer systems using experimental surface tension and density data together with computer simulation data: specifically, the first and second moments of the bond and angle distributions obtained from atomistic molecular dynamics (MD) simulations. Building the model required the introduction of a density-based atomistic-to-coarse-grained mapping in order to obtain a physically realistic description of the system. This model gave values for the bilayer membrane area expansion modulus and power-law scaling of the hydrophobic core thickness that were in excellent agreement with experiment. Izvekov and Voth[4] have developed a force-matching method in which the coarse graining of an interparticle force field is constructed directly from the forces obtained from an atomistic MD trajectory. The procedure has been demonstrated for water and methanol,[4] nanoparticles,[5] and phospholipids.[6]

The motivation for making these connections to the molecular details is to obtain more realistic models, in an effort to describe actual systems. The question that arises naturally, then, is how faithful these models are to the molecular system they are supposed to represent. There are several ways to answer this question, and in doing so attempt to validate the results of a coarse-grained or mesoscale study.

Mesoscale methods are employed in situations where treating the full molecular detail is too costly, so validation attempts are feasible only if limited molecular simulations are required. The final configuration from a CG simulation (e.g. of a self-assembly study) could be restored to its full molecular detail, and the resulting system could be simulated to see if this configuration is at least locally stable. A one-dimensional reaction coordinate identified from CG simulations could be targeted for a free-energy simulation in full molecular detail, and the free energies and conformations could be compared along the pathway. Such validations of CG results are almost nonexistent in the literature. The algorithm presented here is a tool that allows for the molecular detail to be recovered in a straightforward manner. It is hoped that this tool will encourage people to attempt to validate, or at least compare and contrast, their mesoscopic results with limited molecular simulations.

Another use of this algorithm is in atomistic simulations of high-molecular-weight polymer melts, which typically combine an atomistic-to-coarse grained and a coarse grained-to-atomistic mapping algorithm to prepare a relaxed initial condition.[7] The long time scales needed to relax a glassy long-chain polymer melt put the construction of an initial condition out of the realm of atomistic simulation. Employing a coarse-grained model speeds up this process by many orders of magnitude, and once the chains are relaxed, the atomistic details can be reintroduced. Then, only a short equilibration simulation is needed to relax the local atomistic degrees of freedom. Recent developments in multiscale simulation methodology also demand such an algorithm. Hybrid dual-resolution MD is the type of multiscale simulation in which our SO(3) optimization technique is particularly advantageous. Recently, a number of groups have made technical advances to couple a molecular system modeled in full atomistic detail (high resolution) with a molecular system modeled at a lower, coarse grain or united atom level (low resolution)[8–10]. Some of these methods allow matter to change resolution dynamically when it crosses a user-defined boundary. Such a change requires algorithms to convert matter from a coarse to a fine level of representation, and vice versa.

The algorithm presented in this chapter uses SO(3) optimization to align molecular fragments corresponding to coarse-grained sites. In other words, rigid library structures, obtained from atomistic simulated annealing simulations, are placed at each coarse-grained site and rotated in place to minimize an energy function. At the end of this procedure, a globally unfrustrated configuration is obtained. However, local relaxation still needs to be performed on the degrees of freedom that were frozen (the degrees of freedom within each fragment). This is accomplished, for instance, by running a short atomistic simulation.

The approach is based on an algorithm developed by Taylor and Kriegman[11] in which a sequence of local parameterizations of the manifold SO(3) is used rather than relying on a single global parameterization such as Euler angles. The problems caused by singularities in a global parameterization are thus avoided. The efficiency of the algorithm comes from two sources, one technical and one conceptual. On the technical side, the use of quaterion arithmetic and the structure of the algorithm make it fast. From a conceptual standpoint, optimization is performed at the coarse-grained level because the molecular fragments corresponding to coarse-grained sites are treated as rigid bodies with no internal degrees of freedom. This means that there are only three degrees of freedom per coarse-grained site to optimize: namely, an element of SO(3). The algorithm is maximally efficient in the sense that the number of variables to optimize is minimal, and the algorithm for this optimization is extremely fast.

The output from this algorithm consists of rigid molecular fragments positioned at the coarse-grained sites, rotated to minimize an energy function consisting of both intra- and intermolecular terms. We ask the question: In what sense is this configuration a representation of the underlying molecular system? If one insisted that the atomistic configuration be a realization of a particular ensemble [e.g., constant volume and temperature (NVT)], the configuration generated by the algorithm would not be satisfactory. Rather than attempting to provide a mapping algorithm that generates an equilibrated atomistic configuration, we have taken a more pragmatic approach. The mapping algorithm described here quickly generates a globally stable atomistic configuration that further requires very localized relaxation and equilibration. In the context of duel-resolution multiscale simulations, this localized equilibration is inherent in the simulation methodology in the form of a healing region (see the following section) and hence is not required of the mapping algorithm. For wholesale mapping, in which a coarse-grained configuration is mapping entirely to its atomic detail, a short atomistic equilibration simulation needs to be performed to both relax the system locally and to equilibrate it in a particular ensemble. For the latter case, this two-step procedure is expected to be competitive with one-step algorithms.

The remainder of the chapter is organized as follows. First we provide some detail about hybrid dual-resolution methodology, in order to give the reader an appreciation of the utility of the SO(3) algorithm in such a context. Second, the mathematics behind the SO(3) optimization algorithm is introduced. Next, the algorithm is further developed for application to molecular systems. Following this, bulk liquid dodecane is used as an illustrative example.

1. HYBRID COARSE GRAINED–FULLY ATOMISTIC MOLECULAR DYNAMICS

In what follows we briefly describe our implementation of a hybrid MD scheme to model a particular region of a system in atomistic detail while treating everything outside this region at the computational less demanding CG resolution. Further technical details will be published elsewhere (manuscript in preparation). Using the hybrid CG/FA-MD technique, the molecules within a relatively small spatial region interact with each other at the FA level through an atomistic force field. The molecules outside this FA region are treated instead at a CG level, in which several connected atoms are lumped into CG interaction sites (e.g., chemical functional groups of about 10 atoms) that interact with each other through a force field designed for CG MD.

It would be cumbersome to have to develop a third "mixed" force field to describe interactions between atoms and CG sites, and fortunately, this is not necessary. Instead, the interaction sites of the molecules in the FA region (i.e., the atom positions) can easily be mapped onto the CG representation of the molecules, after which the mixed FA–CG interactions can be modeled using the same CG force field as that used for CG–CG interactions. As an illustration, consider a system of methane molecules, partially treated atomistically with the carbon and the four hydrogens as the interaction sites and outside this FA region treated at the united atom level using a single interaction site per CH_4 molecule (see Figure 1). To map the atomistic representation onto the CG representation, we simply take the center of mass of the atoms of each molecule as the position of the CG site. In fact, we can maintain access to both FA and CG representations of the entire system throughout the simulation, by setting the system up in the FA representation and mapping it onto the CG representation by computing centers of mass. Molecules in the FA region are then evolved in their FA representation, while the molecules outside the FA region are evolved in their CG representation. For the latter CG molecules, the extra atomistic degrees of freedom are kept frozen and follow their CG site in a static manner. For the mixed interaction between two molecules in the FA and CG regions respectively, we only need to recompute the center of mass of the molecule in the FA region to know its CG site position and calculate the CG–CG interaction. Then, for the molecule in the FA region, the force resulting from this interaction is distributed in a mass-weighted fashion over its atoms. In this way the molecules in the FA region "feel" the molecules in the outer CG region in the averaged CG manner, and similarly, the molecules in the CG region "feel" the molecules in the FA region as CG molecules.

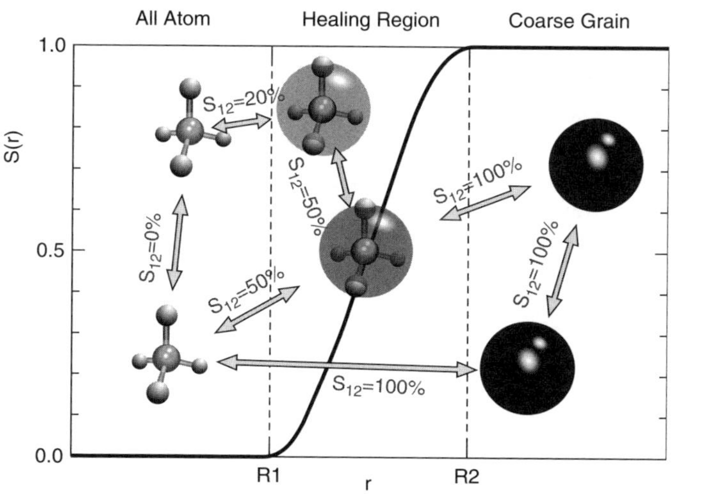

Figure 1 Schematic representation of methane in the atomistic region (at the left), the coarse grain region (at the right), and the connecting healing region (in the middle). The solid black line shows the value of the switching function for a CG site as a function of its position, $S(r)$. The fractional CG character of each pair interaction equals the maximum $S(r)$ among the two interacting CG sites (illustrated in red). (*See insert for color representation of figure*.)

Since the entire system is represented simultaneously in the FA and CG levels of detail throughout the simulation, evaluation of the forces on each molecule and the evolution of each molecule can be chosen to proceed at either of the two resolutions and, moreover, can switch from one resolution to the other whenever necessary. The latter has to take place, of course, when a molecule crosses over from the FA region to the CG region, or vice versa. In practice, however, using a "hard" FA–CG border between the regions causes problems. The atoms of a previously CG molecule can have any orientation and are likely to overlap with other atoms in the FA region when the relatively soft CG interaction is suddenly changed into the atomistic interactions, leading to too-high forces and velocities. To overcome this problem, an intermediate *healing region* (also called an *interface region* or *switching region*) is introduced in between the FA and CG spatial regions. Molecules in this healing region switch in a smooth fashion from one resolution into the other by treating their interaction as partly FA and partly CG. Praprotnik et al.,[8] for example, use force scaling in the healing region so that the total intermolecular force acting between the centers of mass of molecules α and β equals

$$F_{\alpha\beta} = [1 - S(X_\alpha)S(X_\beta)]F_{\alpha\beta}^{\mathrm{CG}} + S(X_\alpha)S(X_\beta)\sum_{i\alpha,j\beta} F_{i\alpha,j\beta}^{\mathrm{FA}} \tag{1}$$

where i and j are the atoms belonging to CG sites α and β, respectively, and $S(X_\alpha)$ is a scaling function that switches smoothly from 0 (in the FA region) to 1 (in the CG region) as a function of the position X_α in the healing region. Note that the interaction between two molecules that find themselves both halfway into the healing region is treated one-fourth atomistically and three-fourths CG.

Our approach is similar, but instead of force scaling, we scale the potentials. The main advantage of scaling the potential is that it allows an estimate of the change in total energy due to the change in potential when a molecule moves through the healing region:

$$\begin{aligned} V &= \sum_{\alpha\beta} V_{\alpha\beta} + \sum_{i\alpha,j\alpha} V^{\text{FA}}_{i\alpha,j\alpha} + V^{\text{inh}} \\ V_{\alpha\beta} &= S_{\alpha\beta} V^{\text{CG}}_{\alpha\beta} + (1 + S_{\alpha\beta}) \sum_{i\alpha,j\beta,\alpha\neq\beta} V^{\text{FA}}_{i\alpha,j\beta} \end{aligned} \tag{2}$$

The total potential is divided into the pair potentials spanning more than one CG site, $V_{\alpha\beta}$, the internal FA pair potentials (bonds and angles) within each CG site, $V_{i\alpha,j\beta}$, and a remainder term, V^{inh}, that accounts for the excess inherent potential energy of CG molecules that is "integrated out" when moving from the atomistic to the CG representation. The interactions spanning two or more CG sites consist of sums of scaled FA–FA and CG–CG potentials (note the similarity with the first equation), while the second term of an internal FA interaction within a single CG site is not scaled in the healing region, but instead, is frozen when moving into the CG region.

For our methane example, this means that the intramolecular bonds and angles evolve atomistically in the healing region, while the intermolecular van der Waals and electrostatic interactions are treated partly FA and partly CG, using the scaling factors $S(r)$. The scaling factor is a polynomial that goes smoothly from zero to unity over a distance from R_1 to R_2, respectively. R_1 and R_2 can, for example, be taken to be the distance from a particular molecule that requires the atomistic treatment, so that a spherical FA region with radius R_1 is defined surrounding the molecule, dressed by a skin of thickness R_2–R_1 defining the healing region, and with a CG region outside the sphere of radius R_2. Alternatively, a rectangular spatial partitioning can be defined, for example, the FA region can be an infinite slab with thickness $2 \times R_1$, flanked on both sides by a healing region of thickness R_2–R_1, with the CG region outside this. Figure 2 shows these two types of partitioning for a box of 1000 methane molecules.

Figure 1 illustrates the scaling of the interactions between pairs of methane molecules in different regions. Each molecule has a switching function value depending on its position, r, being zero in the FA region, unity in the CG region, and an intermediate value in the healing region as shown by the solid black line in Figure 1. Each pair interaction is scaled with the maximum scaling factor among the two interacting CG sites:

$$S_{\alpha\beta} = \max[S(r_\alpha), S(r_\beta)] \tag{3}$$

Figure 2 Two examples of the spatial partitioning of a periodic unit cell of 1000 methane molecules into an atomistic region, a coarse-grained region, and an intermediate healing region. The transparency of the dark blue CG spheres reflects the values of the switching function $S(r)$. On the left is a rectangular partitioning, and on the right is a spherical partitioning. (*See insert for color representation of figure.*)

Although the healing region overcomes the problem of bad overlaps by switching the atomistic degrees of freedom on in a smooth manner, nevertheless the excess potential energy is converted into kinetic energy that has to be removed (e.g., by using a thermostat). In particular, coupling the atomistic degrees of freedom to a thermostat is important for randomly oriented molecules moving from the CG region toward the FA region, for two reasons: (1) allowing the internal degrees of freedom to thermalize into the desired ensemble, and (2) to remove the heat produced during reorientation of the internal degrees of freedom from the energetically unfavorable random orientations assumed in the CG region. However, by using the SO(3) optimization scheme, bad overlaps and energetically unfavorable random orientations in general are largely avoided at the cost of a relatively small overhead, thus reducing the amount of energy needed to be removed from the system. Moreover, boundary effects are expected to be reduced significantly, allowing for a smaller healing region, which reduces the number of interactions that have to be calculated in both representations.

The second important benefit of repairing the internal random orientations by applying the SO(3) optimization scheme for the molecules in CG region is the expected improvement of structural continuity at the boundary for certain systems. Intuitively, this is most easily understood by considering a box of liquid water with the typical hydrogen-bonding network in the FA region and a spherical molecular representation in the CG region. Without the SO(3) optimization, FA water molecules close to the boundary with the CG region cannot seamlessly continue the H-bonded network and will behave as if at an interface to avoid having dangling hydrogens.

2. SO(3) OPTIMIZATION ALGORITHM

The algorithm outlined below is intended to minimize a real-valued objective function E: SO(3) $\to R$ defined on the set of rotation matrices

$$R \in \mathrm{SO}(3) \equiv \{R \in R^{3\times 3} : R^t R = I, \det(R) = 1\} \tag{4}$$

At every point R_0 on the manifold SO(3) we construct a continuous, differentiable mapping between a neighborhood of R_0 on the manifold and an open set in R^3:

$$R(\omega) = R_0 \exp J(\omega), \qquad \omega \in R^3, \quad \|\omega\| < \pi \tag{5}$$

where the skew-symmetric operator J: $R^3 \to$ SO(3) is defined as

$$J(\omega) = \begin{bmatrix} 0 & -\omega_z & \omega_y \\ \omega_z & 0 & -\omega_x \\ -\omega_y & \omega_x & 0 \end{bmatrix} \tag{6}$$

$R(\omega)$ can be computed using the Rodrigues formula, although we will not need to do this. The objective (energy) function can be expanded to quadratic order about R_0 as

$$E(R(\omega)) = E(R_0) + g^t\omega + \omega^t H \omega \tag{7}$$

where g and H are the gradient and the Hessian of the function, respectively, evaluated at the point $\omega = 0$, which corresponds to the rotation matrix R_0. The conjugate gradient incremental step is

$$\omega_s = -H^{-1}g \tag{8}$$

This incremental step determines the new rotation matrix as follows:

$$R = R_0 \exp J(\omega_s) \tag{9}$$

The incremental step must lie within the range of the local parameterization (i.e., $\|\omega_s\| < \pi$). The updating step can be made computationally efficient by representing the rotations by unit quaternions. The relationship between SO(3) and the group of unit quaternions Sp(1) is

$$q = (\cos\theta, \hat{\omega}\sin\theta), \qquad \theta = \frac{\|\omega\|}{2} \tag{10}$$

The incremental step corresponds to the quaternion

$$q_s = \left(\cos\frac{\theta}{2}, \omega\frac{\sin(\theta/2)}{\theta}\right) \qquad \text{where} \quad \theta = \|\omega\| \tag{11}$$

With the rotation R_0 expressed as the unit quaternion q_0, the product of the two rotations, which gives the new rotation matrix, is given by the quaterion multiplication $q_0 q_s$. It has been shown that this algorithm exhibits quadratic convergence provided that the starting point is sufficiently close to a minimum.[12]

3. ALGORITHM SPECIALIZED TO MOLECULAR SYSTEMS

To apply the algorithm outlined in Section 2 to molecular systems, two things are needed. First, an objective function must be chosen that imbues the algorithm with chemical meaning. This function will provide a measure of the potential energy of the molecular configuration associated with a given rotation matrix. Second, the algorithm must be extended to many coupled SO(3) optimizations so that the molecular system is simultaneously and concertedly optimized over all the coarse grain centers. This multibody extension is in fact straightforward and does not incur any significant computational cost aside from the necessary linear scaling with the number of centers. The nature of the multibody aspect of the algorithm will become clear in what follows.

Let us now address the choice of an energy function. Only interactions between atoms belonging to different coarse grain units need be considered. This is because the intraunit degrees of freedom are frozen; the molecular fragment corresponding to a CG site is taken from a library structure, which may be obtained from a simulated annealing FA MD simulation or some other method. The SO(3) algorithm is designed to find the optimal rotational orientation of each of these fragments, where the center of mass of each fragment is constrained to lie at the location of the CG site representing it, and where no internal relaxation of the intrafragment degrees of freedom is allowed. This idea is shown schematically in Figure 3.

We wish to take the functional form and the parameters of all of the terms contributing to the energy function from an underlying atomistic force field. There is no need to invent new potential energy terms when we have well-parameterized terms at our disposal. The first contribution to the energy function is a bonded term arising from the unconnected bonds in the molecular fragments. This function is harmonic in the interatom distance and is expressed as (see Figure 4)

$$E(R_1, R_2) = \frac{k}{2}2\left(\|r + R_2 v - R_1 u\| - d_0\right)^2 \tag{12}$$

where R_1 and R_2 are the rotation matrices corresponding to coarse-grained units 1 and 2, located at positions COM1 and COM2, respectively. The vector from COM1 to COM2 is denoted r, u represents the vector from COM1 to the atom in coarse grain unit 1 involved in the bond, and v represents the vector from COM2 to the atom in coarse grain unit 2 at the other end of the bond. The bond has an equilibrium distance of d_0 and a force constant of k. To perform optimization, the gradient must be evaluated. There are six gradient terms associated with

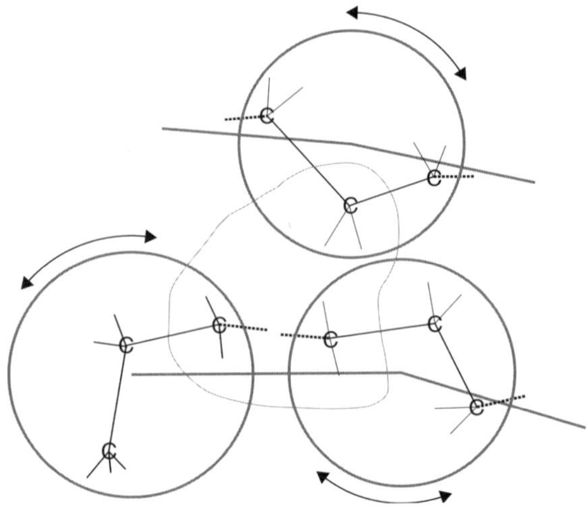

Figure 3 Schematic of the SO(3) optimization algorithm. Atomic fragments are rotated about the centers of mass of the coarse grain units representing them to align the unconnected bonds between fragments, as well as to achieve a low-energy state for the other intra- and intermolecular interactions at the atomistic level.

this energy function, three for rotation matrix R_1 and three for R_2. For the ω_x component of R_1, the gradient is

$$\frac{\partial}{\partial \omega_{x1}} E = -k \frac{\|r + R_2 v - R_1 u\| - d_0}{\|r + R_2 v - R_1 u\|} (r + R_2 v - R_1 u) \cdot \frac{\partial}{\partial \omega_{x1}} R_1 u \tag{13}$$

with

$$\frac{\partial}{\partial \omega_{x1}} R_1 u = R_0^1 J(\hat{x}) u \tag{14}$$

where R_0^1 denotes the initial rotation matrix for coarse-grained center 1. This last result is computationally important because it means that the J operator only ever acts on three elements, $\hat{x}$, $\hat{y}$, and $\hat{z}$, where

$$J(\hat{x}) = \begin{bmatrix} 0 & 0 & 0 \\ 0 & 0 & -1 \\ 0 & 1 & 0 \end{bmatrix}, J(\hat{y}) = \begin{bmatrix} 0 & 0 & 1 \\ 0 & 0 & 0 \\ -1 & 0 & 0 \end{bmatrix}, J(\hat{z}) = \begin{bmatrix} 0 & -1 & 0 \\ 1 & 0 & 0 \\ 0 & 0 & 0 \end{bmatrix} \tag{15}$$

The atoms in the fragment library that have unconnected bonds can be flagged to help automate the optimization procedure.

A bend term in the potential energy function has the form

$$E(R_1, R_2) = \frac{k}{2} 2(\theta - \theta_0)^2 \tag{16}$$

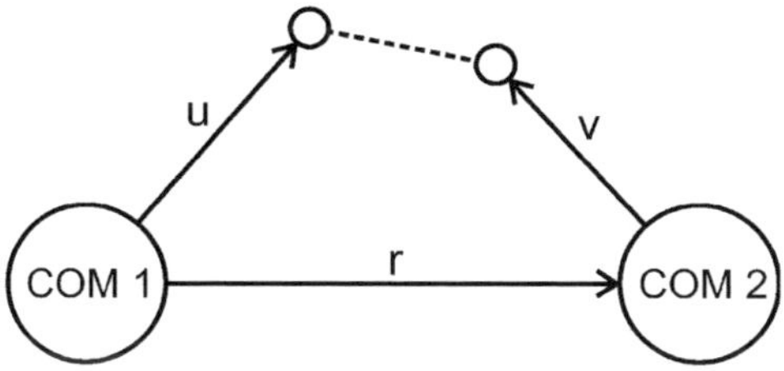

Figure 4 Bond (dashed line) between atomistic fragments corresponding to two neighboring coarse-grained sites. Atom u in fragment 1, centered at coarse-grained site COM1, interacts with atom v, centered at coarse-grained site COM2, through a bonded term in the objective energy function.

where (see Figure 5)

$$\theta = \arccos \frac{(R_1 u' - R_1 u) \cdot (r + R_2 v - R_1 u)}{\|R_1 u' - R_1 u\| \|r + R_2 v - R_1 u\|} \tag{17}$$

For concreteness we have assumed that two of the atoms involved in the bend are located within fragment 1, with the remaining atom in fragment 2. The gradient is straightforward to evaluate by successive applications of the chain rule. The remaining intramolecular potential energy terms are treated in a similar manner and their gradients are again evaluated by the chain rule. For example, a torsional contribution, assuming for concreteness that two of the atoms are in fragment 1 and two are in fragment 2, is constructed as follows. In analogy with Figure 5, we denote the position of the four atoms with respect to their fragment centers of mass as u, u', v, and v' and form the four vectors $b_1 = u'$, $b_2 = u$, $b_3 = r + v$, and $b_4 = r + v'$. We then define three new vectors $a_1 = b_1 - b_2$, $a_2 = b_3 - b_2$, and $a_3 = b_4 - b_2$, from which the torsion angle is formed using the cross product as

$$\tau = \arccos \frac{(a_1 \times a_2) \cdot (a_2 \times a_3)}{\|a_1 \times a_2\| \|a_2 \times a_3\|} \tag{18}$$

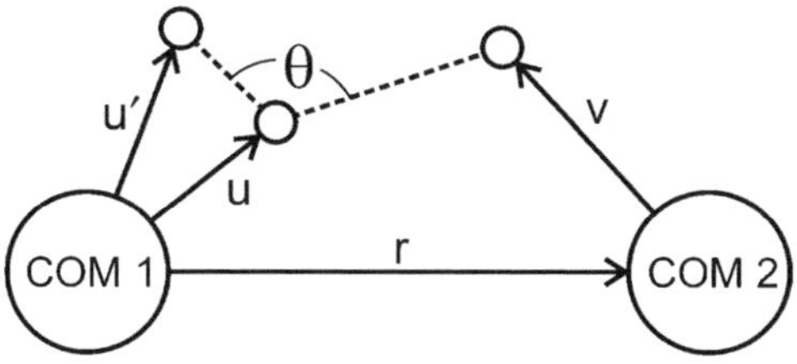

Figure 5 Bend (labeled θ) between atomistic fragments corresponding to two neighboring coarse-grained sites. Atoms u and u′ in fragment 1, centered at coarse-grained site COM1, interact with atom v, centered at coarse-grained site COM2, through a bend term in the objective energy function.

Intermolecular and intramolecular interactions between atoms separated by more than three bonds are critical for condensed-phase systems and must be included in the energy function. Since the crucial contribution of these non-bonded terms is of an excluded volume nature, we take the repulsive part of the Lennard-Jones contribution into the energy function. The multibody nature of the minimization procedure proceeds as follows. The potential energy function is written as a sum over all the coarse-grained sites, with separate terms arising from bonds, bends, torsions, 1-4s, and Lennard-Jones interactions. The gradient of this function is evaluated with respect to each degree of freedom: namely, the three numbers ω_x, ω_y, and ω_z for each coarse-grained site. This gradient is used to decide on a global incremental update step in which all the rotation matrices are changed simultaneous. According to Section 2, we are supposed to evaluate the Hessian as well as the gradient to compute the update step. However, by employing the Fletcher–Reeves–Polak–Ribiere version of the conjugate gradient algorithm, only the gradient is needed.[13]

4. ALGORITHM SPECIALIZED TO LIQUID DODECANE

We now proceed to demonstrate how the SO(3) optimization algorithm works on a specific condensed-phase system: namely, bulk liquid dodecane. The coarse-grained representation we adopt for dodecane has four units,[14] with each unit composed of three carbon atoms and their associated hydrogen atoms. Two of these units include six hydrogen atoms, and two of them include seven hydrogen atoms. A distinction is made between these cases, giving two distinct types of coarse-grained units, middle (M) and terminal (T). An equilibrated fully atomistic bulk dodecane system at 300 K and 1 bar is subjected to simulated annealing runs in which the temperature is reduced to zero. From this final configuration, one atomic fragment corresponding to M and one corresponding to T are selected, and their atomic coordinates are entered in the fragment library after rigid translation to place each fragment center of mass at the origin.

The energy landscape becomes rougher and the SO(3) optimization problem becomes more and more overdetermined as more terms are included in the energy function. We choose the energy function for liquid dodecane to consist of one bond, four bends, four torsions, and four 1-4s per "join" between intramolecular CG sites and all Lennard-Jones repulsions between hydrogen atoms. The potential energy terms corresponding to these contributions are taken from the CHARMM atomistic force field. This selection is shown schematically in Figure 6.

Single Configuration

A single configuration is selected from an equilibrated CG bulk liquid dodecane simulation taken from ref. 14. There are 1050 dodecane molecules in the simulation unit cell at a density of 0.74 g/mL. The goal of the SO(3) optimization algorithm is to replace the coarse-grained configuration with its atomistic

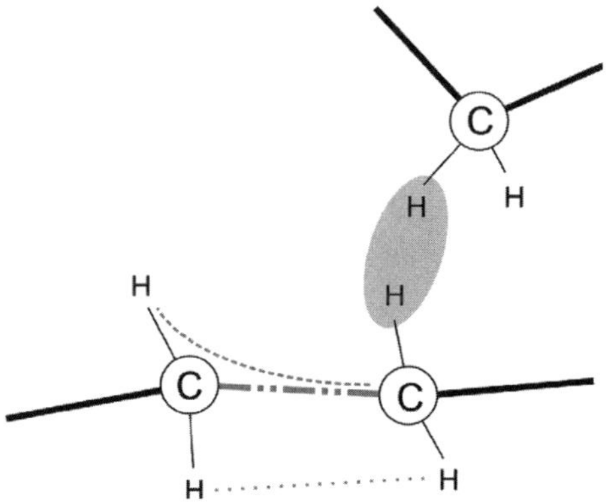

Figure 6 A portion of Figure 3 is expanded to show some of the terms in the objective energy function. Bonds, bends, torsions, and 1-4s are included, involving the two carbon atoms belonging to different coarse-grained units. Lennard-Jones repulsions between all pairs of hydrogen atoms are also included.

counterpart in a low-energy state. Initially, every CG site is replaced with its fragment library structure, placed so that the center of mass of the fragment is at the CG site location. The minimization steps rotate these fragments about their centers of mass to align the unconnected bonds in the fragments with each other, to remove overlapping nonbonded contacts, and in general to find a favorable energy state for the atomistic system. Since the initial condition is far from a local minimum, the conjugate gradient method is of little use; a steepest descent method is performed in stages as follows. Local minima are plentiful when the full potential energy function is used, and with a bad initial condition a nonbonded contact could override an unconnected bond in terms of directing the motion of the atomistic fragments. To avoid this problem, the energy landscape is refined in two stages. Initially, only the bond and bend terms are included. Optimizing on this smooth energy landscape ensures that all the bonds are connected properly. After convergence, the torsions, 1-4s, and Lennard-Jones interactions are turned on and the optimization algorithm is run again. The final configuration is analyzed with the full atomistic force field to see how the potential energy is distributed. Using the equipartition theorem, we would expect to have an energy of $\frac{1}{2}k_{\mathrm{B}}T$ per degree of freedom. Solving for T gives the following results for the intramolecular degrees of freedom:

$$\begin{array}{llll} \text{Bonds}: & T = 294\,\text{K} & \text{Torsions}: T = 75\,\text{K} \\ \text{Bends}: & T = 1125\,\text{K} & 1-4\text{s}: T = 97\,\text{K} \end{array}$$

The bends are too energetic, due to an overdetermined set of equations, but these relax immediately upon running a thermostated simulation.

Incremental Conjugate Gradient Optimization

Once this single configuration has been optimized, we have the option of performing incremental updates on the subsequent configurations under the coarse grain dynamics. Since the CG time step is much larger than the time step used at the atomistic level, the optimized atomistic fragments from the previous CG configuration are no longer appropriate for the updated CG coordinates. However, the conjugate gradient algorithm should work well because the new fragment orientations are not expected to be very different from those optimized to the previous CG coordinates. The conjugate gradient algorithm is used with an arbitrary but very small convergence tolerance for 100 consecutive CG coordinates from a CG MD trajectory. The results are summarized in Figures 7 and 8. Figure 7 shows the objective energy function value plotted for each instance it is evaluated during the conjugate gradient SO(3) optimization procedure.

There are two features worth noticing. First, the conjugate gradient algorithm quickly brings the atomic fragments into line with each new set of CG coordinates. Second, the baseline energy is not constant but varies rather smoothly with each new set of CG coordinates. This occurs because, for example, the CG bonds are flexible (they have a harmonic form) so that the FA bond lengths of the unconnected fragment bonds, with the center of mass of the atomic fragments fixed to the CG sites, cannot always achieve their minimum energy value. Figure 8 shows a histogram of the number of conjugate gradient steps needed to converge the objective function for each of the 100 sets of CG coordinates. It is seen that roughly 10 steps are needed on average. The final configuration is analyzed further by applying the equipartition theorem to it against the full

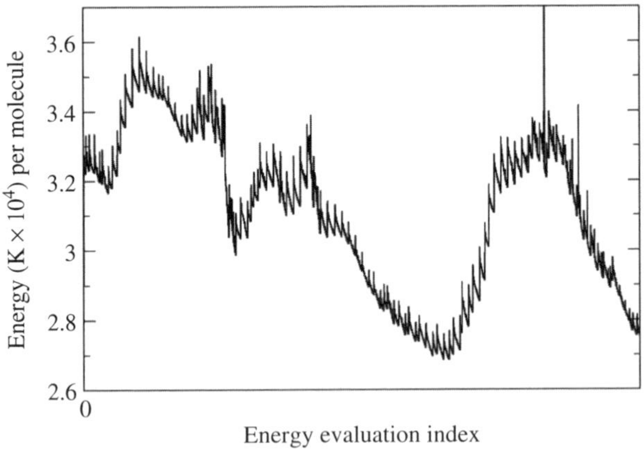

Figure 7 The objective energy function is plotted during the operation of the SO(3) conjugate gradient algorithm acting on 100 consecutive coordinates from a coarse-grained molecular dynamics trajectory of 1050 dodecane molecules at a density of 0.74 g/mL, a temperature of 300 K, and a pressure of 1 bar.

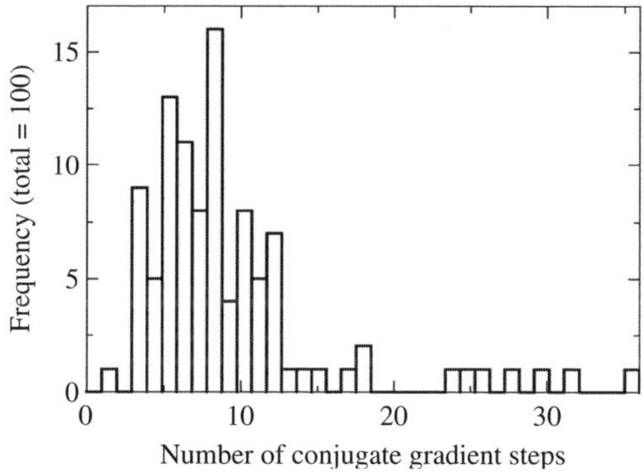

Figure 8 Histogram of the number of conjugate gradient steps needed for the convergence shown in Figure 7 for 100 consecutive coordinates from a coarse-grained molecular dynamics trajectory. The convergence tolerance is set to be an arbitrary, but very small, fixed number.

atomistic force field. The results are similar to those shown earlier in this section.

$$\text{Bonds}: \quad T = 316\,\text{K} \qquad \text{Torsions}: T = 79\,\text{K}$$
$$\text{Bends}: \quad T = 1002\,\text{K} \qquad 1-4\text{s}: T = 247\,\text{K}$$

5. CONCLUSIONS

The coarse grained-to-atomistic mapping algorithm described in this chapter, which is based on SO(3) optimization, is seen to be a mathematically sound, algorithmically efficient, and practical method for recovering the atomistic detail given a reduced level of representation of a molecular system. This algorithm is expected to be of assistance in efforts to validate coarse grain models, a direction that will become increasingly important if coarse-grained simulations are intended to model specific chemical systems. Furthermore, this algorithm will play a central role in our efforts to develop hybrid dual-resolution multiscale simulation methodology, in which atomistic and coarse-grained degrees of freedom coexist in the simulation cell.

REFERENCES

1. Nielsen, S. O.; Lopez, C. F.; Srinivas, G.; Klein, M. L. *J. Phys. Condens. Matter*, 2004, 16: R481.
2. Ortiz, V.; Nielsen, S. O.; Klein, M. L.; Discher, D. E. *J. Polym. Sci. B Polym. Phys.*, 2006, in press.

3. Ortiz, V.; Nielsen, S. O.; Discher, D. E.; Klein, M. L.; Lipowsky, R.; Shillcock, J. *J. Phys. Chem. B*, 2005, 109: 17708.
4. Izvekov, S.; Voth, G. A. *J. Chem. Phys.*, 2005, 123: 134105.
5. Izvekov, S.; Violi, A.; Voth. G. A. *J. Phys. Chem. B*,2005, 109: 17019.
6. Izvekov, S.; Voth, G. A. *J. Phys. Chem. B*,2005, 109: 2469.
7. Queyroy, S.; Neyertz, S.; Brown, D.; Müller-Plathe, F. *Macromolecules*, 2004, 37: 7338.
8. Praprotnik, M.; Delle Site, L.; Kremer, K. *J. Chem. Phys.*, 2005, 123: 224106.
9. Neri, M.; Anselmi, C.; Cascella, M.; Maritan, A.; Carloni, P. *Phys. Rev. Lett.*, 2005, 95: 218102.
10. Abrams, C. F. *J. Chem. Phys.*, 2005, 123: 234101.
11. Taylor, C. J.; Kriegman, D. J. *Minimization on the Lie Group SO(3) and Related Manifolds*, Tech. Rep. 9405, Center for Systems Science, Department of Electrical Engineering, Yale University, New Haven, CT, April 1994.
12. Smith, S. Geometric optimization methods for adaptive filtering, Ph.D. dissertation, Division of Applied Sciences, Harvard University, Cambridge MA, September 1993.
13. Press, W. H.; Teukolsky, S. A.; Vetterling, W. T.; Flannery, B. P. *Numerical Recipes*, Cambridge University Press: New York, 1992.
14. Nielsen, S. O.; Lopez, C. F.; Srinivas, G.; Klein, M. L. *J. Chem. Phys.*, 2003, 119: 7043.

6

MICROSCOPIC INSIGHTS INTO THE DYNAMICS OF PROTEIN–SOLVENT MIXTURES

Taner E. Dirama and Gustavo A. Carri
Department of Polymer Science and The Maurice Morton Institute of Polymer Science, The University of Akron, Akron, Ohio

Proteins are vital molecules for human beings because they perform a variety of critical functions, ranging from the catalysis of chemical reactions needed by the metabolism to the defense of the immune system against bacteria and viruses. In instances where the body is unable to produce the needed proteins by the amount required to carry out their functions, proteins can be introduced in the body externally. The potential of using proteins in pharmaceutical applications has long been acknowledged, but only recently this idea has gained substantial momentum, thanks to advances in genomics research. As developing genetically engineered, protein-based pharmaceuticals is a great technological challenge, due to the difficulty in maintaining the chemical and physical stability of the formulation, protein-based drugs are usually applied through injection as opposed to oral use.[1] However, there is a considerable interest, driven mainly by ease of use and economic considerations, in the pharmaceutical industry to improve the formulations and develop novel ones that stabilize the biological molecules in a dry form.[2] The most common method in creating protein-based pharmaceuticals in the dry form is through lyophilization, where the protein solution is first frozen and, afterward, the frozen solid is dried under vacuum conditions. The long-time stability of proteins in the resulting, largely moisture-free protein–biopreservative mixture is challenged by physical as well as chemical factors. While chemical reactions constitute a major degradation pathway such as oxidation of amino

Multiscale Simulation Methods for Nanomaterials, Edited by Richard B. Ross and Sanat Mohanty

acids, protein can also lose its function by conformational denaturation due to its interactions with the surrounding medium. The conformational denaturation may have thermodynamic or dynamic origins. Experimental evidence points to the importance of dynamics: neutron scattering experiments have suggested that there is an inverse relationship between the thermal stability and the dynamics of a protein powder.[3] A significant number of studies are currently in progress to unveil the relationship between protein dynamics and protein stability. However, the question of how the dynamics affects the stability of the protein does not yet have a clear answer. Naturally, understanding the molecular origins of protein stability and denaturation is a first-class problem in biological physics.[4,5]

While the denaturation of proteins occurs over a very wide time window, short-time-scale protein motions play a fundamental role in the long-time dynamics because they could be precursors to large-scale conformational rearrangements.[2] First, it is known that the rapid and local atomic fluctuations in the native conformation of a protein play a key role in protein functions that involve large-scale motions.[6] Additionally, it has been shown that these fluctuations serve as a "lubricant" for large-scale conformational rearrangements such as perturbation of the average structure by the binding of ligands.[7,8] This idea can also be extrapolated to the large-scale protein motions (e.g., those required for the conformational denaturation of the protein). Namely, suppressing the atomic motions in the pico-nanosecond time range slows down the motions of the protein at longer time scales and, consequently, improves the conformational stability of the protein. Indeed, this hypothesis has been supported by recent studies that combine the analysis of the short-time dynamics and the measurements of enzymatic activity.[2,9] These arguments suggest that a study of short-time dynamics in the pico to nanosecond range has the potential of revealing crucial information, which, in turn, relates to the conformational denaturation of proteins. Our main objective here is to shed some light on the effect of the dynamics of the surrounding medium on the protein dynamics using results obtained from atomistic molecular dynamics simulations.

Trehalose and Glycerol as Biopreservation Agents While combinations of various ingredients are used in formulating the host environment for biological agents, today simple sugars and sugar alcohols are used as principal constituents.[2] A series of carbohydrates have been examined and the effectiveness of their glasses in biopreservation has been illustrated.[10–12] The basic molecular mechanisms that render carbohydrates effective are still a subject of discussion. Yet there appear to be two mechanisms that come forward when explaining the protective effect of carbohydrates on freeze-dried proteins. First, they can form hydrogen bonds with the protein surface, thus serving as water substitutes. Second, they form glasses at room temperature and their glass transition temperature (T_g) has been correlated with the stability of the hosted protein.[13–15]

The selection of carbohydrates as candidates for biopreserving agents originated from observations done in living organisms. For instance, it has been

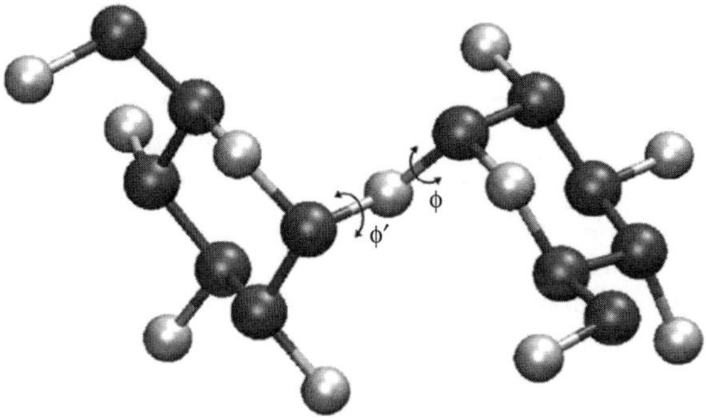

Figure 1 Trehalose (α-D-glucopyranosyl α-D-glucopyranoside).

observed that trehalose, a disaccharide, (see Figure 1) is present in large concentrations in cells undergoing anhydrobiosis.[16] This finding is followed by a number of studies that focused mainly on understanding the molecular mechanisms responsible for the enhanced effectiveness of trehalose in the biopreservation of living organisms.[17–20] Some of the explanations are as follows. Trehalose, with four hydroxyl sites in each ring, is capable of making hydrogen bonds with the polar and charged groups of proteins that are normally hydrated. This results in the creation of an environment that is similar, in terms of the local dynamics of the protein, to the one seen by the protein in aqueous solutions.[16,21–24] Additionally, trehalose is in the glassy state up to a distinctively high temperature of 390 K. These reasons motivated us to select trehalose as a model solvent for studying how glassy solvents suppress the dynamics of proteins.

Although it is a small triol, glycerol has proven to be very efficient in stabilizing the native state of proteins and preserving the activity of enzymes at low temperatures, and has become one of the most widely used cryopreservatives.[25] Additionally, it is liquid at temperatures as low as 190 K. Thus, it is a good model system to study the dynamics and compare it to glassy trehalose.

1. MODELING AND SIMULATION PROTOCOL

We begin this section with a brief and general description of the molecular dynamics (MD) simulation method followed by a discussion about the validation of force fields employed in this work. The section concludes with a presentation of the details of the simulation protocol.

Brief Description of Molecular Dynamics Simulations

The idea behind the MD method was introduced over 50 years ago to study the interactions of hard spheres.[26] However, nearly two decades passed until the first MD simulation of a realistic system was performed for the simulation of liquid water.[27] Simulation of the bovine pancreatic trypsin inhibitor (BPTI) was the first simulation of a protein system.[28] Today, MD simulation has become a conventional method for the study of solvated proteins and protein–DNA complexes addressing a variety of problems. MD simulation techniques are widely used in conjunction with experimental procedures such as x-ray crystallography and nuclear magnetic resonance structure determination.

MD simulations are commonly employed to compute the equilibrium and transport properties of many-body systems.[29] It is similar to experiments in the sense that the measured quantities are subject to statistical noise since the properties evolve with certain time intervals. Successive configurations of the system are generated by solving Newton's equations of motion, which are solved using finite-difference methods such as the leapfrog Verlet and Beeman's algorithms.[29] The integration is broken down into small time intervals (i.e., δt), usually on the order of a femtosecond. The total force applied on a particle at time t is calculated as the vectorial sum of its interactions with other particles. The acceleration on the particle is calculated from the force and it is combined with the positions and velocities at time t in order to calculate the positions and velocities at time $t + \delta t$. The main assumption in this estimation is that the force remains constant during each time step. Therefore, the selection of the time interval is crucial for a successful MD simulation.[30]

The MD simulation algorithms begin by reading in the parameters that specify the conditions of the simulation run and those that describe the system. Subsequently, the velocities of the particles are usually assigned randomly, but they have to satisfy that the combined momentum must have zero magnitude. Then the forces on all particles are calculated based on the potential function with force-field parameters.[30]

Validation of the Force Field Selected

The success of classical molecular dynamics simulations is strongly dependent on the soundness of the potential function and its empirically determined constants (i.e., force field) for a given problem. The efficacy of the given potential and the force field should be verified prior to reaching conclusions from the output of the simulation. In this section we validate the force fields employed in modeling the solvents (glycerol and trehalose) and the protein (hen egg white lysozyme). Further validation of the force field and simulation protocol is presented in the subsequent sections.

The solvents of the simulation work presented in succeeding sections were modeled using the general AMBER force field (GAFF).[31] GAFF is an extension of the AMBER force field, which is designed specifically for amino acids and nucleic acids, with the purpose of encompassing a wider variety of organic molecules. The performance of the force field parameters has been tested for three different cases.[31] In the first test case, 74 crystallographic structures were compared to the GAFF minimized structures. It is noteworthy to point out that all of these test structures are different from the molecules used in the parameter development of the force field. GAFF produced root-mean-square deviation

(MSD) of atomic displacement of 0.26 Å, which is comparable to that of the Tripos 5.2 force field (0.25 Å) and better than those of MMFF 94 and CHARMm (0.47 and 0.44 Å, respectively). In the second test case, the gas-phase minimizations of 22 nucleic acid base pairs were carried out. The MSD of displacements and relative intermolecular energies were comparable to results from Parm99 force field, which was parameterized to these base pairs. In the third test case the relative energies of 71 conformational pairs that were used in development of the Parm99 force field were calculated. The MSD error in relative energies (compared to experiment) is about 0.5 kcal/mol. These test cases show that GAFF can be applied successfully to a wide range of molecules.[31]

We performed additional validation studies of the GAFF force field by reproducing experimental crystal lattice parameters for glycerol and trehalose. The glycerol crystal lattice has been created by replication of the orthorhombic unit cell (space group $P2_12_12_1$, $Z = 4$). The experimental lattice parameters for glycerol are $a = 7.00$ Å, $b = 9.96$ Å, and $c = 6.29$ Å.[32] The simulation box was created by replicating the unit cell six times in each direction. The resulting system consisted of 864 glycerol molecules. After energy minimization the system was equilibrated initially in NPT simulation for 50 ps followed by a 100-ps data collection run at 10 K. The same procedure was applied as the temperature was raised by 10 K each time. At 200 K, where experimental data were reported (198 K), the system had undergone 2.8 ns of simulation. The analysis of the radial distribution function has resulted in lattice constants $a = 7.04$ Å, $b = 9.69$ Å, and $c = 6.59$ Å. The error margin of 3% suggests that the general AMBER force field (GAFF)[32] used for glycerol is able to reproduce an important structural property, in good agreement with experiments. The same analysis for the trehalose dihydrate crystal lattice has been carried out. The results were also in a good agreement with experimental lattice parameters of $a = 12.23$ Å, $b = 17.89$ Å, and $c = 7.59$ Å.[33]

One of the most common physical properties used in evaluating the success of a force field by reproducing experimental quantities is the heat of vaporization (ΔH_{vap}). Chelli et al.[34] used an AMBER-type force field developed by Cornell et al.[35] The GAFF force field is an improved version of this force field which emerged with the purpose of being able to model general organic molecules successfully. An experimental value of ΔH_{vap} for glycerol is 76.1 kj/mol. This MD simulation study reported a ΔH_{vap} value of 73.0 kj/mol, which shows good quantitative agreement with experiment.[35] The authors of ref. 35 have also calculated the cohesive energy density of the glycerol crystal at the melting point. The simulation estimated that $E_{coh} = 98.3$ kj/mol, whereas the experimental value was $E_{coh} = 94.8$ kj/mol. Additional calculations of molar volume of glycerol were carried out. Comparison with experiments showed that the simulation results were in a very good agreement with experiments, especially at low temperatures. Up to 300 K the agreement is within 1%, whereas at 400 K this error margin increases to 3%. These validation results of ref. 35 along with our comparison of the crystal lattice parameters for glycerol demonstrate that AMBER-type force fields are capable of modeling this molecule with very high accuracy.

The main challenge in modeling carbohydrates is related to the interplay between the intermolecular and intramolecular forces. Accurately modeling of these interactions is a major task for force field development.[37] Glennon and Merz analyzed the torsional motions of carbohydrates using methanol as a model. The experimental rotational barrier of methanol was reported to be 1.09 kcal/mol, while an MD simulation study based on an AMBER force field found a value of 1.04, which is in excellent agreement with experiment.[38] Another model molecule that has been used to test an AMBER force field is α-D-glucose. In line with the experimental crystal structure[39] the orientation of the hydroxymethyl group was shown to have a minimal effect on the ring structure. Specifically, the different conformations of the hydroxymethyl group in glucose (i.e., GG and GT orientations) resulted in nearly the same bond and dihedral angles in the ring.

The disaccharide α,α-trehalose is composed of two glucopyranosyl units linked together by a glycosidic oxygen bridge between their anomeric carbon atoms. The rotation around the glycosidic oxygen and these carbon atoms largely determines the conformation of trehalose (see ϕ and ϕ' in Figure 1). Therefore, the study of conformations of these dihedral angles is imperative for the conformation of the molecule as a whole. Bonanno et al.[40] compared these dihedral angles as obtained by experiments to the results of MD simulations employing various force fields. Their evaluation showed that the experimental data measured by Duda and Stevens[41] using optical rotatory dispersion (ORD) ($\phi = 60$ and $\phi' = 60$) are mostly compatible with the MD simulation results of Sakurai et al.[42] ($\phi = 54 \pm 13.6$ and $\phi' = 54 \pm 12.9$), who implemented an AMBER force field for the modeling of trehalose. The aforementioned validation analyses of the GAFF force field demonstrate that it can be used successfully to model both glycerol and trehalose. Further validation of the force field is shown in subsequent sections.

The model protein lysozyme was modeled using a Parm99 force field, which is a slightly improved version of Parm94 force field.[35] This AMBER force field is one of the most preferred force fields in the modeling of amino acids and nucleic acids. It has been tested against quantum mechanical calculations as well as experimental results. Such test cases include conformational energies of large fundamental units of nucleic acids (deoxyadenosine nucleoside), the normal-mode frequencies of the four nucleic acid bases and solvation free energies of a representative set of molecules. These quantities, along with others, showed that the AMBER force field can model biological molecules (i.e., amino and nucleic acids) and the results are in good agreement with experiment and quantum mechanical calculations.

One of the research areas where the AMBER force field has been used is for the study of protein docking. For example, Sotriffer et al.[43] used the AMBER force field to study protein–ligand interactions by computational docking. Given the structures of the protein and the ligand, the objective was to determine the structure of the resulting complex. Three complexes were used as new test cases to analyze the ability of the method to reproduce experimental results. The x-ray structures could be reconstituted, and the corresponding solutions were ranked with the best energy score in all cases.[43]

Simulation of Lysozyme in Glycerol and Trehalose

The AMBER 8.0 molecular dynamics package[44] with ff99[45] (a common AMBER force field for proteins) was used in this work to model lysozyme. The structure of hen egg white lysozyme was obtained from the Protein Data Bank (193L). Using the most probable ionization states of lysozyme at neutral pH results in a net charge of + 8. To satisfy a zero net charge, mobile counter ions were added. Namely, 8 Cl^- ions were placed randomly around the protein. The Lennard-Jones parameters for the interactions between different atom types were derived from those of same atom types using standard Lorentz–Berthelot combination rules.

Rectangular parallelepiped periodic boundary conditions were used. Long-range electrostatic interactions were calculated using the particle-mesh Ewald (PME) method, while van der Waals interactions were calculated using the Lennard-Jones 6–12 potential. The cutoff distance for nonbonded van der Waals interactions was set to 8 Å. However, in the case of electrostatic interactions, this cutoff is used for the evaluation of Ewald's standard direct sum; corrections are taken into account via the reciprocal sum. The hydrogen bonds in this all-atom potential function are represented by a balance between electrostatic and van der Waals interactions. Random initial velocities were assigned to all atoms after minimization of the initial structure. The equations of motion were integrated using the leapfrog Verlet algorithm with a step size of 1 fs. Constant temperature and pressure were satisfied by a weak coupling algorithm.[46]

For the simulation of lysozyme in glycerol, a simulation box of glycerol was preequilibrated at 300 K and the energy-minimized structure of lysozyme was placed at the center. The glycerol molecules within a proximity of 2 Å from the surface of lysozyme were removed from the simulation box. The resulting protein–solvent mixture with 1118 glycerol molecules was then equilibrated, first under constant-volume conditions (NVT) for 50 ps and then under isobaric conditions (NPT) for 300 ps at 500 K and 0.1 MPa. During this step of the equilibration, harmonic restraints were put on the protein atoms. Following an additional equilibration at 300 K for 300 ps, where the restraints were removed, the data collection run was performed in isobaric conditions for 2 ns and coordinate sets were saved for every 0.1-ps interval for subsequent analysis. For the simulations at 250, 200, and 150 K, the system was first annealed to 50 K below the current temperature at a cooling rate of 0.1 K/ps followed by equilibration in the NPT ensemble at this temperature before the data collection run.

For the simulation of lysozyme in trehalose, a simulation box of trehalose was preequilibrated at 300 K and the energy-minimized structure of lysozyme was placed at the center. The trehalose molecules within 2 Å from the surface of lysozyme were removed from the simulation box. The resulting protein–solvent mixture with 283 trehalose molecules was then equilibrated first under constant volume and temperature conditions for 50 ps and then under isothermal–isobaric conditions for 300 ps at 500 K and 0.1 MPa. Afterward, the simulation box was annealed to 300 K at a rate of 0.1 K/ps. During this stage of the equilibration, harmonic restraints were put on the protein atoms to maintain the native structure. Following an additional equilibration at 300 K for 2.7 ns where the restraints were removed, the data collection run was performed in isothermal–isobaric conditions for 2 ns. For the simulations at 250, 200, and 150 K, the system was first annealed to 50 K below the current temperature at a cooling rate of 0.1 K/ps followed by an equilibration in the NPT ensemble at this temperature before the data collection run. At temperatures higher than 300 K, the system was equilibrated at the higher temperature followed by a data collection run.

2. SOLVENT-INDUCED SUPPRESSION OF PROTEIN DYNAMICS

Proteins show a resemblance to glassy systems[48] in the sense that they are complex systems with many conformational substates related to local minima of the potential energy surface.[49] The molecular motions occurring within substates close to the global energy minimum are crucial for protein function. As discussed earlier, the rapid and local atomic fluctuations in the native conformation of a protein are important in protein function, which involves large-scale motions.[6] These rapid and local atomic fluctuations serve as a "lubricant" for large-scale conformational rearrangements.[7,8]

The idea that rapid and local atomic fluctuations are related to larger-scale conformational motions can be elaborated using the concept of a hierarchical arrangement of substates.[50] According to this concept, the substates in tier 0 describe a small number of different conformations, each corresponding to different functions. Each of these tier 0 substates can assume a large number of tier 1 statistical substates which perform the same function at varying rates. Since the molecular motions in tier 0 occur as a result of successive transitions among substates in tier 1, Fenimore et al. suggested that freezing out these transitions would prevent occurrence of the motions in tier 0.[51] This idea could be extrapolated to large-scale protein motions whose characteristic time scales exceed those of the transitions required for protein function and lead to the conformational denaturation of the protein. Namely, suppressing the atomic motions in the pico- to nanosecond time range may prevent the motions of the protein at longer time scales and consequently, may help improve the conformational stability of the protein. Indeed, this idea has been supported by recent studies that combine analysis of the short-time dynamics and the measurements of enzymatic activity.[2,9]

The solvent surrounding the protein could play an essential role in the activation of the fluctuations among the substates.[48] In fact, it has been reported that there is a correlation between the structural fluctuations of the protein and the thermal motion of water.[52] Specifically, it has been proposed that these fluctuations are promoted by solvent water molecules through a hydrogen-bonding network in bulk water which occurs on the same picosecond time scale as the conformational fluctuations.[53] In addition, Paciaroni and collaborators have recently found a direct correlation between the logarithm of the solvent viscosity and the mean-square displacements of the protein hydrogen atoms for lysozyme in water and in glycerol–water and glucose–water mixtures.[54] This experimental correlation further supports our previous statement. In this section we discuss how different solvents influence the dynamics of the protein and how their molecular properties correlate with their biopreservation abilities. Specifically, we show how trehalose and glycerol control the dynamics of the model protein using insights obtained from molecular dynamics simulations in the pico- to nanosecond time window.

It appears that glycerol is a better biopreserving agent than trehalose at low temperatures, while the reverse is valid at high temperatures.[55] Indeed, low-temperature Raman spectroscopy of lysozyme in glycerol and trehalose has revealed that the former is a better biopreserving agent (i.e., it suppresses the dynamics of the protein more effectively) than the latter at low temperatures, even though trehalose is deeper in the glassy state.[56] However, trehalose is known to be more effective than glycerol at room temperature and has been reported to be the most effective bioprotecting agent by means of functional recovery.[56] These conclusions have been verified further by studies of geminate CO rebinding to myoglobin after flash photolysis in trehalose and glycerol environments, which showed significant differences in terms of reaction kinetics.[57,58] Namely, at low temperatures (< 200 K), glycerol inhibits the reaction kinetics more effectively.[57] These counterintuitive findings in the dynamics of the protein could be related to differences in protein–solvent interactions.[55] Here we focus on identifying and studying these interactions and their influence on protein dynamics.

To demonstrate the effect of solvent on the dynamics of the protein in the 1-ns time window, we compared the dynamics of lysozyme in pure glycerol and in pure trehalose. For this purpose we utilized the incoherent intermediate scattering function of the hydrogen atoms in the protein:

$$S(q,t) = \frac{1}{N}\left\langle \sum_{j=1}^{N} e^{i\mathbf{q}\bullet[\mathbf{R}_j(t)-\mathbf{R}_j(0)]} \right\rangle \tag{1}$$

where $\mathbf{q}$ is the scattering wave vector, t is time in nanoseconds, N is the total number of hydrogen atoms, $\mathbf{R}_j(t)$ is the position of the jth hydrogen atom at time t, and the angle brackets indicate average over time origins. This function carries information about the structural relaxation of the protein.

At 300 K, the dynamics of lysozyme in glycerol is faster than in trehalose, as shown by the faster decay of $S(q,t)$ with time in Figure 2*a*. The physical origin of this result is the liquid nature of glycerol, which is above its glass transition temperature T_g of 192 K,[59] and the glassy behavior of trehalose, whose T_g value is 388 K.[60] At 300 K, glycerol allows fluctuations among the conformational states of the protein beyond the harmoniclike motions, whereas trehalose prevents such anharmonic motions constraining the dynamics.[61] However, at lower temperatures the dynamic behavior of lysozyme reverses. Indeed, at 200 K lysozyme becomes slightly slower (in our time window) in glycerol than in trehalose, although the latter is *substantially deep in the glassy state while the former is in the liquid phase*. This is shown by a slower decay of $S(q,t)$ as a function of time in Figure 2*a*. Another interpretation of the dynamics of lysozyme is presented in the inset of Figure 2*a*, where the ratio of the mean-square displacement

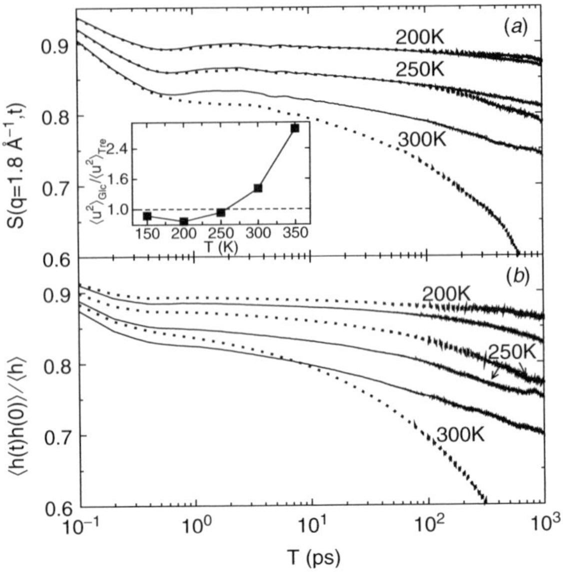

Figure 2 (a) Incoherent intermediate scattering function of hydrogen atoms (without methyl group hydrogens) in lysozyme immersed in trehalose (continuous lines) and glycerol (dotted lines) at three temperatures. The inset shows the ratio of $\langle u^2 \rangle$ for lysozyme hydrogen atoms in glycerol and trehalose as a function of temperature. Averaging of $\langle u^2 \rangle$ (t) was carried out over 1 ns. (b) Hydrogen-bond correlation function for lysozyme–trehalose (continuous lines) and lysozyme–glycerol (dotted lines) systems at three temperatures.

$\langle u^2 \rangle$ of lysozyme in glycerol to that in trehalose at 1-ns resolution is shown. Clearly, glycerol provides a better suppression of protein dynamics compared to trehalose at low temperatures. This result agrees with experiments done on the same systems and was attributed to the superior effectiveness of glycerol in the preservation of biological agents at low temperatures.[62]

The dynamics of proteins are influenced greatly by hydrogen bonds.[63] Since the solvents under consideration are capable of forming hydrogen bonds, it is expected that the dynamics of such protein–solvent systems will be influenced by the formation of hydrogen bonds at the protein–solvent interface. Consequently, we evaluated the protein–solvent hydrogen-bonding network using geometric criteria based on the distance between donor and acceptor oxygens and the angle formed by the donor oxygen, the acceptor hydrogen, and the acceptor oxygen.[64] For this purpose, we computed the hydrogen-bond correlation function,

$$c(t) = \frac{\langle h(t)h(0) \rangle}{\langle h \rangle} \tag{2}$$

where $h(t)$ is the hydrogen-bond population operator, which is equal to 1 when a donor–acceptor pair satisfies the hydrogen-bond criteria at time t and it is zero otherwise. $c(t)$ yields the probability that a random donor–acceptor pair that is hydrogen bonded at time zero is still bonded at time t. Thus, the decay of $c(t)$ measures the effectiveness of the hydrogen-bond network between protein and solvent molecules.

Figure 2*b* shows $c(t)$ for lysozyme in pure glycerol and in pure trehalose at different temperatures. The initial decay corresponds to the rotation and libration of solvent molecules, which lead to hydrogen bonds with lifetimes shorter than 1 ps. Therefore, they do not affect the dynamics of the protein in a significant manner. However, the secondary decay observed at longer times characterizes long-living hydrogen bonds between the protein and the solvent. Since the time scales for the long-living hydrogen bonds are comparable to those that correspond to the structural relaxation of the protein [e.g., relaxation of $S(q,t)$], the secondary decay in the correlation function has important implications on the dynamics of the protein. Comparison of the hydrogen-bond correlation function with $S(q,t)$ shows a clear correlation between both functions. At low temperatures, the effectiveness of the hydrogen-bonding network between lysozyme and glycerol is more robust than the one between lysozyme and trehalose (i.e., the hydrogen-bond correlation function for the latter decays faster than for the former at long times even though trehalose is a glass while glycerol is in the liquid state, whereas the opposite is true at 300 K). At intermediate temperatures (250 K), the strengths of the hydrogen, bonding networks for both systems are similar, in line with the similar dynamics displayed by lysozyme in both solvents (Figure 2*a*).

The correspondence between the analysis of hydrogen bonds and the structural relaxation of the protein suggests that the hydrogen-bonding network at the protein–solvent interface is responsible for the enhanced effectiveness of glycerol as a bioprotecting agent at low temperatures. Consequently, the answer to the question of why glycerol is a better biopreserving agent (i.e., suppresses the protein dynamics more effectively) than trehalose at low temperature is that lowering the temperature strengthens the intermolecular hydrogen bonds between lysozyme and glycerol at a faster rate than with trehalose. These different temperature dependencies lead to a crossing at a temperature close to 250 K.

Based on the above-mentioned arguments we infer that the pico- to nanosecond dynamics of proteins correlate with their conformational stabilities. Therefore, our findings provide a molecular explanation for the enhanced ability of glycerol in biopreservation applications at low temperatures. However, the large difference between the time scales accessible to MD simulations and those involved in the denaturation of proteins does not permit us to draw a definite correspondence between the pico- to nanosecond dynamics and the denaturation of proteins.

3. PATHWAYS LEADING TO SUPPRESSION OF THE PROTEIN DYNAMICS AND THE ROLE OF SOLVENTS

As illustrated above, solvents have a profound effect on protein dynamics. The next question to be answered is: What are the mechanisms through which the two molecules interact with each other that lead to this eminence of the solvent over the protein? We dwell on the following ideas that we introduce below: the protein dynamic transition and the coupling of protein and solvent dynamics. Atomistic molecular dynamics simulations are particularly valuable in clarifying the molecular mechanisms for both processes. First, we reproduce the protein dynamic transition and dynamic coupling between the solvent and the polymer and then seek the details of the molecular mechanisms behind these processes.

Protein Dynamic Transition

The biological function of a protein decreases with decreasing temperature, a phenomenon possibly due to the suppression of the transitions between conformational substates.[65] Similar to the α- relaxation in glassy systems, proteins exhibit a relaxation process that governs the large-scale motions.[48] This marks the separation of two types of molecular motions. At low temperature, the protein undergoes small-amplitude harmoniclike motions, and at high temperature, large-amplitude anharmonic motions dominate the dynamics. The change in dynamic behavior from harmoniclike to anharmonic motions is commonly called the *dynamic transition*.[66] This transition, observed by experimental[48] as well as MD simulation methods,[67] promotes the flexible motions in the protein that are widely accepted to be required for biological function.[65] Additionally, correlations have been found between the presence of anharmonic motions and protein activity.[68,69] These findings demonstrate that the dynamic transition is an important process that affects the protein function.

The protein dynamic transition observed routinely at $T \sim 200$ to 230 K for proteins in water was suggested to originate from the solvent glass transition.[70] According to this scenario the solvent molecules slow down and trap the protein molecules in long-living conformations such that the conformational motions necessary for function are strongly hindered. This scenario was supported by the findings that the conformational relaxation of a protein can be suppressed by highly viscous solvents even at room temperature.[17] MD simulations by Vitkup et al.[70] showed that the magnitude of the protein fluctuations is determined largely by solvent viscosity. Based on these results it was concluded that the protein motion below the dynamic transition is inhibited mainly by high solvent viscosity.[17,48,67]

One of the most widely studied properties of protein–solvent systems is the dynamic transition temperature, T_d, which is the temperature at which the conformational component, using the terminology of Fenimore et al., of the atomic mean-square displacement of the atoms $\langle u^2 \rangle$ in the protein departs from zero.[51] We start the presentation of the results of the MD simulations with a verification of the simulation protocol by reproducing T_d. We choose the protein hydrogens to compute T_d and other dynamic quantities, such as the mean-square displacement, $\langle u^2 \rangle$, for two reasons. First, they are distributed throughout both the protein and solvent molecules, thus making it possible to probe the dynamics of all molecules consistently. Second, experimental data from incoherent neutron spectroscopy is dominated by hydrogen atoms, and thus the hydrogens are a useful probe that allow validation of the simulation methodology. Consequently, the results can be compared with experimental data directly. T_d can be estimated from the simulation data as the intercept between two straight lines: the first line is a fit of $\langle u^2 \rangle$ as a function of temperature at low temperatures, and the second line is the same fit but at high temperatures. $\langle u^2 \rangle$ was derived from the MD trajectories through the relationship

$$\langle u^2 \rangle = \langle (r(t + t_0) - r(t_0))^2 \rangle \tag{3}$$

where $r(t_0)$ and $r(t + t_0)$ are the coordinates of atoms at reference time t_0 and after time t. Brackets represent averaging over the hydrogen atoms and the reference time. The MSD for each temperature was obtained after averaging over 800 ps using multiple time origins.

Lysozyme in Glycerol Figure 3 shows a plot of $\langle u^2 \rangle$ for the hydrogen atoms in lysozyme as a function of temperature. The two regimes traditionally observed in protein–solvent systems are clearly visible in Figure 3. At low temperatures, $\langle u^2 \rangle$ increases linearly with increasing temperature up to ~250 K. Above this temperature the dynamics are largely nonlinear. The low-temperature behavior shown in Figure 3 is typical of a harmonic oscillator. Thus, in this regime the

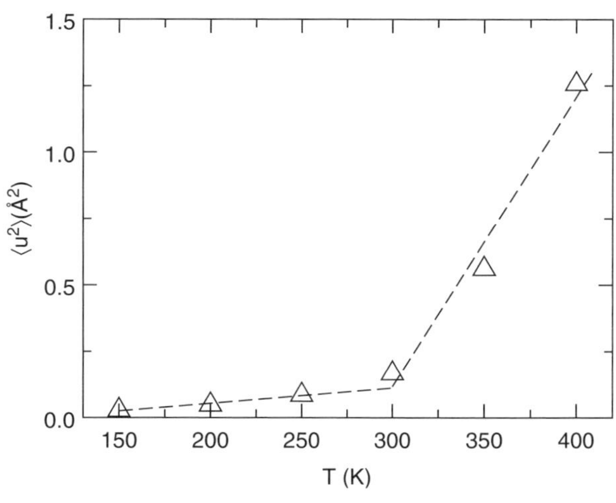

Figure 3 Temperature dependence of the mean-square displacement of hydrogen atoms in lysozyme immersed in glycerol. (Reprinted with permission from T. E. Dirama et al., *J. Chem. Phys*., 2005, 122: 244910. Copyright © 2005, American Institute of Physics.)

atoms in the protein can be visualized as vibrating in a harmoniclike potential around their equilibrium positions. At high temperatures, the deviation of $\langle u^2 \rangle$ from linear low-temperature behavior becomes clear and is due to the activated transitions between different substates. These transitions give rise to anharmonic motions. The intercept between both straight lines occurs around 300 K, implying that T_d for the system, which contains 12% of lysozyme by weight, is located close to 300 K. Tsai et al.[71] have reported values of T_d for 80% and 50% (by weight) lysozyme in glycerol to be 330 K and 270 K, respectively. Our estimate agrees with these results on a quantitative level, providing some experimental support to the simulation protocol and the force field employed in this study.

Lysozyme in Trehalose Figure 4 shows $\langle u^2 \rangle$ for the hydrogen atoms in lysozyme as a function of temperature. $\langle u^2 \rangle$ was obtained after averaging over 800 ps of the 2-ns data set using multiple time origins. A straight line was fitted to the first four data points at low temperatures. In this region $\langle u^2 \rangle$ increases linearly with temperature up to 300 K, whereas above this temperature, $\langle u^2 \rangle$ escalates nonlinearly, indicating that there are two distinct dynamic regimes, as has been observed experimentally in various solvated proteins.[19,20,71] In the trehalose–lysozyme system studied in this work, the onset of anharmonic behavior was found to be ~ 350 K, which is below the glass transition temperature of trehalose. Previous neutron scattering[19] and Mössbauer spectroscopy[20] measurements of myoglobin (MbCO) at similar time resolutions did not show a clear indication of the dynamic transition because $\langle u^2 \rangle$ data were reported for temperatures below 300 K.

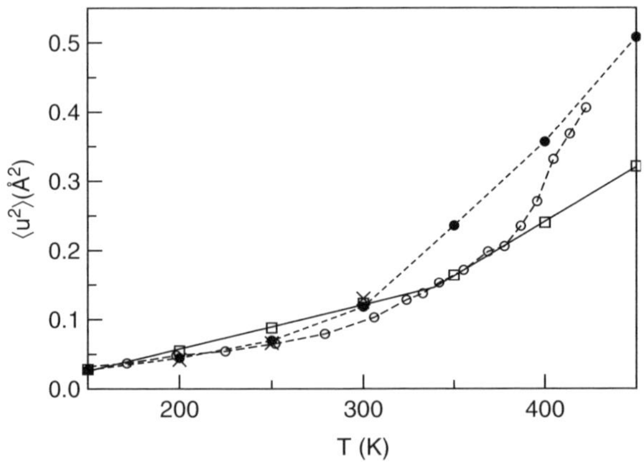

Figure 4 Temperature dependence of the mean-square displacement of hydrogen atoms in lysozyme immersed in trehalose (□) and in the surrounding trehalose molecules (•) and (O). Experimental data for pure trehalose from ref. 9. The continuous lines are linear fits to the low- and high-temperature behavior of lysozyme. Mean-square displacement values were calculated for 0.8 ns of the simulated trajectory. (Reprinted with permission from ref. 61. Copyright © 2006, American Institute of Physics.)

Figure 4 also shows $\langle u^2 \rangle$ for the hydrogen atoms in trehalose and experimental data for pure trehalose (empty circles).[9] $\langle u^2 \rangle$ for trehalose obtained from the simulation displays a change in dynamics at temperatures close to 300 K, which is below the glass transition temperature. This change in dynamics of a dehydrated trehalose glass agrees with the experimental results (see Figure 4),[9] thus also validating the simulation protocol for trehalose. The analysis of the experimental data for pure trehalose shows three temperature regimes. The first regime is below 300 K, the second regime is between 300 and 390 K, and the last regime is above 390 K. The experimental T_g value of trehalose is ~390 K. Since the change in the temperature dependence of $\langle u^2 \rangle$ of trehalose around 300 K occurs deep in the glassy state, it should be related to a secondary relaxation and not to a translational diffusion of trehalose molecules. Indeed, it is known that pure trehalose has secondary relaxations active in the glassy state.[72] According to dielectric studies,[72] this secondary relaxation process is strongly stretched and should reach the nanosecond time window at temperatures $T \sim 330$ to 390 K. This agrees well with the results of the simulations (Figure 4) and neutron scattering studies.[9]

The onset of the protein dynamic transition in trehalose occurs over a much broader temperature range than that observed in hydrated protein powders or proteins in glycerol.[19,20,71] This onset of anharmonic behavior of the protein occurs above the 300 K transition temperature of trehalose, suggesting that some softening of the glass is required before the protein can undergo dynamic transition. In other words, the T_d of lysozyme being above the onset of this secondary relaxation implies that for large solvent molecules, a change in conformation is enough to break the hydrogen bonds between the protein and the solvent, thus allowing the protein to begin to undergo dynamic transition. This mechanism is different for small solvent molecules such as water.

Previous experimental studies have not reported a protein dynamic transition in trehalose, possibly due to the temperature range studied, which is limited by the thermal degradation of trehalose at high temperatures. Thus, the fact that these data can be collected by simulations at high temperatures which are experimentally inaccessible enabled us to detect the dynamic transition for the lysozyme–trehalose system. Furthermore, it was found that the onset of the anharmonic motions inherent in the protein dynamic transition are observed far below the T_g of trehalose, indicating that the relaxation of the protein in such a glass is extended over a much wider temperature range than an hydrated amorphous powder or proteins in glycerol.

Coupling of Protein and Solvent Dynamics

It is well known that the temperature of the dynamic transition is affected greatly by the surrounding solvent.[48] The role of solvent has been demonstrated in many experimental studies on different proteins, such as lysozyme and myoglobin.[51,73–76] The strong influence of the solvent on the protein dynamics has been described as the protein being a "slave" of the solvent or coupling of protein and solvent dynamics.[62,76,77] Moreover, MD simulations of myoglobin in water[70] found that solvent mobility is a crucial parameter in the determination of the atomic fluctuations in the protein, which, in turn, determine T_d. Other MD simulation studies (e.g., ribonuclease A in water[78] and copper plastocyanin in water[79]) have provided additional support for the strong coupling of protein and solvent dynamics.

The ability of certain chemical agents utilized in biopreservation applications could be correlated with this dynamic coupling. Namely, the protein becomes "slaved" to the solvent molecules and its dynamics is suppressed, resulting in retardation of the denaturation process. The α-relaxation of the solvent has generally been regarded as the most significant process influencing protein dynamics. However, Caliskan et al.[62] have suggested that the fast conformational fluctuations of glass-forming systems that usually occur on a picosecond time scale may also influence protein dynamics. Using Raman and neutron spectroscopy, they showed that on a picosecond time scale, the solvent controls the dynamics of the protein through coupling in both low-frequency vibrations and relaxations of the protein.

In this work, the incoherent intermediate scattering function, $S(q,t)$, and dynamic structure factor, $S(q,\nu)$, for the (pure) solvent and the protein atoms were used to probe the existence and the extent of the dynamic coupling. Here q is the scattering wave vector, t is time, and ν is frequency. A comparison between $S(q,t)$ of the pure solvent and, separately, $S(q,t)$ of the protein in the solvent gives some insight into the coupling of their dynamics within the accessible time window of 1 ns. A comparison of the frequency of the low-energy vibration mode, the *boson peak* in $S(q,\nu)$, yields additional information about the collective molecular vibrations in the solvent and collective vibrations of residues in the protein.

First, we investigate $S(q,t)$. Figure 5 shows the results for lysozyme in glycerol and pure glycerol. At 150 and 200 K the time dependence of $S(q,t)$ for lysozyme and glycerol is approximately the same over the entire time window studied (except for a prefactor), indicating that up to 1 ns there is strong coupling between the dynamics of glycerol and lysozyme. This coupling could be a consequence of both types of molecules being trapped in a glassy state at very low temperatures as well as molecular interactions between the two. This result implies that practically all the modes with characteristic time scales between 1 ps and 1 ns are strongly coupled. On the other hand, $S(q,t)$ for lysozyme and glycerol differ at 250 K; it decreases in glycerol faster than in lysozyme. Therefore, at these temperatures and for the time window mentioned above, lysozyme does not show the same dynamics as pure glycerol. Only some molecular motions in lysozyme and glycerol are coupled.

A similar analysis for lysozyme embedded in trehalose was performed where $S(q,t)$ was computed in two different ways. First, $S(q,t)$ was computed considering all hydrogen atoms, and second, those hydrogens belonging to methyl groups were excluded. The methyl groups are unique in the sense that their motions in protein powders are activated at temperatures as low as 100 K[80] while the rest

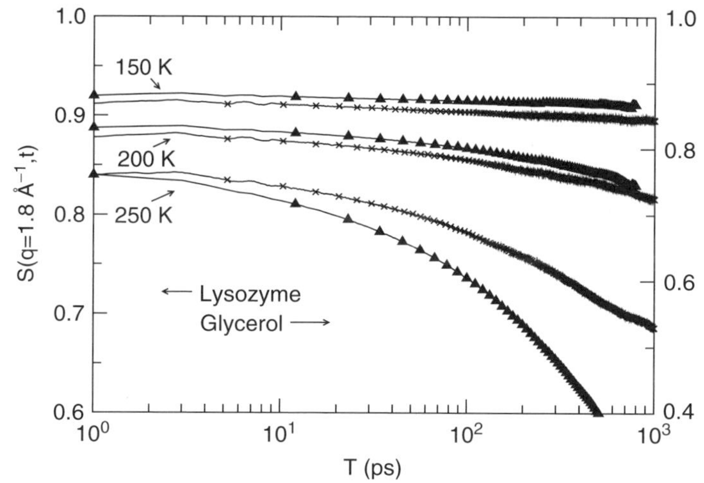

Figure 5 Incoherent intermediate scattering function of the hydrogen atoms in lysozyme (×) and pure glycerol molecules (▲) at 150, 200, and 250 K. (Reprinted with permission from T. E. Dirama et al., *J. Chem. Phys.*, 2005, 122: 244910. Copyright © 2005, American Institute of Physics.)

of the protein has predominantly harmonic vibrations in the nanosecond time scale. Thus, the methyl group hydrogens have distinct dynamic behavior and their incorporation into the calculation of the dynamic observables may complicate interpretation of the structural relaxation of the protein.[80] The emerging hypothesis is that the structural relaxation of the protein is related to the activation of modes involving large-scale motions of secondary structural elements. Since the dynamics of methyl protons occur on this moving framework, the inclusion of their contribution to the relaxation process may overwhelm or complicate the analysis of the underlying physical processes.

Figure 6 illustrates how the dynamics of lysozyme relate to the dynamics of pure trehalose. At 200 K (Figure 6*a*), when the dynamics of methyl group hydrogens are neglected, the $S(q,t)$ values are very similar for lysozyme and trehalose indicating a strong correlation in the structural relaxation of lysozyme and trehalose in the 1-ns time window. This result implies that on average, practically all the modes with characteristic time scales between 1 ps and 1 ns are strongly coupled. The dynamic coupling at this low temperature could be due to the fact that the system is deep in the glassy state, where protein atoms could be trapped in the cage created by the surrounding solvent molecules. At higher temperatures, as shown in Figure 6*b*, $S(q,t)$ for trehalose decays faster than $S(q,t)$ for lysozyme. This faster decay may be a consequence of the secondary relaxation process that is activated at 300 K.

To further evaluate the conformity between experiments and the simulations, the dynamic structure factor $S(q,\nu)$ obtained from the MD simulations are compared to experimental neutron scattering data for lysozyme.[62] $S(q,\nu)$ was

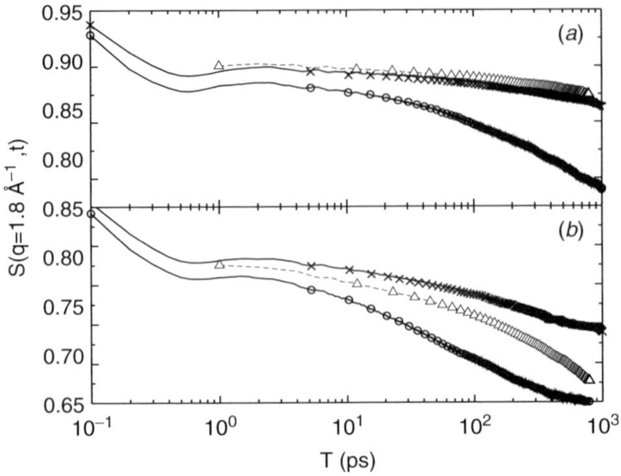

Figure 6 Incoherent intermediate scattering function of the hydrogen atoms in lysozyme: with (○) and without (×) methyl hydrogen atoms and pure trehalose molecules (Δ) at (a) 200 K and (b) 300 K. (Reprinted with permission from ref. 61. Copyright © 2006, American Institute of Physics.)

calculated from $S(q,t)$ by Fourier transformation:

$$S_{\mathrm{inc}}(q,\nu) = \frac{1}{2\pi}\int_{-\infty}^{+\infty} e^{-2i\nu t} S_{\mathrm{inc}}(q,t)\,dt \tag{4}$$

after multiplication with a Gaussian function that takes into account the experimental resolution of the spectrometer. For the case of lysozyme in glycerol, the resolution was set to 300 μeV, which corresponds to a Gaussian function with a full-width at half-maximum (FWHM) value of 700 μeV. Higher resolutions (200 and 100 μeV) were also tried; however, a reduction in the value of this parameter led to some small oscillations on top of the curves shown in Figures 7 and 8. Similar oscillations have also been observed in simulations of other proteins, such as azurin.[81] The origin of these oscillations is the lack of enough structural inhomogeneity, which can be resolved by using not a single lysozyme but a cluster of many lysozymes (e.g., four or five proteins in the simulation) or by running various MD simulations where the glassy phases are prepared differently and averaging the results.[82] The main interest, however, is in the behavior of $S(q,\nu)$ in the vicinity of the boson peak ($\sim$2 to 6 meV) and not in the low-frequency (quasielastic) region. Thus, a resolution of 300 μeV is a reasonable choice because it removes the small oscillations, allowing us to resolve the boson peak clearly. For lysozyme in trehalose, a FWHM of 500 μeV appeared to be low enough to avoid the oscillations. The price paid for the low resolution is a disagreement at low frequencies (quasielastic spectrum). In addition, the experimental data in ref. 62 were obtained after averaging over

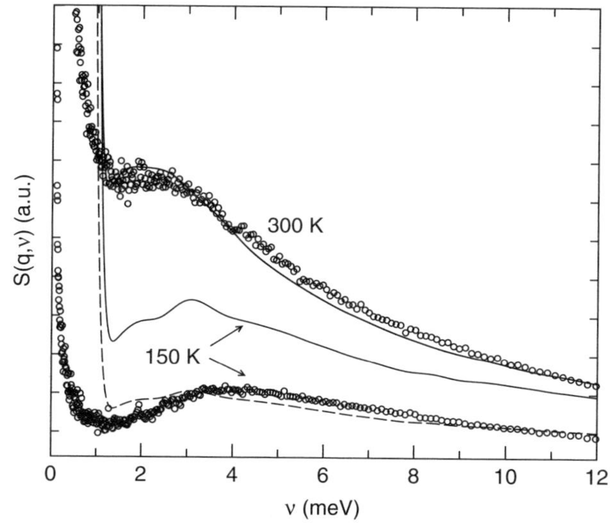

Figure 7 Dynamic structure factor for lysozyme in glycerol from MD simulation (lines) and neutron scattering spectra (symbols) from Caliskan et al.[62] The experimental data were multiplied by the frequency and Bose factor. The upper and bottom curves/symbols correspond to temperatures of 300 and 150 K, respectively. The meaning of the continuous and dashed lines is explained in the text. (Reprinted with permission from T. E. Dirama et al., *J. Chem. Phys.*, 2005, 122: 244910. Copyright © 2005, American Institute of Physics.)

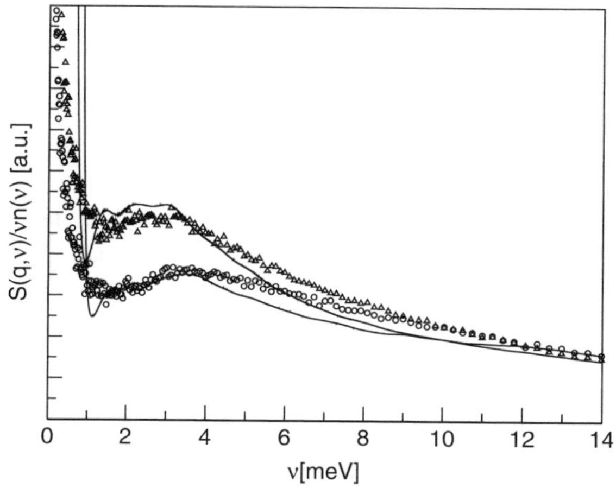

Figure 8 Dynamic structure factor for lysozyme in trehalose from MD simulation (lines) and neutron scattering spectra (symbols) from Caliskan et al.[62] The upper and bottom curves/symbols correspond to temperatures of 300 and 150 K, respectively. $n(\nu) = [\exp(h\nu/kT) - 1]^{-1}$ is the Bose factor. (Reprinted with permission from ref. 61. Copyright © 2006, American Institute of Physics.)

values of q ranging from 0.25 to 1.5 Å^{-1}, so $S(q,t)$ was averaged over 20 randomly chosen wave vectors with moduli between the aforementioned values. Finally, numerical inaccuracies appeared in the Fourier transformation at 150 K. Since the origin of these inaccuracies were traced to the truncation of $S(q,t)$, the intermediate scattering function was extrapolated at times beyond 1 ns using a stretched exponential decay with parameters estimated from the fit of the simulation data obtained from the last 997 ps. The Fourier transformation shown in Eq. (4) was applied to this extended function, where the first 1 ns of data was taken from the MD simulation and the fitted stretched exponential was used after 1 ns. The data at 300 K did not need this correction because $S(q,t)$ decayed fast enough during the 1-ns time window.

Figure 7 shows two quantitative comparisons of the simulation and experimental results for lysozyme in glycerol. The simulation data obtained at 300 K (line) were scaled and shifted vertically to maximize the quantitative agreement between both results. This is justified due to the arbitrary units used in the experimental $S(q,\nu)$. The results of the simulations reproduce well the experimental data in the frequency range of interest (Figure 7): The presence of an inelastic boson peak and an increase in the quasielastic scattering intensity (QES) with decreasing frequency are captured by the simulations. The simulation data obtained at 150 K were treated in two different ways. First, the same scaling and vertical shifts used with the 300 K data were employed with the data obtained at 150 K. This result is shown as a continuous line in Figure 7. Second, the vertical scaling and shift were optimized to obtain the best quantitative agreement with the experimental data (dashed line). As expected, the first treatment of the simulation data leads to a curve above the experimental data. The origin of this vertical displacement is the different cooling rates used in typical MD simulations (about 1 K/ps) and experimental studies. However, when the second approach is used, the simulation study reproduces the experimental data quite well in the frequency window (1 meV, 10 meV) of interest. Yet the maximum of the boson peak in the simulation appears at slightly lower frequency (the difference is about 0.6 meV). This has also been reported for other systems, many of which were studied using different force fields and MD simulation packages (see ref. 82 and references therein). The difference in the frequency of the boson peak has been rationalized in terms of the softness of the potential force field. However, another, perhaps more physical origin of the disagreement might be found in the preparation of the system at low temperatures. In particular, it is known experimentally that systems quenched into the glassy state have boson peaks with lower frequencies and higher amplitudes than those of systems annealed into the glass.[84] Thus, systems studied using MD simulations will always have the boson peak shifted to lower frequencies compared with experimental results. Another possible origin of the discrepancy between simulation and experiment might be the concentration of lysozyme: It was 50% by weight in the experimental studies, whereas it was 12% in this study.

When we studied the case of trehalose as a solvent, good quantitative agreement was also observed between the simulations and the experimental data at

150 and 300 K, shown in Figure 8. The presence, location, and temperature dependence of the inelastic boson peak is captured by the simulations. Reproducing $S(q, \nu)$ enabled us to study the coupling of the protein and the solvent dynamics in the frequency domain. For this purpose, only one value, $q = 1.8$ Å^{-1}, was employed.[81] The resulting $S(q, \nu)$ curves for lysozyme–trehalose and lysozyme–glycerol (not shown) systems showed the boson peak clearly. The only systems that did not show a boson peak were pure glycerol and lysozyme at high temperatures, where the quasielastic contribution overwhelmed the contribution from the boson peak. The physical origin of this peak for glassy systems as well as proteins is still a subject of discussion.[85] However, for the case of proteins, Tarek and Tobias[86] showed that the boson peak vibrations involve the protein as a whole. Strong similarities in the spectral shape of $S(q, \nu)$ for lysozyme and solvents were observed as discussed below. The frequencies of the boson peak for lysozyme and solvents were very similar.

A comparison of the solvent and protein dynamics as quantified by $S(q, \nu)$ for the case of lysozyme in trehalose is shown in Figure 9 at temperatures equal to 200, 300, and 400 K. Note that the spectral shapes of $S(q, \nu)$ for lysozyme and pure trehalose follow one another very closely. The boson peak is observed clearly in both systems at 200 and 300 K, while at 400 K, it appears as a shoulder due to the intensity of QES, which increases with temperature and overlaps with the contribution arising from inelastic scattering. QES is due to the relaxation-like

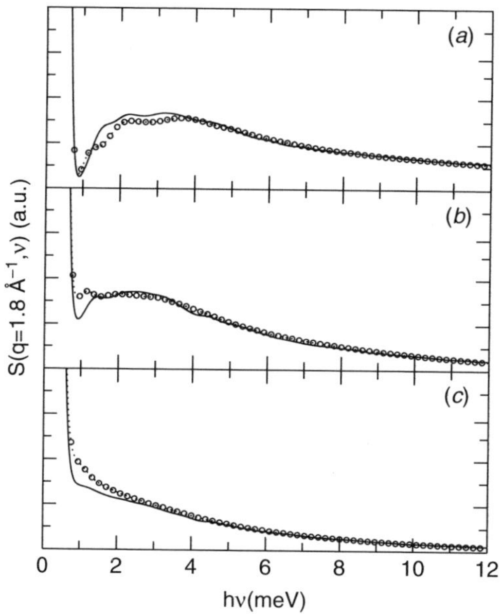

Figure 9 Dynamic structure factor for pure trehalose (O) and lysozyme in trehalose (continuous line) at (a) 200 K, (b) 300 K, and (c) 400 K. (Reprinted with permission from ref. 61. Copyright © 2006, American Institute of Physics.)

dynamics, such as overdamped vibrations or activated processes.[88] The frequency of the boson peak (ν_{BP}) was plotted as a function of temperature for both systems in Figure 10. ν_{BP} was extracted from the simulation data using the extrapolation formula

$$S(q,\nu) = \frac{A\nu_0}{\nu_0^2 + \nu^2} + B\exp\left\{-\frac{[\ln(\nu/\nu_{BP})]^2}{2[\ln(W/\nu_{BP})]^2}\right\} \quad (5)$$

which comprises two terms: The first approximates the quasielastic scattering (QES) part with a Lorentzian function of width ν_0 and height A/ν_0, whereas the second term (lognormal function) fits the boson peak of width W.[87] A very similar temperature dependence of the frequencies of the boson peak for lysozyme and pure trehalose from 150 to 450 K is evident, implying that the low-frequency collective vibrations are coupled, in agreement with results from Raman spectroscopy.[62] The same type of analyses for lysozyme-glycerol system resulted in similar conclusions. This result, in conjunction with the analysis of $S(q,t)$, implies that the simulation protocol is also capable of reproducing the coupling of the protein and solvent dynamics for both solvents.

Molecular Origins of Dynamic Coupling

Perhaps the key question to ask about the protein–solvent system at this point is how the coupling of the solvent and protein dynamics occurs. Since the protein–solvent interactions occur at the protein–solvent interface, this region carries crucial information regarding the microscopic origins of the coupling between the protein and solvent dynamics. We discuss this region by dividing

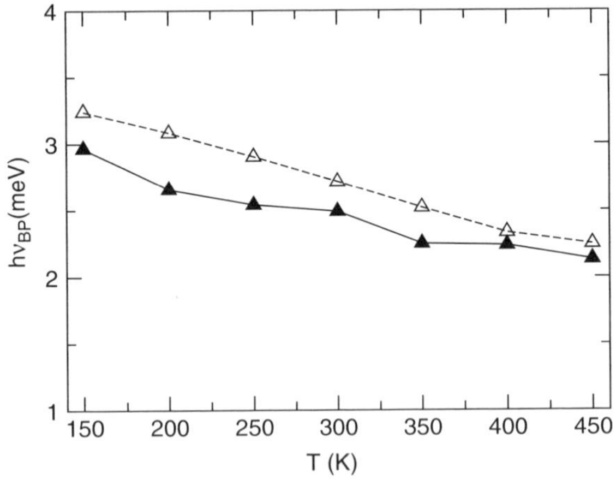

Figure 10 Temperature dependence of the frequency of the boson peak from MD simulation. Lysozyme in trehalose (▲) and pure trehalose (Δ). (Reprinted with permission from ref. 61. Copyright © 2006, American Institute of Physics.)

the subject into three parts: the dynamic heterogeneity in the protein, the influence of the protein on the solvent dynamics, and the hydrogen-bonding interactions.

Surface Versus Core Atoms of the Protein The protein atoms comprising the protein side of the interface are expected to be affected by the solvent molecules more strongly than the core atoms. Therefore, the dynamics of the surface and core atoms of the protein are likely to be influenced by the solvent differently. In an attempt to enhance an understanding of the surface protein dynamics, we compared the dynamics of the hydrogen atoms that are on the protein surface to the dynamics in the core of the protein. The hydrogen atoms in the protein that are the closest ones to any hydrogen atom in any solvent molecule were defined as surface hydrogen atoms, and the rest were considered as core atoms. In other words, the distances between a particular hydrogen atom in a solvent molecule and all hydrogen atoms in the protein were calculated. The hydrogen atom in the protein that was closest to the solvent hydrogen atom was considered to be on the surface of the protein. This method was repeated for all the hydrogen atoms in all the solvent molecules present in the system. This provided us with a list of those protein hydrogens that are closest to the solvent: that is, the surface hydrogens. This definition found 440 hydrogens on the surface of the protein and 537 in the core.

Figure 11 shows $S(q,t)$ for glycerol and the surface and core hydrogen atoms of lysozyme. For all the temperatures studied, $S(q, t)$ shows a first fast decay in the subpicosecond regime. However, the behavior of the dynamics at long times is substantially different. For example, at temperatures below T_d (150 and 200 K), $S(q,t)$ for both the surface and the core atoms does not show any indication of a second decay at long times and remains approximately constant (note the scale). It is interesting to notice that the curve for the core atoms is below the curve for the surface atoms, indicating that the atoms in the core of the protein are more mobile than the atoms on the surface. This result implies that glycerol reduces the size of the cage around the surface residues. However, at 250 and 300 K, $S(q,t)$ for the surface atoms starts to follow the decay observed for glycerol and crosses the curves that correspond to the core atoms. This implies that the protein atoms on the surface become more mobile than the core atoms at long enough times. The core atoms also follow the decay in $S(q,t)$ of glycerol, but the effect is less pronounced than for the surface atoms. Thus, it can be stated that their dynamics are more shielded from the effects of the solvent than those of the surface atoms. However, the figure clearly shows the effect of the solvent properties on the core residues. This result suggests that the dynamic coupling between the protein and the solvent is translated through the surface of the protein. Basically, the influence of the solvent dynamics is conveyed to the surface atoms by means of hydrogen-bond interactions (see below) and, afterward, the surface atoms translate that effect onto the core atoms through intramolecular interactions. This inference is supported further by MD simulations of Walser and van Gunsteren.[89]

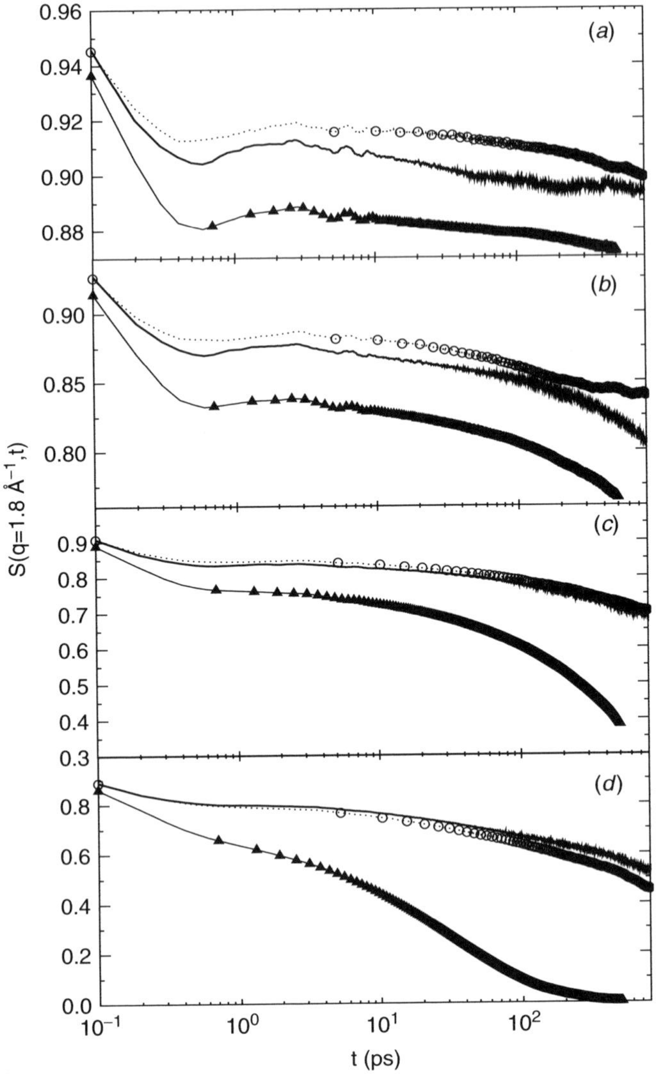

Figure 11 Incoherent intermediate scattering function for glycerol (▲) and surface (○) and core (continuous line) hydrogen atoms in lysozyme at (a) 150, (b) 200, (c) 250, and (d) 300 K. (Reprinted with permission form T. E. Dirama et al., *J. Chem. Phys*., 2005, 122: 244910. Copyright © 2005, American Institute of Physics.)

For the lysozyme–trehalose system let us first analyze the dynamics of the protein and the solvent molecules before focusing on the dynamics of the surface and core separately. At 200 K (Figure 12*a*), the atoms of trehalose are trapped in smaller cages than the protein atoms, as manifested by higher values of $S(q,t)$ at short times. Additionally, the secondary relaxation in trehalose is

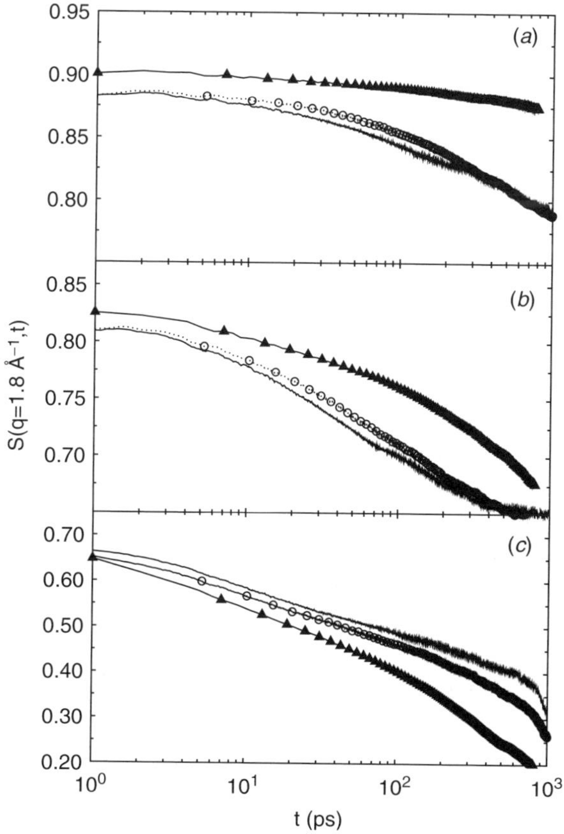

Figure 12 Incoherent intermediate scattering function for trehalose (▲) and surface (○) and core (continuous line) hydrogen atoms in lysozyme at (a) 200 K, (b) 300 K, and (c) 400 K. (Reprinted with permission from ref. 61. Copyright © 2006, American Institute of Physics.)

slower than in lysozyme. At 300 K, trehalose shows a secondary decay similar to that of lysozyme. An important observation can be made at this point: At 200 K, when the dynamics of trehalose are relatively slower than those of lysozyme, the dynamics of protein atoms on the surface are somewhat slower than the core atoms, whereas at 400 K (Figure 12*c*), when trehalose is more mobile than lysozyme, the protein surface has relatively faster dynamics than the core. At intermediate temperatures (i.e., 300 K), when the dynamics of the solvent approach that of the protein, the surface and the core protein atoms show nearly the same dynamic behavior. These results indicate that the surface of the protein is more sensitive than the interior to the dynamics of the solvent.

Effect of the Protein on the Solvent The influence of the solvent on the dynamics of the protein is well established. However, few studies have been done to investigate the effect of the protein on the dynamics of the surrounding solvent.[90–92] Yet it has been shown[91] that the dynamics of water molecules near the surface of the protein are more restricted than in bulk water. The restricted mobility of water near the protein surface has been attributed to three factors: the decrease of the dimensionality of the space at the interface, solute surface roughness, and solvent structuring.[91]

To explore if this result is applicable to the solvents under consideration, the profile of $\langle u^2 \rangle$ for the hydrogen atoms in the solvent molecules as a function of the distance from the surface of lysozyme was calculated for different temperatures. This distance was computed following the work of Pettitt and co-workers.[91] Average positions of solvent hydrogen atoms were sorted into five shells with respect to the distance from their nearest protein atom. The first shell comprised the hydrogens within a distance of 4.5 Å from the protein surface; the following four shells were created with thicknesses of 2.5 Å from the surface of the protein. $\langle u^2 \rangle$ was calculated for each shell and the results at 300 K are shown in the insets of Figures 13 and 14 for glycerol and trehalose, respectively.

We observed that the dynamics of glycerol are suppressed near the surface of the protein for all the temperatures studied (Figure 13). $\langle u^2 \rangle$ increases with increasing distance from the surface until it plateaus around 10 to 15 Å and reaches the bulk value. The magnitude of $\langle u^2 \rangle$ near the surface can differ from that in the bulk by a factor as large as 2 at 300 K. Additional analysis of $S(q,t)$ was done for the five shells around the protein at 300 K. The plot clearly shows that

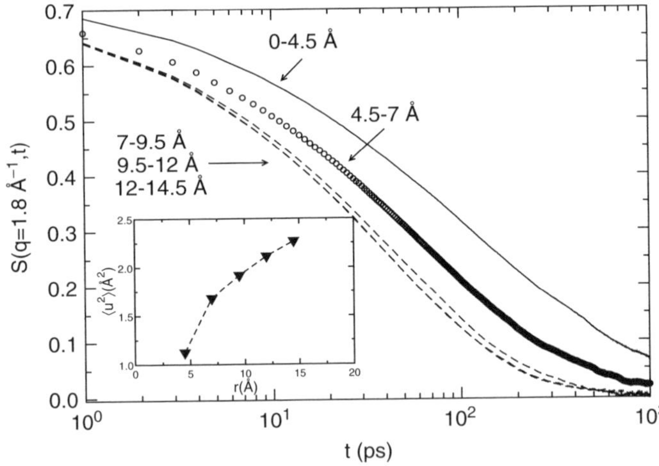

Figure 13 Incoherent intermediate scattering function for glycerol molecules within 0 to 4.5 Å (solid line), 4.5 to 7 Å (circles), and larger distances (dashed lines) from the surface of lysozyme at 300 K. The inset shows the mean-square displacement of the hydrogen atoms in glycerol as a function of the distance from the surface of lysozyme at 300 K. (Reprinted with permission from T. E. Dirama et al., *J. Chem. Phys*., 2005, 122: 244910. Copyright © 2005, American Institute of Physics.)

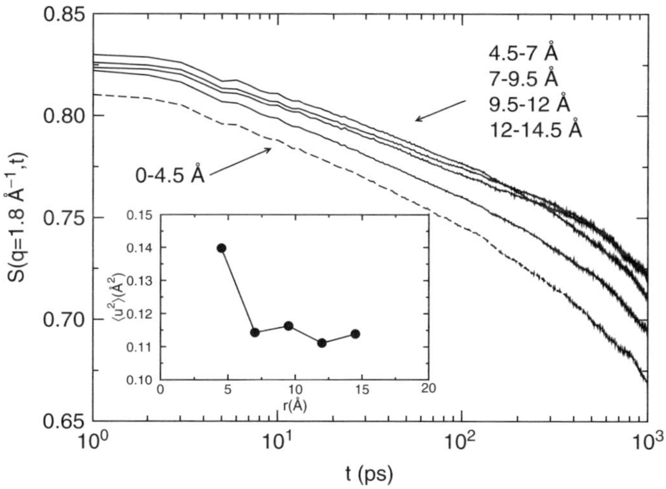

Figure 14 Incoherent intermediate scattering function for trehalose molecules within 0 to 4.5 Å (dashed line) and larger distances (continuous lines) from the surface of lysozyme at 300 K. The inset shows the mean-square displacement of the hydrogen atoms in trehalose as a function of the distance from the surface of lysozyme at 300 K. (Reprinted with permission from ref. 61. Copyright © 2006, American Institute of Physics.)

the farther the solvent molecules are from the surface, the faster $S(q,t)$ decays. This result shows that the dynamics of the glycerol molecules in close proximity to the protein surface are affected significantly by the presence of the protein. However, the effect of the protein on the dynamics of the solvent vanishes for distances longer than ~10 Å. This is observed in the rate of decay of the curves and in the amplitude of $\langle u^2 \rangle$, which remains virtually unchanged beyond 10 Å. Similar behaviors were observed at temperatures 150, 200, and 250 K.

The restricted mobility near the protein was suggested to be a universal behavior for protein–solvent systems.[91] To evaluate the validity of this observation in the case of trehalose, let us look at Figure 14. The initial decay of $S(q,t)$ is larger for the first shell of solvent than for the other shells, suggesting that trehalose is more mobile near the protein surface than in the bulk. This is confirmed by $\langle u^2 \rangle$, shown in the inset; $\langle u^2 \rangle$ increases from the bulk value as one approaches the protein surface. This result contrasts the findings of previous studies of protein–solvent systems, where it was reported that the dynamics of solvent molecules should be suppressed near the protein surface. To clarify this observation and understand the relevance of this unexpected behavior on the dynamic coupling, $\langle u^2 \rangle$ was plotted as a function of the distance from the protein surface for a wide range of temperatures in Figure 15. At temperatures below 450 K, $\langle u^2 \rangle$ decreases with increasing distance. This implies that trehalose molecules near the surface are more mobile than trehalose in the bulk. This behavior contradicts previous results on protein–glycerol and protein–water systems. However, as the temperature is increased from 300 to 450 K, (Figure 15*a* to *c*), the difference

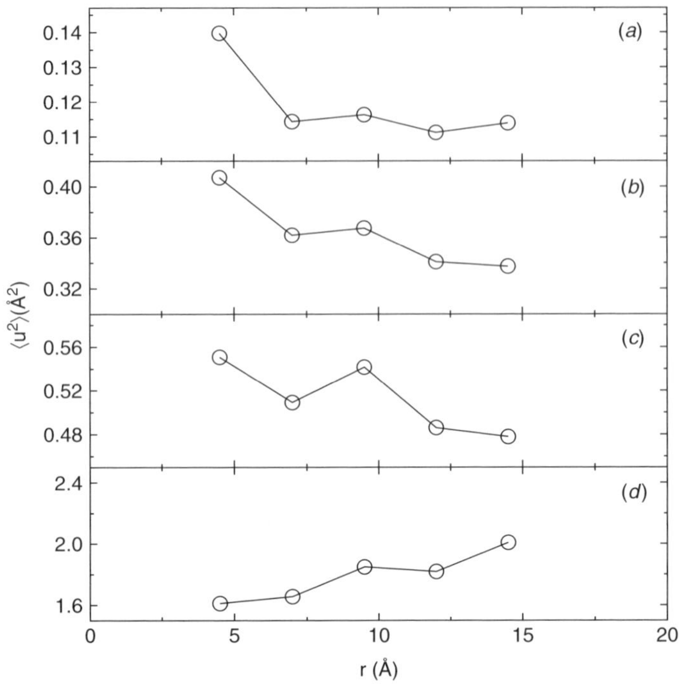

Figure 15 Mean-square displacement of the hydrogen atoms in trehalose as a function of the distance from the surface of lysozyme at (a) 300, (b) 400, (c) 450, and (d) 550 K. (Reprinted with permission from ref. 61. Copyright © 2006, American Institute of Physics.)

in the bulk-to-surface value of $\langle u^2 \rangle$ decreases from about 27% to 15%, suggesting that the effect becomes less pronounced. At 550 K the behavior reverses; trehalose becomes more mobile in the bulk than near the surface. Note that as predicted earlier by MD simulations, the glass transition temperature for trehalose is 456 K. Thus, at temperatures below and near the T_g of trehalose, the protein is more mobile (as measured by $\langle u^2 \rangle$). Under this condition, to achieve dynamical coupling, the dynamics of trehalose molecules near the surface adapt to the dynamics of the protein (Figure 15). When temperature is above T_g, trehalose becomes more mobile than the protein (Figure 15*d*). This forces trehalose to reduce its dynamics gradually as it approaches the surface of the protein.

The data shown in Figure 13 can be interpreted from a different perspective using the concept of local viscosity of the medium surrounding the protein. It is intuitively clear that the viscosity of a fluid should increase monotonically with decreasing MSD of the molecules. This is clearly the case for a simple fluid that satisfies the Stokes–Einstein relationship

$$\eta = \frac{k_b T}{6\pi r D} \tag{6}$$

where η is the viscosity of the liquid and D is the diffusion coefficient, which can be expressed in terms of the atomic MSD as

$$D = \lim_{t\to\infty} \frac{\sum_{n}[x(t) - x(0)]^2}{6tn} \tag{7}$$

where t is time, x is atomic position of the center of mass, and n is the number of molecules. η can be computed from Eqs. (6) and (7) from MSD data obtained from MD simulations. However, the linear dependence of the MSD on time must be satisfied.[49,93] Similarly, a monotonic dependence of the viscosity on the MSD has been reported for melts of glass-forming polymers and proteins in glassy solvents.[54,94] Following this line of reasoning, an interesting observation can be made about the data shown in Figure 13. The suppression of the dynamics of glycerol near the surface of the protein can also be interpreted as the protein being immersed in an environment of higher viscosity than the bulk viscosity of the solvent. Thus, the protein sees an effective local viscosity higher than the viscosity of the bulk solvent.

On the other hand, using the same concept, the enhanced dynamics of trehalose near the surface of the protein (Figure 14) can be viewed as the protein being surrounded by a solvent with a lower viscosity than the bulk viscosity. Such a situation implies that the protein sees an effective local viscosity that is lower than the viscosity of bulk solvent. In fact, this explains why lysozyme is more mobile than trehalose at 200 K, as illustrated by a faster decay of $S(q,t)$ for the protein atoms in Figure 12a. Despite 200 K being lower than the T_g value of trehalose, trehalose molecules near the surface are more mobile than trehalose in the bulk, thus facilitating the conformational changes of the protein, which would otherwise, not be viable in an environment with bulk viscosity.

Hydrogen-Bonding Network Among the various types of interactions in molecular systems, hydrogen bonds are particularly important in the dynamics of proteins.[64] Proteins contain many polar groups in the backbone as well as in the side groups that can form intramolecular and intermolecular hydrogen bonds. Some previous studies on protein–water[48,78,79] mixtures suggest that the hydrogen-bonding network plays an important role in the dynamic behavior of these types of systems. Thus, the behavior of the hydrogen bonds present in the current system were studied using a geometric criterion based on the distance between the donor and acceptor oxygen atoms and the angle formed by the donor oxygen, the acceptor hydrogen, and the acceptor oxygen atoms. The cutoff distance between oxygen atoms was set to 3.4 Å, which is about the location of the minimum after the first peak of the radial distribution function; the cutoff for the angle was set to 120°. Using this geometric criterion, the hydrogen-bonding network was characterized using the hydrogen-bond correlation function, $c(t)$, which represents the probability that a hydrogen bond formed originally at $t = 0$ between a randomly chosen donor–acceptor pair exists at time t.

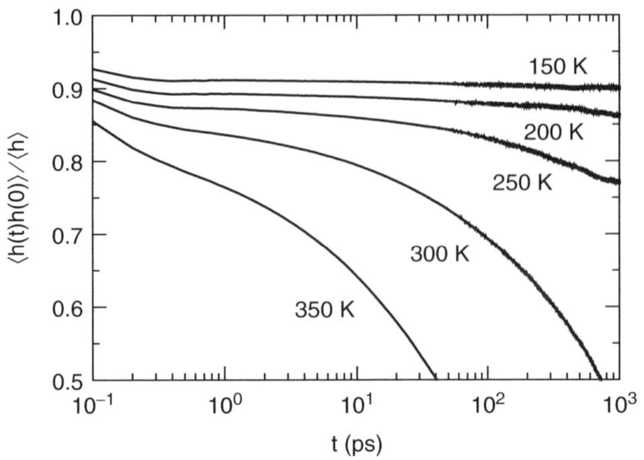

Figure 16 Hydrogen-bond correlation function for the hydrogen bonds between glycerol and lysozyme at five temperatures. (Reprinted with permission from T. E. Dirama et al., *J. Chem. Phys*., 2005, 122: 244910. Copyright © 2005, American Institute of Physics.)

Figure 16 shows the hydrogen-bond correlation functions for the hydrogen bonds between glycerol and lysozyme for five temperatures: 150, 200, 250, 300, and 350 K. At 150 and 200 K, the correlation functions show an initial decay for time scales shorter than 1 ps and then remain approximately constant for the time window explored in this study (1 ns). At 250 K, the hydrogen-bond correlation function shows the initial decay in the subpicosecond regime and the beginning of a second decay at times close to 1 ns. At higher temperatures, the hydrogen-bond correlation function decays to small values within the time window studied in this work. The existence of two decays, one in the subpicosecond regime and the other one at long times, shows the existence of two types of hydrogen bonds: fast and slow.[78] The fast hydrogen bonds correspond to rotation and libration of the solvent molecules and affect the fast motions of the protein. The dynamics of the slow hydrogen bonds are important because the structural relaxation of the protein (i.e., changes in the conformational substates) requires relaxation of the protein–solvent hydrogen-bonding network. This is achieved via solvent translational displacement.[78] Therefore, long-living hydrogen bonds have an effect on the dynamics of the protein, thus affecting T_d and other properties. In addition, Tarek and Tobias[78] have demonstrated that the hydrogen-bonding network relaxation time correlated with the dynamics of the protein as opposed to the fast hydrogen-bonding lifetime, which did not exhibit such a relationship. Therefore, the focus will be on the slow hydrogen bonds.

Figure 17 shows a semilogarithmic plot of the average lifetime (τ_R) of the slow hydrogen bonds as a function of temperature. The dashed line corresponds to 1 ns, which is the time window of the simulations. The inset is a plot of the same data but as a function of T^{-1} (Arrhenius form). τ_R is the relaxation time of the slow hydrogen bonds and was extracted from the data as follows. A

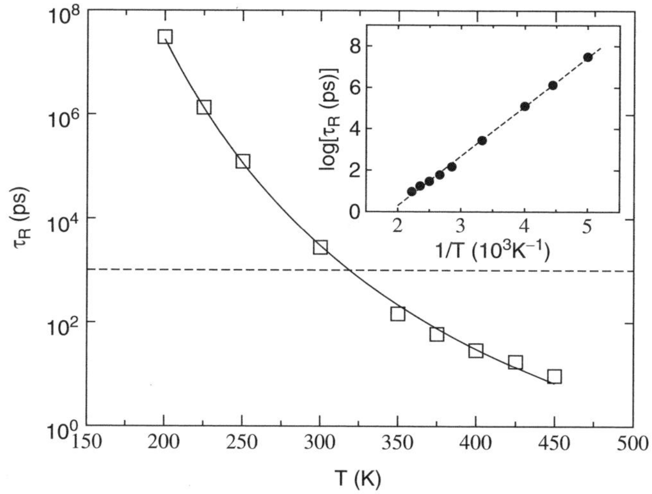

Figure 17 Relaxation time of the hydrogen-bond network (τ_R) as a function of temperature for the lysozyme–glycerol system. The inset shows the same data but presented as log(τ_R) versus T^{-1} (Arrhenius form). The horizontal (dashed) line represents the resolution function of the MD simulation. Since the data points above this line were obtained by stretched exponential fitting, these data points are susceptible to larger errors. (Reprinted with permission from T. E. Dirama et al., *J. Chem. Phys.*, 2005, 122: 244910. Copyright © 2005, American Institute of Physics.)

stretched exponential function was fitted to the data collected during the last 997 ps and τ_R was extracted from the fit. Figure 17 shows that the values of τ_R vary from the picosecond time scale at high temperatures to the microsecond time scale at low temperatures. At temperatures close to 300 K the dynamics of the hydrogen-bonding network enters the time window accessible to the MD study. A comparison of Figures 3 and 17 suggests a clear correlation between the lifetime of the slow hydrogen bonds and the dynamics of the protein. Namely, when the dynamics of the hydrogen-bonding network enter the time window of the simulation study, $\langle u^2 \rangle$ increases rapidly, indicating the presence of the dynamic transition. The inset shows that the behavior of the relaxation time follows an Arrhenius law ($E_a = 46$ kj/mol), implying that there is only one type of dynamic process. In other words, the dynamic transition does not involve a transition between two different relaxation processes. This result is in agreement with a recent study by Fenimore et al.[51]

There is a remarkable similarity between $S(q,t)$ without methyl groups and the hydrogen-bond correlation function (Figures 5 and 11 versus Figure 16). In particular, the effect of temperature on both functions is practically the same (e.g., both functions show very similar decays at long times). It is interesting to note that the hydrogen-bond interactions that determine the hydrogen-bond correlation function occur within a shell of thickness 3.4 Å around the protein, while $S(q,t)$ is determined by the dynamics of the whole protein.

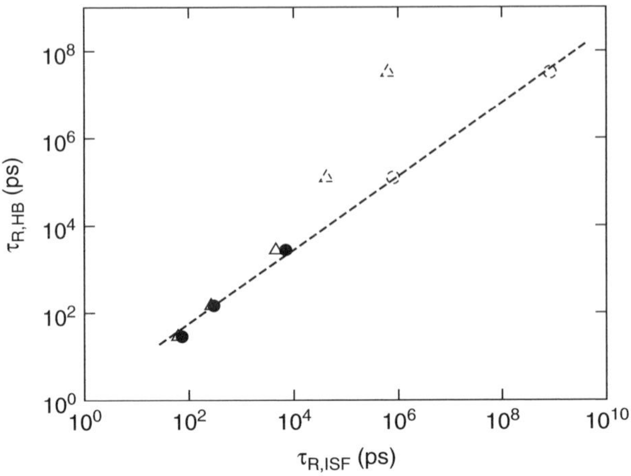

Figure 18 Plot of the relaxation time for the hydrogen bonds between the protein and glycerol as a function of the relaxation time of $S(q, t)$ of lysozyme considering the methyl group hydrogen atoms (Δ) and neglecting them (•). The data points with dashed symbols were obtained after extrapolation of the simulation data for several orders of magnitude. Therefore, these particular data points are susceptible to larger errors. (Reprinted with permission from T. E. Dirama et al., *J. Chem. Phys.*, 2005, 122: 244910. Copyright © 2005, American Institute of Physics.)

One way to quantify the extent of similarity between $S(q,t)$ and $c(t)$ is to compare their relaxation times. Figure 18 shows a plot of the relaxation time for the slow hydrogen bonds ($\tau_{R,\mathrm{HB}}$) as a function of the relaxation time of the protein obtained from $S(q,t)$ ($\tau_{R,\mathrm{ISF}}$) with and without the contributions from methyl groups for the lysozyme–glycerol system. It was emphasized in ref. 80 that methyl group rotation in lysozyme appears in the ps–ns time window at $T \sim 100$ K and gives significant contribution to the dynamic structure factor at higher temperatures. Thus, particular attention is paid to the methyl group contribution in the simulations. When the methyl hydrogens are neglected, $\tau_{R,\mathrm{HB}}$ and $\tau_{R,\mathrm{ISF}}$ show a power law relationship with exponent and prefactor equal to 0.86 and 0.81, respectively. The correlation coefficient for these data is 0.999. On the other hand, this simple relationship breaks down when methyl groups are included, especially at low temperatures (200 and 250 K). These are the temperatures where the dynamical transition has not been reached yet while the methyl group rotations are active. In addition, note that the methyl group rotations do not contribute in a significant manner to structural relaxations in the protein. Therefore, it is expected that the hydrogen-bond relaxation time correlates with the relaxation time of $S(q,t)$ when the motions of methyl groups are omitted. The same correlation between relaxation times of $S(q,t)$ and $c(t)$ has been observed for the lysozyme–trehalose system. These findings show that the relaxation of the slow hydrogen bonds on the surface of the protein determines the structural relaxation of the protein, at least in the time window accessible to this MD simulation study. In fact, based on these findings it can be argued that the hydrogen-bonding network is a major factor controlling the dynamics of the protein.

4. CONCLUSIONS

We have presented results about the dynamics of lysozyme in pure glycerol and trehalose over a wide temperature range using analyses based on fully atomistic molecular dynamic simulations. We have provided evidence that shows clearly that the dynamics of the solvents have a profound effect on the protein dynamics. Different solvents induce their effects differently on the protein dynamics. For example, lysozyme showed relatively dissimilar fast dynamics depending on the surrounding solvent medium and the temperature. After demonstrating the effect of solvent on the protein dynamics, we focused on understanding the origins behind it. One of the indicators of this effect, the protein dynamic transition, was reproduced successfully by our simulation study. The MD simulation results have provided further insight into the molecular-level mechanisms and pointed to the importance of the dynamic coupling of the protein and solvent dynamics.

The dynamics of the interface between the solvent and the protein were found to play a fundamental role in the evolution of the dynamics of the protein. In particular, we provided evidence that the dynamics of the hydrogen-bonding network between the protein and the first shell of solvent control the structural relaxation of the protein as a whole. This was clearly shown by the power law relationship displayed by the structural relaxation time of the protein and the relaxation time of the slow hydrogen bonds between the protein and the solvent. Moreover, this study, based on lysozyme–glycerol and lysozyme–trehalose systems, suggested a molecular-level mechanism that leads to coupling between the dynamics of the protein and those of the solvent. First, the hydrogen bonds between the solvent and the surface atoms of the protein couple the dynamics of the surface of the protein to those of the solvent. This coupling is propagated into the core atoms via intramolecular interactions (i.e., van der Waals, electrostatic, bonded, etc.). However, this propagation shields the core atoms, and their dynamics are not affected by the solvent as strongly as are those of the surface atoms.

We also found that the temperature dependence of the hydrogen-bond relaxation time for the lysozyme–glycerol system follows an Arrhenius law, leading to the conclusion that during the dynamic transition nothing special happens to the dynamics; there is only one dynamic process in the system above and below the dynamic transition temperature. Simply, the relaxation process enters the accessible time window and this leads to strong increase in MSD. This finding provides some insight into the debate about the physical phenomenon that leads to the dynamic transition.

The glycerol dynamics in proximity to the protein surface were shown to be strongly suppressed, in line with other small molecules, such as water. This result was corroborated with calculations of the incoherent intermediate scattering function and interpreted using the concept of viscosity. The protein sees a viscosity higher than that of the bulk solvent. Unlike water and glycerol, we observed enhanced dynamics of trehalose close to the protein surface.

An interesting finding about the dynamic transition of lysozyme in the trehalose matrix, which implies the dynamic coupling, is that the dynamic transition starts at temperatures below the glass transition temperature of the solvent. The activation of some secondary relaxation in the solvent appeared to facilitate sufficient mobility for the dynamic transition of the protein. This suggests that translational diffusion of solvent molecules is not the only molecular mechanism that can activate the onset of anharmonic dynamics in the protein. When combined with the secondary relaxation of trehalose at 300 K, the enhanced dynamics near the protein surface facilitates the molecular motions of the protein to such a degree that the initiation of anharmonic motions in the protein occurs far below the glass transition temperature of the solvent.

Acknowledgments

This material is based on work supported by the National Science Foundation under grant CHE-0132278. Acknowledgment is also made to the Ohio Board of Regents, Action Fund (grant R566) for financial support.

REFERENCES

1. Wang, W. *Int. J. Pharm*., 2000, 203: 1–2.
2. Cicerone, M. T.; Tellington, A.; Trost, L.; Sokolov, A. *Bioprocess. Int*., 2003, 1: 36–47.
3. Tsai, A. M.; Udovic, T. J.; Neumann D.A. *Biophys. J*., 2001, 81: 2339–2343.
4. Magliery, T. J.; Regan, L. *Eur. J. Biochem*., 2004, 271: 1595–1608.
5. Tollinger, M.; Crowhurst, K. A.; Kay, L. E.; Forman-Kay, J. D. *Proc. Natl. Acad. Sci. USA* 100: 4545–4550.
6. Creighton, T. E., Ed. *Protein Folding*, W.H. Freeman, New York, 1992, p. 129.
7. Brooks, C. L., III; Karplus, M.; Pettitt, B. M. *In Proteins: A Theoretical Perspective of Dynamics, Structure and Thermodynamics*, Advances in Chemical Physics, Vol. 71, Prigogine, I.; Rice, S., Eds., Wiley, New York, 1988.
8. Dunn, R. V.; Réat, V.; Finney, J.; Ferrand, M.; Smith, J. C.; Daniel, R. M. *Biochem. J*., 2000, 346: 355–358.
9. Cicerone, M. T.; Soles, C. L. *Biophys. J*., 2004, 86: 3836–3845.
10. Mouradian, R.; Womersley, C.; Crowe, L. M.; Crowe, J. H. *Biochim. Biophys. Acta*, 1984, 778: 615–617.
11. Carpenter, J. F.; Crowe, J. H. *Biochemistry*, 1989, 28: 3916–3922.
12. Carpenter, J. F.; Prestrelski, S. J.; Arakawa, T. *Arch. Biochem. Biophys*., 1993, 303: 456–464.
13. Green, J. L.; Angell, C. A. *J. Phys. Chem*., 1989, 9: 2880–2882.
14. Bell, L. N.; Hageman, M. J.; Muraoka, L. M. *J. Pharm. Sci*., 1995, 84: 707–712.
15. Duddu, S. P.; Zhang, G.; Dal Monte, P. R. *Pharm. Res*., 1997, 14: 596–600.

16. Crowe, J. H.; Crowe, L. M.; Chapman, D. *Science*, 1984, 223: 701–703.
17. Hagen, S. J.; Hofricter, J.; Eaton, W. A. *Science*, 1995, 269: 959–962.
18. Gottfried, D. S.; Peterson, E. S.; Sheikh, A. G.; Wang, J.; Yang, M.; Friedman, J. M. *J. Phys. Chem.*, 1996, 100: 12034–12042.
19. Cordone, L.; Galajda, P.; Vitrano, E.; Gassmann, A.; Ostermann, A.; Parak F. *Eur. Biophys. J.*, 1998, 27: 173–176.
20. Cordone, L.; Ferrand, M.; Vitrano, E.; Zaccai, G. *Biophys. J.*, 1999, 76: 1043–1047.
21. Roser, B. *Trends Food Sci. Technol.*, 1991, 2: 166–169.
22. Tanaka, K.; Takeda, T.; Miyajima, K. *Chem. Pharm. Bull.*, 1991, 39: 1091–1094.
23. Allison, S. D.; Chang, B.; Randolph, T. W.; Carpenter, J. F. *Arch. Biochem. Biophys.*, 1999, 365: 289–298.
24. Cleland, J. L.; Lam, X.; Kendrick, B.; Yang, J.; Yang, T.; Overcashier D.; Brooks D.; Hsu C.; Carpenter J.F. *J. Pharm. Sci.*, 2001, 90: 310–321.
25. Gekko, K.; Timasheff, S. N. *Biochemistry*, 1981, 20: 4667–4676.
26. Alder, B. J.; Wainwright, T. E. *J. Chem. Phys.*, 1957, 27: 1208–1209.
27. Stillinger, F. H.; Rahman, A. J. *Chem. Phys.*, 1974, 60: 1545–1557.
28. McCammon, J. A.; Gelin, B. R.; Karplus, M. *Nature (London)*, 1977, 267: 585–590.
29. Frenkel, D.; Smit, B. *Understanding Molecular Simulation*, Academic Press, San Diego, CA, Boston, 1996.
30. Leach, A. R. *Molecular Modeling: Principles and Applications*, Prentice Hall, Upper Saddle River, NJ, 2001.
31. Wang, J.; Wolf, R. M.; Caldwell, J. W.; Kollman, P. A.; Case, D. A. *J. Comput. Chem.*, 2004, 25: 1157–1174.
32. van Koningsveld, H. *Rec. Trav. Chim.*, 1968, 87: 243–254.
33. Brown, G. M.; Rohrer, D. C.; Berking, B.; Beevers, C. A.; Gould, R. O.; Simpson, R. *Acta Cryst. B*, 1972, 28: 3145–3158.
34. Chelli, R.; Procacci, P.; Cardini, G.; Della Valle, R. G.; Califano, S. *Phys. Chem. Chem. Phys.*, 1999, 1: 871–877.
35. Cornell, W. D.; Cieplak, P.; Bayly, C. I.; Gould, I. R.; Merz, K. M.; Ferguson, D. M.; Spellmeyer, D. C.; Fox, T.; Caldwell, J. W.; Kollman, P. A. *J. Am. Chem. Soc.*, 1995, 117: 5179–5197.
36. Lide, D. R., Ed. *Handbook of Chemistry and Physics*, CRC Press, Boca Raton, FL, 1997.
37. Glennon, T. M.; Merz, K. M., Jr. *J. Mol. Struct. (Theochem)*, 1997, 395: 157–171.
38. Shull, H.; Hall, G. G. *Nature (London)*, 1959, 184: 1559–1560.
39. Brown, G. M.; Levy, H. A. *Acta Cryst. B* 1979, 35: 656–659.
40. Bonanno, G.; Noto, R.; Fornilli, S. L. *J. Chem. Soc. Faraday Trans.*, 1998, 94: 2755–2762.
41. Duda, C. A.; Stevens, E. S. *J. Am. Chem. Soc.*, 1990, 112: 7406–7407.
42. Sakurai, M.; Murata, M.; Inoue, Y.; Hino, A.; Kobayashi, S. *Bull. Chem. Soc. Jpn.*, 1997, 70: 847–858.
43. Sotriffer, C. A.; Flader, W.; Winger, R. H.; Rode, B. M.; Liedl, K. R.; Varga, J. M. *Methods*, 2000, 20: 280–291.

44. Case, D. A.; Pearlman, D. A.; Caldwell, J. W.; Cheatham, T. E., III; Wang, J.; Ross, W. S.; Simmerling, C. L.; et al. *AMBER 7*, University of California, San Francisco, CA, 2002.
45. Wang, J.; Cieplak, P.; Kollman, P. A. *J. Comput. Chem*., 2000, 21: 1049–1074.
46. Berendsen, H. J. C.; Postma, J. P. M.; van Gunsteren, W. F.; DiNola, A.; Haak, J. R. *J. Chem. Phys*., 1984, 81: 3684–3690.
47. Dirama, T. E.; Carri, G. A.; Sokolov, A. P. *J. Chem. Phys*., 2005, 122: 114505.1–114505.8.
48. Iben, I. E. T.; Braunstein, D.; Doster, W.; Frauenfelder, H.; Hong, M. K.; Johnson, J. B.; Luck, S.; et al. *Phys. Rev. Lett*., 1989, 62: 1916–1919.
49. Bizzarri, A. R.; Cannistraro, S. *J. Phys. Chem. B*, 2002, 106: 6617–6633.
50. Ansari, A.; Berendzen, J.; Bowne, S. F.; Frauenfelder, H.; Iben, I. E. T.; Sauke, T. B.; Shyamsunder, E.; Young, R. D. *Proc. Natl. Acad. Sci. USA*, 1985, 82: 5000–5004.
51. Fenimore, P. W.; Frauenfelder, H.; McMahon, B. H.; Young, R. D. *Proc. Natl. Acad. Sci. USA*, 2004, 101: 14408–14413.
52. Doster, W.; Bachleitner, A.; Dunau, R.; Hiebl, M.; Lüscher, E. *Biophys. J*., 1986, 50: 213–219.
53. Barron, L. D.; Hecht, L.; Wilson, G. *Biochemistry*, 1997, 36: 13143–13147.
54. Cornicchi, E.; Onori, G.; Paciaroni, A. *Phys. Rev. Lett*., 2005, 95: 158104–158104. 4.
55. Caliskan, G.; Kisliuk, A.; Tsai, A.; Soles, C.; Sokolov, A. P. *J. Chem. Phys*., 2003, 118: 4230–4236.
56. Uritani, M.; Takai, M.; Yoshinaga, K. *J. Biochem*., 1995, 117: 774–779.
57. Doster, W.; Kleinert, T.; Post, F.; Settles, M.; Gregory, R. B. *Protein–Solvent Interactions*, Marcel Dekker, New York, 1993, p. 375.
58. Sastry, G. M.; Agmon, N. *Biochemistry*, 1997, 36: 7097–7108.
59. Xu, L.; Hu, X.; Lin, R. *J. Solution Chem*., 2003, 32: 363–370.
60. ZangZang, J.; Zografi, G. *J. Pharm. Sci*., 2001, 90: 1375–1385.
61. Dirama, T. E.; Curtis, J. E.; Carri, G. A.; Sokolov, A. P. *J. Chem. Phys*., 2006, 124, 034901.1–034901.8.
62. Caliskan, G.; Mechtani, D.; Roh, J. H.; Kisliuk, A.; Sokolov, A. P.; Azzam, S.; Cicerone, M. T.; Lin-Gibson, S.; Peral, I. *J. Chem. Phys*., 2004, 121: 1978.
63. Doster, W.; Settles, M. In *Hydration Processes in Biology*, Bellissent-Funel, M. C., Ed., Nato Science Series A: Life Science, Vol. 305, IOS Press, Berlin, 1999, pp. 177–191.
64. Mezei, M.; Beveridge, J. *J. Chem. Phys*., 1981, 74: 622–632.
65. Frauenfelder, H.; Petsko, G. A.; Tsernoglou, D. *Nature*, 1979, 280: 558–563.
66. Doster, W.; Cusack, S.; Petry, W. *Nature*, 1989, 337: 754–756.
67. Arcangeli, C.; Bizzarri, A. R.; Cannistraro, S. *Chem. Phys. Lett*., 1998, 291: 7–14.
68. Ferrand, M.; Dianoux, A. J.; Petry, W.; Zaccai, G. *Proc. Natl. Acad. Sci. USA*, 1993, 90, 9668–9672.
69. Rasmussen, B. F.; Stock, A. M.; Ringe, D.; Petsko, G. A. *Nature*, 1992, 357: 423–424.
70. Vitkup, D.; Ringe, D.; Petsko, G. A.; Karplus, M. *Nat. Struct. Biol*., 2000, 7: 34–38.
71. Tsai, A. M.; Neumann, D. A.; Bell, L. N. *Biophys. J*., 2000, 79: 2728–2732.

72. De Gusseme, A.; Carpentier, L.; Willart, J. F.; Descamps, M. *J. Phys. Chem. B*, 2003, 107: 10879–10886.
73. Paciaroni, A.; Cinelli, S.; Onori, G. *Biophys. J.*, 2002, 83: 1157–1164.
74. Frauenfelder, H.; Fenimore, P. W.; McMahon, B. H. *Biophys. Chem.*, 2002, 98: 35–48.
75. Beece, D.; Eisenstein, L.; Frauenfelder, H.; Good, D.; Marden, M. C.; Reinisch, L.; Reynolds, A. H.; Sorensen, L. B.; Yue, K. T. *Biochemistry*, 1980, 19: 5147–5157.
76. Fenimore, P. W.; Frauenfelder, H.; McMahon, B. H.; Parak, F. G. *Proc. Natl. Acad. Sci. USA*, 2002, 99: 16047–16051.
77. Caliskan, G.; Kisliuk, A.; Sokolov, A. P. *J. Non-Cryst. Solids*, 2002, 307–310: 868–874.
78. Tarek, M.; Tobias, D. J. *Phys. Rev. Lett.*, 2002, 88: 138101.1–138101.4.
79. Bizzarri, A. R.; Wang, C. X.; Chen, W. Z.; Cannistraro, S. *Chem. Phys.*, 1995, 201: 463–472.
80. Roh, J. H.; Novikov, V. N.; Gregory, R. B.; Curtis, J. E.; Chowdhuri, Z.; Sokolov, A. P. *Phys. Rev. Lett.*, 2005, 95: 038101.1–038101.4.
81. Bizzarri, A. R. *J. Phys. Condens. Matter*, 2004, 16: R83–R110.
82. Smith, J. C. *Q. Rev. Biophys.*, 1991, 24: 227–291.
83. Paciaroni, A.; Stroppolo, M. E.; Arcangeli, C.; Bizzarri, A. R.; Desideri, A.; Cannistraro, S. *Eur. Biophys. J*, 1999, 28: 447–456.
84. Isakov, S. L.; Ishmaev, S. N.; Malinovsky, V. K.; Novikov, V. N.; Parshin, P. P.; Popov, S. N.; Sokolov, A. P.; Zemlyanov, M. G. *Solid State Commun.*, 1993, 86: 123–127. Malinovsky, V. K.; Sokolov, A. P. *Solid State Commun.*, 1986, 57: 757–761.
85. Sokolov, A. P. *J. Phys. Condens. Matter*, 1999, 11: A213–A218.
86. Tarek, M.; Tobias, D. J. *J. Chem. Phys.*, 2001, 115: 1607–1612.
87. Pócsik, I.; Koós, M. *Solid State Commun.*, 1990, 74: 1253–1256.
88. Bee, M. Quasielastic neutron scattering: principles and applications, in *Solid State Chemistry, Biology and Material Science*, Adam Hilger, Philadelphia, PA, 1988.
89. Walser, R.; van Gunsteren, W. F. *Proteins Funct Genet.*, 2001, 42: 414–421.
90. Phillips, G. N.; Pettitt, B. M. *Protein Sci.*, 1995, 4: 149–158.
91. Makarov, V. A.; Feig, M.; Andrews, B. K.; Pettitt, M. *Biophys. J.*, 1998, 75: 150–158.
92. Merzel, F.; Smith, J. C. *Proc. Natl. Acad. Sci. USA*, 2002, 99: 5378–5383.
93. Bizzarri, A. R.; Rocchi, C.; Cannistraro, S. *Chem. Phys. Lett.*, 1996, 263: 559–566.
94. Starr, F. W.; Sastry, S.; Douglas, J. F.; Glotzer, S. C. *Phys. Rev. Lett.*, 2002, 89: 125501.1–125501.4.

7

MESOSCALE SIMULATIONS OF SURFACE-MODIFIED NANOSPHERES IN SOLVENTS

SANAT MOHANTY
Corporate Materials Research Laboratory, 3M Company, St. Paul, Minnesota

Over the last few years, nanoparticles have begun to find extensive industrial use in polymers and glasses to impart specific properties to composites. These include strength of various composites used in dental fillers, in adhesives, and in other consumer products, as well as optical properties of a number of ceramic and polymer based materials used to manipulate electromagnetic waves. These nanoparticles can change the mechanical,[1,2] thermal, and interfacial properties of the polymers quite significantly. Nanoparticles can also be engineered to change permeability,[3] viscosity, adhesive properties,[4] accessible surface area or flame-retardant properties of the materials and potentially find use in next-generation catalysts,[5,6] selective membranes,[7] and photonic bandgap materials.[8] In addition, nanoparticle-based fillers have the additional advantage of nanoscale mixing with the polymer, leading to better mechanical optimization of properties than micrometer-sized fillers. Nanoparticles thus find themselves fulfilling a critical need in design of materials. Yet the ability to predict the behavior of nanoparticles in solvents and polymers of interest is rather limited, to say the least. In the absence of such predictive abilities, formulation of materials with nanoparticles remains an expensive and inefficient empirical process.

Nanoparticles need to be mixed with the polymer of interest. Usually, they are first mixed with another solvent, which might then be polymerized or mixed with the polymer. Often, silica (or other) nanoparticles are incompatible with the polymer of interest or with the solvent. To better engineer these materials, the

Multiscale Simulation Methods for Nanomaterials, Edited by Richard B. Ross and Sanat Mohanty

nanoparticles may be chemically modified by attaching short organic groups to the surface of the particles. Depending on the functionality of these short organic groups and the solvents of interest, these nanoparticles may form a clear solution, a cloudy colloid, a gel, or may simply precipitate at the bottom. By controlling the state of the nanoparticles in solution, one can control the properties of the material of interest. Thus, the ability to predict the thermodynamics of surface-modified nanoparticles in different solvents and polymer mixtures is of critical interest to a rapidly growing industrial market. The industrial market could also benefit from prediction of rheological behavior of these systems. It is the former that this study attempts to address.

1. MODELING NANOPARTICLES

The thermodynamics of two-phase systems has been studied extensively in the case of liquid–liquid systems. However, there have been fewer systematic studies of systems where the second phase is liquid crystalline, and even fewer studies on the effect of additive self-assemblies or aggregates in fluid mixtures for real systems of industrial significance. The investigations of solids in fluids have focused largely on the effect of liquids in the presence of stationary solids such as a wall, or a substrate. In this chapter we focus on understanding the behavior of solutions with mobile particles in the nanometer-to-submicrometer size range. Experimental and theoretical studies of surface-modified nanoparticles in solutions have been few, and our understanding of such systems is limited.

One reason for the paucity of modeling efforts to understand this system has been the complexity of the system. For one, the system spans at least two size scales. Whereas interactions between the solvents and the surface modifying functional groups attached to the nanosphere occurs in the angstrom size, interactions between nanospheres occur in the nanometer or even submicrometer sizes. The mean field used to describe long-range solvent interaction with particles is also in that scale. Thus, the model must capture multiple size scales. If the atomistic details of the interactions are included, the system must include at least a few nanospheres, thousands of attached functional groups, and hundreds of thousands of solvent molecules. Clearly, this system is too large to be simulated with a detailed atomistic method. On the other hand, coarse graining the system threatens to bury details of the atomistic interactions. Such studies present significant results about model systems; however, they are of little use to industrial applications in which the behavior of specific chemicals is of interested.

Another set of algorithms that has been reasonably successful has included self-consistent field theories and density functional theories using Flory–Huggins parameters to describe the behavior of nanoparticles in polymeric mixtures.[9] These methods closely replicate experimental behavior but also reflect the weakness of Flory–Huggins theory. Although, they reflect experimental trends and show phase behavior that mimics experimental evidence, their predictive capability depends on the accuracy of the Flory–Huggins parameter. The Flory–Huggins parameter for a functional group that is free in a solvent is rather different from that of the same functional group tethered to a nanoparticle. This parameter does not capture the specificity of interactions between functional groups that are oriented or self-assembled in this fashion. For this reason we do not use the Flory–Huggins approach.

The lack of understanding of the system is also due to the difficulty in characterizing the experimental system. For example, although we may believe that a nanoparticle is surface treated with a certain functional group, we are never sure that the entire surface is covered or that small parts of the surface are bare or covered by other functional groups. Thus, it is important to develop a system that includes multiscales as well details of the atomistic interactions. An algorithm of this nature has been developed in work of Powell et al.[10] By using simple descriptors of energy functions, this work is able to capture the effect of particle–particle, colloid–particle, and colloid–colloid interactions. However, it does not include a good method to understand the effect of different functional groups on the particle surface. Prediction of the effect of various surface-modifying agents is necessary for engineering of nanoparticle-based solutions or composites. Next, we present an algorithm that includes molecular-level interactions as input for development of coarse-grained models of nanospheres. Results from the nanoscale simulations are used to model the dispersion of the nanospheres in various solvents.

2. THE ALGORITHM

It is clear that any model that is to be useful industrially must include the chemistry of the functional groups and the solvent, yet be able to account for the behavior of a number of nanosized particles. Clearly, we are looking at two different size scales: one in the range 1 to 10 Å and the other in the range 30 to 100 nm. The coarse-grained approach described here accounts for both of these scales. The algorithm to coarse-grain the system of nanoparticles in a solvent is similar to the one used to calculate properties of simple atomistic systems. First we calculate the energy of interaction between two surfaces that have functional groups attached to and covering them as they lie in a solvent (Figure 1). Periodic boundary conditions were used in all three directions. Along the x and y directions, periodic boundary conditions prevent edge or artificial surface effects from dominating the calculation. Along the z-direction, periodic boundary condition ensures that the top of the unit cell, with solvent in it, is interacting with the bottom of the unit cell, which has functional groups attached to the surface.

For the purpose of this study, we use the Discover module in the Materials Studio suite of Accelrys. The Compass force field is used to calculate the interaction energies. The energy of interaction is a function of the distance between the two surfaces. In case of charged functional groups, cell size becomes critical. In this system, one can change the height of the inorganic layer, and effectively keep a large cell size even for small gaps, without significantly affecting the simulation times. The energy that results from these simulations includes the interactions

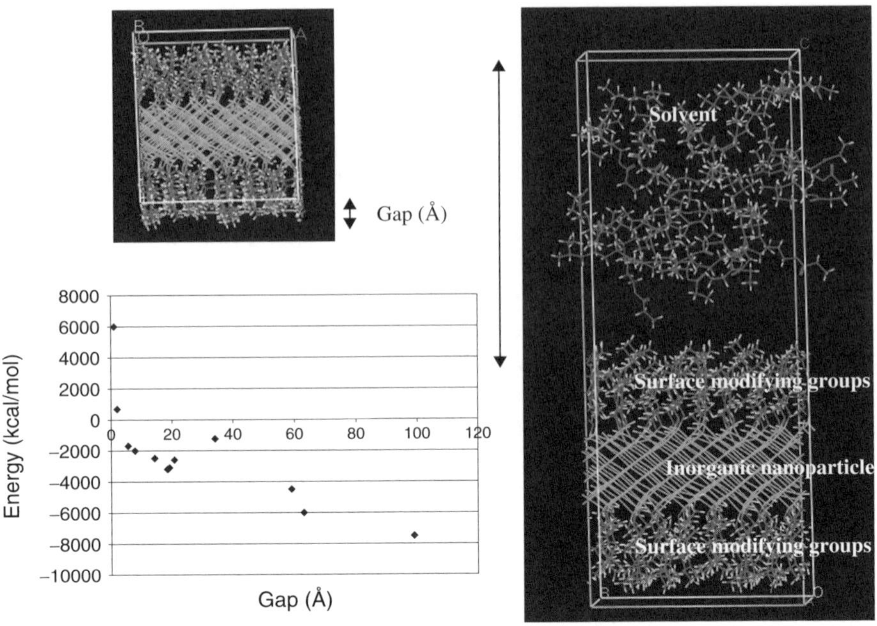

Figure 1 Calculating energy of interaction between two functionalized surfaces that are at different distances when placed in a solvent.

between the solvent molecules as well as the energy of the functional groups on the silica layer. In accounting for interaction between the surfaces, we will exclude the energies of the nanoparticles and the self-energy of the solvent and account for them separately. Thus,

$$E(A,d) = E_{\text{sim}} - E_{\text{solvent}} - E_{\text{nanoparticle}} \tag{1}$$

Here $E(A,d)$ is the energy between two surfaces with an area of overlap A (which is the projection on the second suface of one of the two surfaces that are facing) with the distance of separation between the two surfaces being d: E_{sim} is the energy that results from the calculations; E_{solvent} is the energy of solvent (which lies between the surfaces) by itself; and $E_{\text{nanoparticle}}$ is the energy of the surfaces separated by a large distance (so that there is no surface–surface interaction) in the absence of a solvent. $E(A,d)$ is then normalized with respect to the area of the surfaces, to give

$$E_{s,12}(d) = \frac{E(A,d)}{A} \tag{2}$$

where $E_{s,12}$ is the energy between the two surfaces, normalized for their area of overlap. The interaction energies between two nanoparticles of any size and shape can be obtained from the energies of the two surfaces (although for aspherical shapes, the relative orientation of the two particles will also affect the energy).

Figure 2 Schematic showing how interactions between surfaces can be used to estimate interactions between particles of specific shapes.

In this study we focus only on nanospheres. Two nanospheres can be approximated as a set of infinitesimally thick disks, and the interaction energy between these two spheres can be approximated by integrating over these surfaces as presented schematically in Figure 2:

$$E_{ij}(d) = \int_{d \text{ to } d+D} E_{s,ij}(x)A(x)\,dx \tag{3}$$

where D is the diameter of the sphere and A is the area of overlap (described above) at a gap of x. $E_{s,12}(d)$ is the interaction between two nanoparticles of given shapes and at a distance d in a given solvent medium. Once the energy of interaction between two nanospheres of given size in a solvent is known, the nanosphere–solvent system can be reduced to an interacting many-body system of particles. The physics of interactions is now embedded in the energy functional and the properties of the system can be calculated by simulating this simple many-body system with the complicated energy functional.

E_{solvent} must be added back in a way that weights the concentration of the particles in solvent. We can thus obtain the interaction energy of a mole of particles for a given amount of solvent. Thus, the energy of the solution will include the self-energy of n moles of solvent (E_{solvent}) along with the energy of m moles of the particles and the interactions between the two ($E_{ij}(d)$).

As we shall see, the energy functional one obtains is not a simple one—unlike a Lennard-Jones 12-6 function. The energy functional can be stored as an array of numbers and can be used in Monte Carlo calculations. It is not easily replaced by an analytical expression. Thus, it does not give itself easily to molecular dynamics (MD) (since MD simulations need derivatives of functions). However, it can be approximated by a set of functions fitted to the curve, and this composite function can be used in MD calculations.

Once the interaction energy between two surface-modified nanospheres in a given solvent is known, we can replace the system of surface-modified nanospheres and solvents by particles in a regular Monte Carlo simulation and study the aggregation behavior of these particles. The physics of interaction between these particles is accounted for completely in the energy functional, and the simulation of this system is like the simulation of any simple system of atoms whose interactions are defined by a Lennard-Jones (or some other) interaction potential. This is presented schematically in Figure 3.

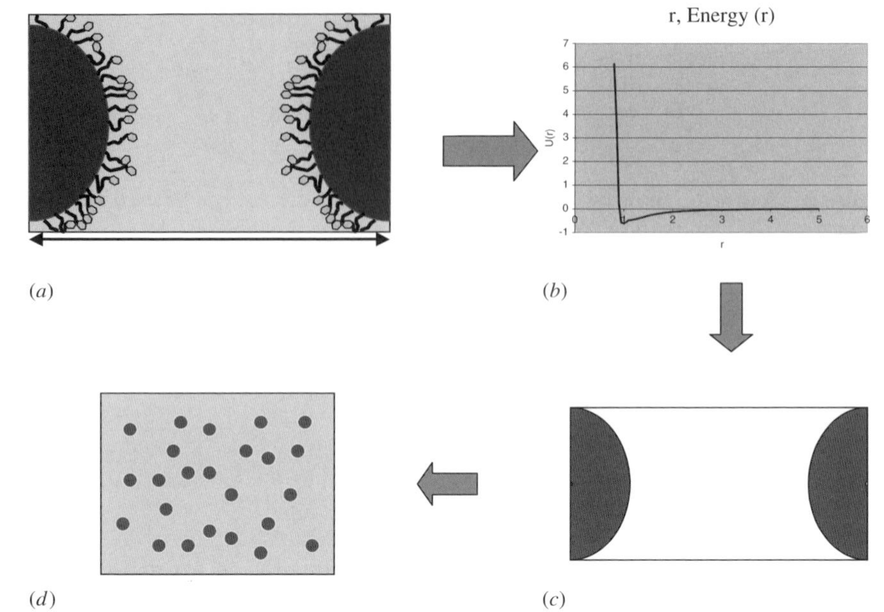

Figure 3 Schematic of a process that will be used to predict the behavior of particles in a solvent, accounting for specific interactions of functionalized groups on the surfaces. (*a*) Interaction between two nanoparticles in a solvent is captured by (*b*) the energy of interaction over changing distance between these particles. (*c*) The atomistic details of the interaction can now be captured in the energy functional and the particles can be coarse-grained. (*d*) Now the nanoparticles in the solvent can be treated as a box of particles with their motion defined by the energy functional.

3. ENERGIES OF INTERACTING SURFACES

The silica surface is covered with functional group A as shown in Figure 1. The solvent used is B. A few surface-modifying groups and solvents have been explored in this study. The modifying groups are all attached to the nanoparticle surface via siloxane groups. Examples of modifying groups include

1. C_2H_4OH (abbreviated here as c2oh) in solvents such as ethanol (c2oh) and water
2. OH in solvents such as water, butane (c4), and hexane (c6)
3. isooctyltrimethylsiloxane (abbreviated as io) in solvents such as methanol, toluene, 3-methyl heptyl acrylate (ioa), and 3-ethyl hexyl acrylate (eha)

The notations in parentheses are used to identify the systems in subsequent graphs and tables. The variances in energy of these systems are typically approximately 5%. In all future discussions, the gap is the separation between the functionalized surfaces as shown in Figure 1.

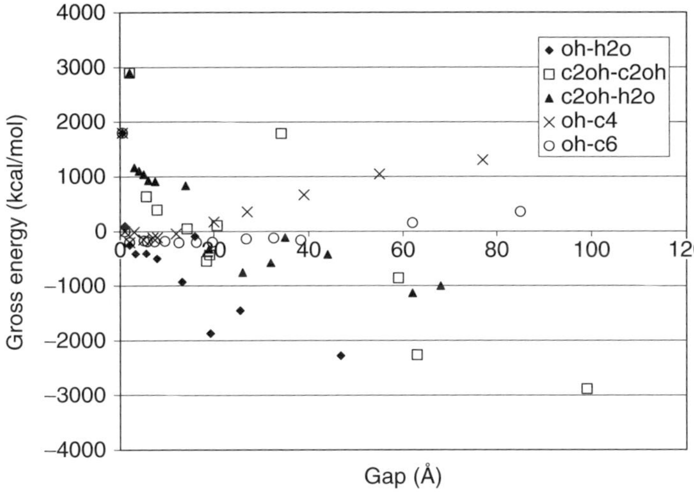

Figure 4 Gross interaction energies between two surfaces modified by OH or c2oh groups in ethanol, water, or alkanes.

Figure 4 shows the gross interaction energy of two surfaces that are surface-modified by short-chain alkanes and alcohols and have a solvent between them. The energies are unusually large since the units are in moles of particles (not moles of solvent). There is very little difference between the gross energies of a surface modified by OH groups and those modified by c2oh groups when placed in water. Surfaces modified by c2oh have different energies when placed in water and in ethanol. This is due to the more attractive self-energy of water. The interactions of the OH- or c2oh-modified surfaces with alkanes as solvents are significantly different. The gross interaction energies are only mildly attractive.

Figure 5 shows the "normalized" interaction energy between the two surfaces. In effect, we have removed certain components of the energies as shown in Eq. (1). For one, we have subtracted the energy of the silica surface along with the surface-modifying functional groups (when the surfaces are separated by large distances). This is to ensure that the size of the silica slab, the energy of the silica itself, or that of the modifiers does not influence the interaction energy. Another component that must be removed from the gross energies shown in Figure 4 relates to the self-energy of the solvent itself. When particles are added to a solvent, the amount of solvent in the system remains the same. Thus, the amount of total solvent does not influence the gap between these particles; the concentration of the particles does. We remove the energy of the solvent from the gross energy of interaction between these surfaces, but the energy of interaction between solvent and surfaces remains. Later, we add the total self-energy of the solvent. Figure 5 shows the normalized interaction energies between these surfaces for some short-chain alkanes and alcohols.

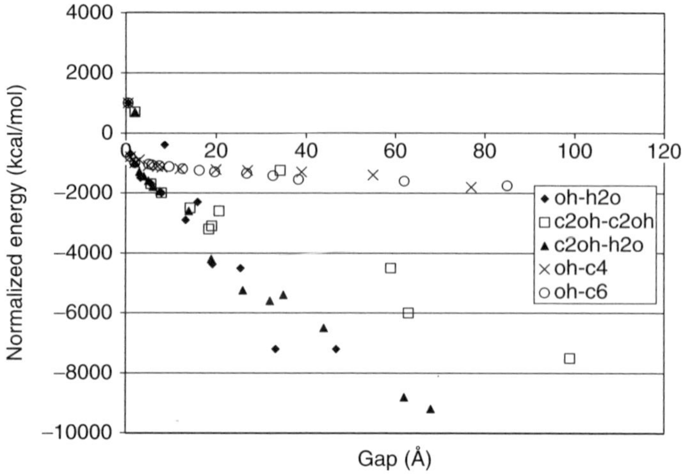

Figure 5 Normalized interaction energies between two surfaces modified by OH or c2oh groups in ethanol, water, or alkanes.

Water has a more attractive self-energy than that of ethanol. When the self-energies of the solvents are removed, there is little difference between OH- and c2oh-modified surfaces in either water or ethanol. c2oh in water is a little less attractive than OH in water. This is probably due to stronger interactions between OH and water than between c2oh and water. c2oh in ethanol has a repulsive peak at small gaps between the surfaces; c2oh in water does not. This is probably due to the difference in ordering between water and ethanol. All of these profiles show a local well at a separation of about 2 nm—it is possible that at smaller separation between the surfaces, the solvent molecules are more ordered (as we discuss later based on evidence of Figures 8 and 9). In addition, when two surfaces modified with c2oh or OH groups are brought closer together, there is an increasing attraction, but when these surfaces are so close that solvent is completely excluded, the attraction between the surface decreases. The surfaces prefer solvent between them. c2oh-modified surfaces seem to have greater attraction for solvent between them than do OH-modified surfaces.

Based on the results from these simulations, it seems that particles that are surface-modified by OH or c2oh groups will be driven (by energy considerations) to have water or ethanol around them. Whether they aggregate or stay dispersed will be influenced by entropic considerations as well. Subsequent simulations that also account for the size of the particles will predict the behavior of these particles. At very large gap sizes, the only interaction of consequence in the normalized energy plots is the energy of interaction between the surface and the solvent. The normalized energies of OH in the alkane groups are repulsive. In addition, the energy of two modified surfaces interacting with each other is lowest when all solvent is removed from between them and they are in proximity to each other. Based on this analysis, one could infer that particles with OH groups will be driven (by energy considerations) to arrange themselves such that there is minimal interaction with the hydrocarbon solvent. They will aggregate and precipitate when they are "large" or aggregate in suspension when they are "small".

Figures 6 and 7 present the gross and normalized energies of surfaces modified with io in ioa, methanol, toluene, and eha. On increasing gaps between the surfaces, the gross energies of all four systems go through a rapid decrease followed by a gradual increase and subsequent plateau. The decrease in energy is associated with a reduction in steric interactions followed by the increasing

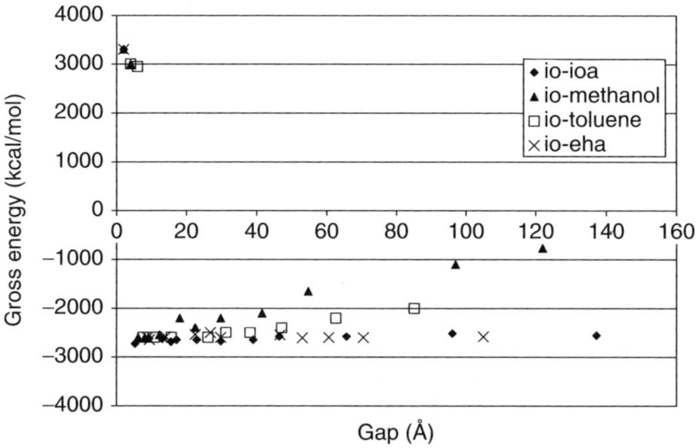

Figure 6 Gross interaction energies between two surfaces modified by io in ioa, eha, toluene, and methanol.

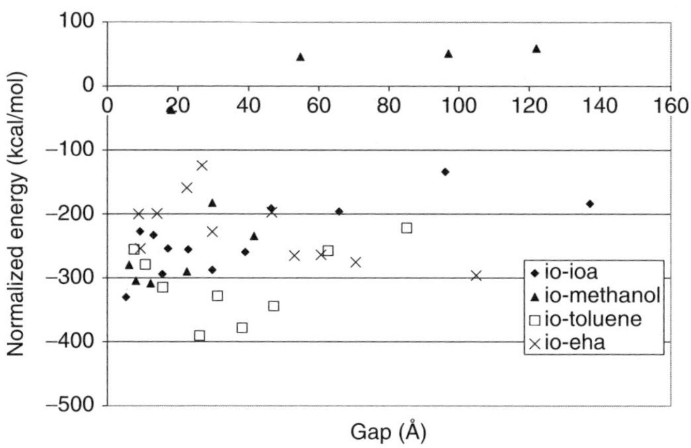

Figure 7 Normalized interaction energies between two surfaces modified by io in ioa, eha, toluene, and methanol.

presence of solvent. The gross energy becomes increasingly less favorable for methanol with increasing gap size. There is much less difference among the other three. When the energies are normalized (Figure 7), methanol with repulsive interaction energies is the least favored solvent. The modified surfaces prefer interactions with the other solvents. There is a significant difference between the energies of the particles in ioa and in eha despite these solvents having very similar molecular structures. The source of this disparity lies in the difference in the cohesive energies of these solvents: eha, with its increased branched structure, has a smaller cohesive energy than ioa and hence is more amenable to interacting with the surfaces. The energy of dissolution depends on competition between how much the solvent likes itself and the interaction energy between the particles and the solvent. Note that the energy of formation of the surface is the same for all of these systems in Figures 6 and 7. Based on these trends, one could infer that energy contributions might make dissolution of these particles difficult in methanol. The other solvents seem more amenable to dissolution.

In addition, at small gaps, there may be ordering of the solvent molecules between the surfaces. Figures 8 and 9 show that when the gap is small, the

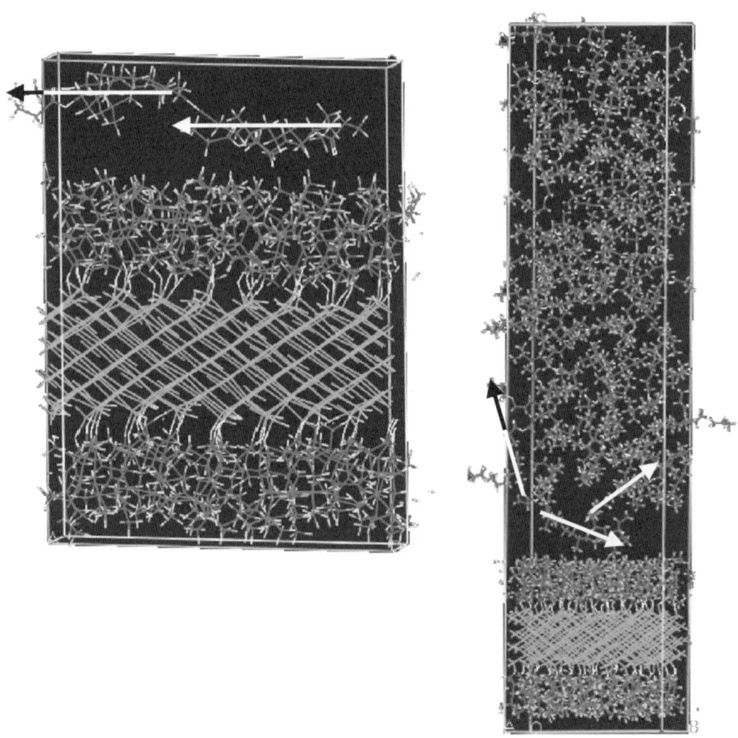

Figure 8 Increased ordering of acrylate molecules with reduced gaps between surfaces. Notice that the vectors representing the chain orientations lie parallel to the surface when the gap is small and are more disordered when the gap is larger.

Figure 9 Increased ordering of methanol molecules with reduced gaps between surfaces. Notice that the vectors representing the chain orientations lie parallel to the surface when the gap is small and are more disordered when the gap is larger.

solvent molecules (eha and methanol, respectively) are aligned along the surfaces. Notice that the vectors representing the chain orientations lie parallel to the surface when the gap is small and are more disordered when the gap is larger. Further, the larger molecules (eha) are significantly more constrained than methanol in the gap. It is likely that this influences the interaction energy of the solvent molecules as well as their entropy. However, at this stage we have not accounted for the effects of entropy.

4. ENERGIES OF INTERACTING NANOSPHERES

Using Eqs. (2) and (3) we calculate the interaction energies between the particles. Figure 10 shows the energy of interaction between 20-nm particles with OH and c2oh functional groups in various solvents at 50% volume fraction. The particles in water and ethanol show increasingly attractive interaction with increasing distance before they flatten out. In fact, c2oh-modified silica in water shows a shallow well. This implies that the particles would prefer to stay in dispersed states although they may associate in higher concentrations. The presence of a well indicates that c2oh in water might be more prone to association. The particles in the alkanes become increasingly repulsive before the energy flattens out at large separation. This indicates their preference to aggregate and perhaps precipitate. These trends were perhaps expected based on the surface interaction energies.

Figure 11 shows the energy of interaction between 20-nm particles with isooctyltrimethylsiloxane groups in various solvents at 50% volume fraction. The energy plot suggests that the particles would find it difficult to disperse in

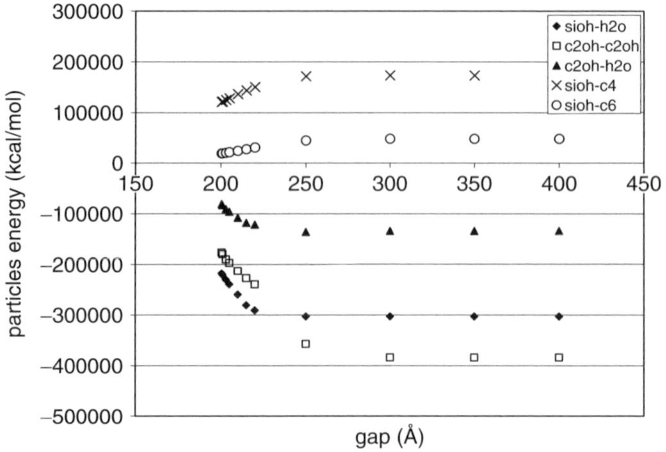

Figure 10 Interaction energies between two spheres modified by OH and c2oh groups in water, ethanol, and alkanes.

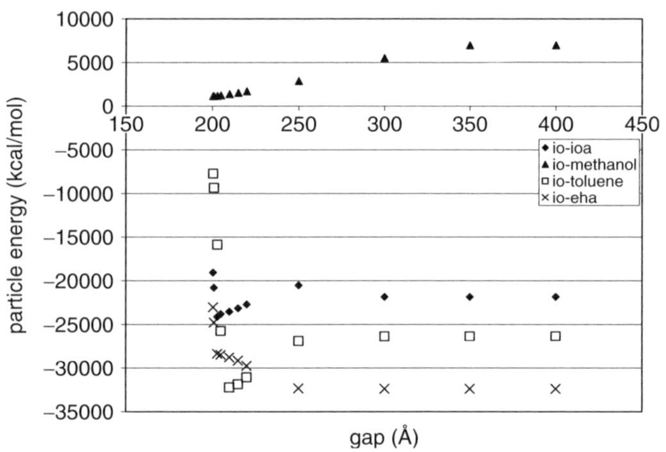

Figure 11 Interaction energies between two spheres modified by io in ioa, eha, toluene, and methanol.

methanol, owing to positive (repulsive) interaction energies, but would disperse in the other solvents. The interaction energies in the other solvents show an energy well, implying that at higher concentrations of these particles, association or gel formation is possible. Experiments conducted on these systems in-house show similar behavior. In addition, it is important to note the significant difference between the energies of these particles in eha and in ioa. This difference is consistent with the difference that was observed in the normalized energy plots in Figure 7. Particles disperse more easily in eha—a trend that is observed experimentally as well.

5. EFFECT OF PARTICLE SIZE

Figure 12 shows the effect of increasing particle size on the energy of interactions for io-functionalized particles in toluene. The strongly repulsive energies of the particles caused by steric interactions at smaller gaps are not shown so that the scales of the plot allow us to focus on the interactions between the particles due to functional groups and solvent. The energies are normalized with respect to the total amount of silica. The plot shows the energy of 1000 mol of 20-nm particles, 64 mol of 40-nm particles, and 1 mol of 200-nm particles at 50 vol %. Clearly, as one increases the particle size, it becomes less soluble. Similar trends are true for the other solvents as well. This trend is also consistent with experimental observations. It is assumed that the particles are small enough that molecular assumptions still hold.

There is another aspect of particle size: Given the shape of the particles, each particle has associated with it certain interstitial volume. At a given value of separation between particles, the solution made up of larger particles has a greater volume percent loading. Conversely, at a given loading, smaller particles are in closer proximity to each other than are larger particles. Two-nanometer particles can be packed as little as 2 nm apart and will only make 6.5% of the solution volume 20-nm particles will make 39.3% of the volume, and 200-nm particles will make up 51.1% of the solution volume at that packing. As the particles get larger, there is less intimate interaction between them. Thus, there is less opportunity for these particles to find minimum energy configurations by associating or gelling. They will either be dispersed or they will precipitate.

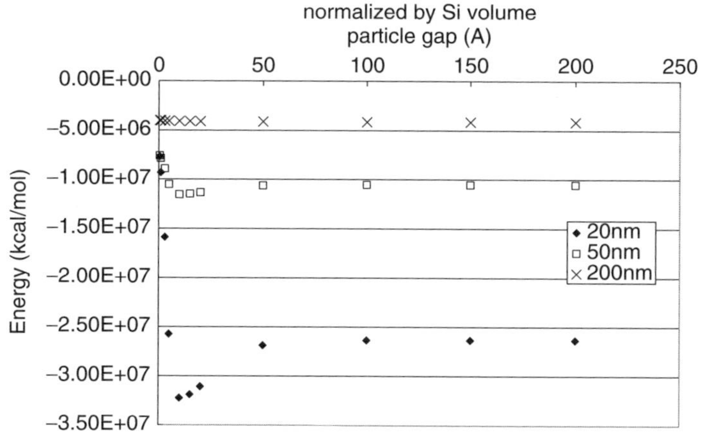

Figure 12 Energy of particles various sizes modified by io in eha.

6. CONCLUSIONS

These simulations present results that are in agreement with experimental trends. They can be used to understand and predict the behavior of systems of interest. In addition, these simulations provide the force fields that can be used in Monte Carlo simulations to predict the behavior of macroscopic systems, such as aggregation of these particles, gelation or dissolution, and possible temperature effects. This algorithm and these force fields show promise to develop into a robust predictive tool for engineering surface-modified nanoparticles and materials and composites made from these nanoparticles.

Acknowledgments

I would like to thank W. J. Schultz for discussions on applications of such models in engineering nanoparticles. I am also grateful to B. U. Kolb, G. Caldwell, and M. Jain for extensive discussions on various aspects of nanoparticle behavior and multiscale modeling.

REFERENCES

1. Kojima, Y.; Usuki, A.; Kawasumi, M.; Okada, A.; Kurauchi, T.; Kamigaito, O.; *J. Polym. Sci. A Polym Chem.*, 1993, 31: 983.
2. Krishnamoorti, R.; Vaia, R. A.; Giannelis, E. P. *Chem Mater.*, 1996, 8: 1728.
3. Yano, K.; Uzuki, A.; Okada, A.; Kurauchi, T.; Kamigaito, O. *J. Polym. Sci. A Polym Chem.*, 1993, 31: 2493.
4. Kolb, B. U.; Baran, J. R., Jr.; Johnson, M. A.; Johnson, G. G.; Lehmann, M. P.; Sokalski, J. S. Foams and plastic foams containing surface-modified nanoparticles and suitable for production of adhesive tapes, PCT Int. Appl., 2002.
5. Makarova, O. V.; Rajh, T.; Thurnauer, M. C.; Martin, A.; Kemme, P. A.; Cropek, D. *Environ. Sci. Technol.* 2000, 34: 4797.
6. Han, M. Y.; Gan, L. M.; Huang, W.; Chew, C. H.; Zou, B. S. *Chem. Lett.*, 1997, 8: 751.
7. Belin, S.; Santos, L.R.B.; Briois, V.; Lusvardi, A.; Santilli, C. V.; Pulcinelli, S. H.; Chartier, T.; Larbot, A. 2003 *Colloids Surfaces A Physicochem. Eng. Aspects*, 216: 195.
8. Ullal, C.; Tripathi, R.; Breiner, T.; Bockstaller, M.; Urbas, A.; Thomas, E. *Polym. Prepr.*, 2002, 43: 544.
9. Ginzburg, V. V.; Balasz, A. C. *Macromolecules*, 1999, 32:, 5681.
10. Powell, C.; Fenwick, N.; Bresme, F.; Quirke, N.; *Colloids Surfaces A Physicochem. Eng. Aspects*, 2002, 206: 241.

8

FIXING INTERATOMIC POTENTIALS USING MULTISCALE MODELING: AD HOC SCHEMES FOR COUPLING ATOMIC AND CONTINUUM SIMULATIONS

CLIFFORD W. PADGETT
Department of Materials Science and Engineering, North Carolina State University, Raleigh, North Carolina

J. DAVID SCHALL
Department of Chemistry, United States Naval Academy, Annapolis, Maryland

J. WESLEY CRILL AND DONALD W. BRENNER
Department of Materials Science and Engineering, North Carolina State University, Raleigh, North Carolina

Many variations on multiscale modeling have been introduced over the last three decades, attempting to overcome the size and time limitations inherent in molecular dynamics simulations. Starting in the 1970s, various thermostating procedures were developed with the intent of introducing temperature control that models interactions with a macroscopic heat bath. Work by Tully and others, for example, used Langevin forces applied to atoms surrounding a "reaction zone" to model heat flow between a surface reaction and bulk phonon modes.[1] Andersen, Hoover, Evans, Berendsen, and others introduced various modified Lagrangians that control temperature (and pressure/stress state) of a molecular simulation.[2] Building on this work, Nosé introduced a classical Lagrangian that maintains not only a

Multiscale Simulation Methods for Nanomaterials, Edited by Richard B. Ross and Sanat Mohanty

desired average temperature, but also reproduces the fluctuations appropriate for a finite system of atoms embedded within a well-defined macroscopic ensemble.[3]

More recently, various schemes have been introduced that couple the dynamics of finite atomic regions with other macroscopic system properties.[4] For example, methods have been introduced that couple finite-element spatial regions to an atomic region through a "handshaking" zone, allowing coupling between long-range stresses and atomic dynamics.[5,6] In a related method, a finite-element mesh is reduced to the atomic scale, where fine resolution is needed, and the constitutive equations that govern the dynamics of these nodes are determined from a molecular simulation.[7] Similarly, bridging scale decomposition has been developed in which a Lagrangian is written involving both overlapping atoms and finite-element nodes.[8] Many similar schemes exist.

Several groups have recently introduced schemes for extending the time scale of simulations. In Voter's hyperdynamics approach, a potential energy hypersurface is defined in which potential energy wells between regions corresponding to rare events are "filled in" while the true potential is used to determine trajectories near transition states associated with the rare events.[9,10] A molecular dynamics simulation is carried out on the hypersurface, and a time *boost factor* is defined that consists of the ratio of the Boltzmann factors of the true and hypersurface potential energies. Between rare event crossings, the scaled time step is large, while the true dynamics are followed in the transition-state region.

Implicit to these schemes is the assumption that the forces and energies used in the atomistic simulation are correct and appropriate. Although the reliability of analytic potentials that enable large-scale atomic simulations (such as the embedded-atom method[11,12]) has improved dramatically over that last two decades, there are properties that an analytic potential will never describe well, regardless of the functional form or parameters chosen. One important example is heat flow in metals, where in most cases thermal conductivity at room temperature is dominated by electronic effects rather than phonon dynamics.

The response of the modeling community to the issue of heat flow in metals has been in large part simply to avoid simulations in which a poor description of thermal conductivity would produce nonphysical results. One exception to this was innovative work by Ivanov and Zhigilei in simulating laser fields interacting with metals.[13,14] In their approach, the electronic and phonon contributions to the dynamics of the system are decoupled and treated using a combination of continuum and atomic dynamics in which continuum equations were solved stepwise on a grid superimposed over the atomic simulation.

In a recent paper we analyzed a simplification of the Zhigilei multiscale modeling approach that takes the form of a thermostat in which velocities in a molecular dynamics simulation are coupled to the numerical solution of the continuum thermal transport equation through ad hoc feedback.[15] A key feature of this work is the use of an experimental thermal diffusivity in the continuum equations that indirectly allows the influence of missing electronic effects on heat flow to be included in a large-scale molecular dynamics simulation without adding a large computational burden. In an extension of this idea of ad hoc atomistic–continuum coupling, we developed a coarse-grained method that incorporates current flow and associated Joule heating into a large-scale atomic simulation,[16] again without explicitly including electronic degrees of freedom.

Our intent in the chapter paper is twofold. The first intent is to present our prior work on continuum–atomistic coupling in the larger context of multiscale modeling and to present a new example simulation of electrically "hot" and "cold" metal–metal contacts that represent an asperity contact in a microelectromechanical system (MEMS) device. The second intent is to propose other possible uses for the ad hoc continuum–atomistic coupling in molecular simulations, in particular incorporating diffusion, dislocation climb, and creep in a molecular dynamics simulation, and coupling accurate kinetic rates for chemical reactivity with mesoscale dynamic simulations.

1. AD HOC COUPLING SCHEME

The concept behind this approach is very straightforward. A concurrent molecular dynamics and continuum simulation is carried out over the spatial range of the atoms, with ad hoc feedback for the common properties of the two simulations. Both calculations are carried out stepwise in time, with the common properties used to set the initial conditions of each step. In the thermostat discussed below, for example, the continuum heat flow equation is solved numerically on a finite-difference grid superimposed on a molecular simulation. The initial temperatures within each grid region are defined from the kinetic energies in the molecular simulation. One (or more) steps of the continuum solution are iterated, and the atomic kinetic energies are scaled to match the prediction of the continuum simulation. The molecular dynamics simulation is moved forward in time one (or more) step(s) and the process is repeated.

There are several advantages to this scheme compared to other multiscale formalisms. First, as discussed above, parameters in the continuum simulation can compensate for deficiencies in the potential function. Second, the scheme is straightforward to implement and can be added over an existing molecular dynamics code. It is also independent of system size and interatomic potential. Similarly, standard methods can be used to solve the continuum equations. Third, the continuum can be extended beyond the atomistic region, thereby providing feedback between macroscopic and atomic scales. Finally, although both the continuum and atomic equations are solved stepwise, they do not necessarily have to have the same time step size if the system is near equilibrium. This implies that multiple time scales can be treated in a single simulation. For example, a continuum equation for impurity diffusion in a solid can be solved with a macroscopic time step and the resulting concentrations used to populate a lattice from which atomic motion is implemented. This type of multitime modeling would be especially important for situations in which some long-time process triggers an atomic-scale event, such as vacancy diffusion leading to dislocation climb and slip. This example is discussed in detail below.

There are also some important potential disadvantages to this scheme. The most critical is that there is no obvious conserved quantity or anything equivalent to a fictitious Lagrangian. Instead the coupling is done on an ad hoc basis. This can make it difficult to identify system conditions and step sizes for which numerical solutions may be unstable. There is also no set procedure for implementing the coupling, and unanticipated issues involving the coupling can arise, as discussed in more detail in the next section. A solution to these and related issues is to test the method thoroughly as implemented for a given system and code.

Continuum Thermostat

In the thermostat implementation, a molecular dynamics simulation system is first coarse-grained into grid regions (Figure 1, dotted lines), and the temperature of each grid region is assigned according to the average kinetic energy of the atoms within that region. The temperature in each grid region is then calculated at the time of the next molecular dynamics time step $t + \Delta t$ by solving the continuum heat equation

$$\frac{\partial \mathrm{T}}{\partial t} = D\frac{\partial^2 T}{\partial R^2} + Q \tag{1}$$

where T is the temperature derived from the atomic simulation, D is the thermal diffusivity given by

$$D = \frac{\lambda}{c_p \rho} \tag{2}$$

and Q is heat generated (converted to a temperature change) in the grid region from an external source such as current flow. In Eq. (2), λ is the thermal conductivity, c_p the heat capacity, and ρ the density of the material. In the simplest

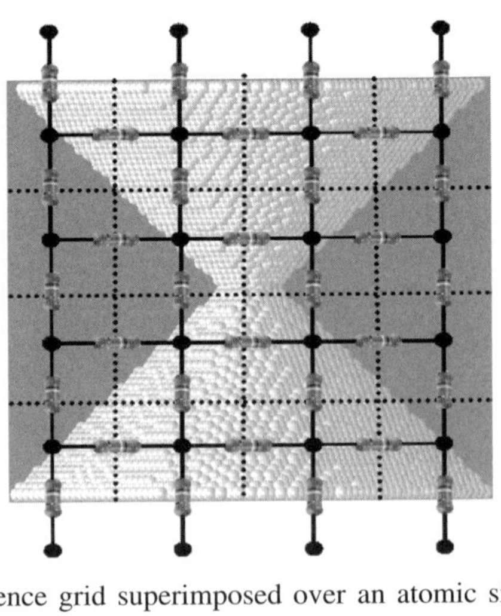

Figure 1 Finite-difference grid superimposed over an atomic simulation (dotted lines) and the resistive network (solid lines).

implementation of this scheme, a Euler's method solution to Eq. (1), the temperature T_{new} for each grid region is given by

$$T_{\text{new}} = T_{\text{old}} + \Delta t D \left(\sum_{\alpha} \frac{T^{\alpha}_{i+1} - 2T^{\alpha}_{i} + T^{\alpha}_{i-1}}{\Delta r^2_{\alpha}} \right) + Q_i \tag{3}$$

where the partial derivative in Eq. (1) has been approximated by a centered finite-divided difference formula, Δr_{α} is the grid spacing in direction α, the sum is over the three Cartesian coordinates, and i specifies the particular grid point. The atomic velocities of the atoms in each grid region are then scaled to match the numerical solution of Eq. (1), and the atoms then move according to the interatomic potential (plus a constraint force described below) using a standard molecular dynamics integrator.

To evaluate the numerical stability and accuracy of this scheme, a series of simulations were carried out describing heat flow from a surface maintained at a constant temperature into a lower-temperature substrate. The first simulation was a strictly molecular dynamics calculation using a standard embedded-atom method potential for silver. In this case, the constant surface temperature was maintained by scaling the velocities of atoms near the surface at each time step. In the second simulation, the continuum equation (1) was solved numerically using the experimental diffusivity for silver. In this case the constant surface temperature was maintained by not updating the grid regions at the surface. The third simulation followed the ad hoc continuum–atomic coupling description above. To validate the results, the temperature profiles from each simulation were compared to the analytic solution to this problem using the experimental thermal diffusivity. As expected, the temperature profiles from the analytic and numerical solutions to Eq. (1) matched at all times. In the strictly molecular dynamics simulations, heat flow from the surface was several orders of magnitude slower than the continuum results due to the much lower thermal diffusivity produced by the analytic potential compared to experiment, again as expected. In the case of the ad hoc coupling, the heat flow into the solid was much faster than the strictly molecular dynamics simulation, but was still slower than the continuum solution. This discrepancy arises because part of the kinetic energy given to the atomic simulation through the velocity scaling is converted to potential energy at each timestep due to the equipartition principle. Over time this results in a "time lag" in the heat flow through the coupling to the continuum part of the calculation.

To compensate for the partial conversion of kinetic to potential energy, the procedure outlined above was modified by adding to the interatomic forces a Hoover frictional force applied locally to each grid region. This friction coefficient for this force is given by

$$\gamma = \frac{-\sum_j F_j V_j}{\sum_j V_j^2 m_j} \tag{4}$$

where the sum is over atoms in a given grid region, and F_j, V_j, and m_j are the total force on atom j from the interatomic potential, the velocity of atom

j, and the mass of atom j, respectively. Note that there is a unique frictional force for each grid region. With this modification the temperature profile for the coupled continuum–atomic modeling agreed with the strictly continuum result. In addition to validating this method, at least for this application, this exercise also demonstrated that the coupling scheme does indeed influence the simulation results (see Figure 2).

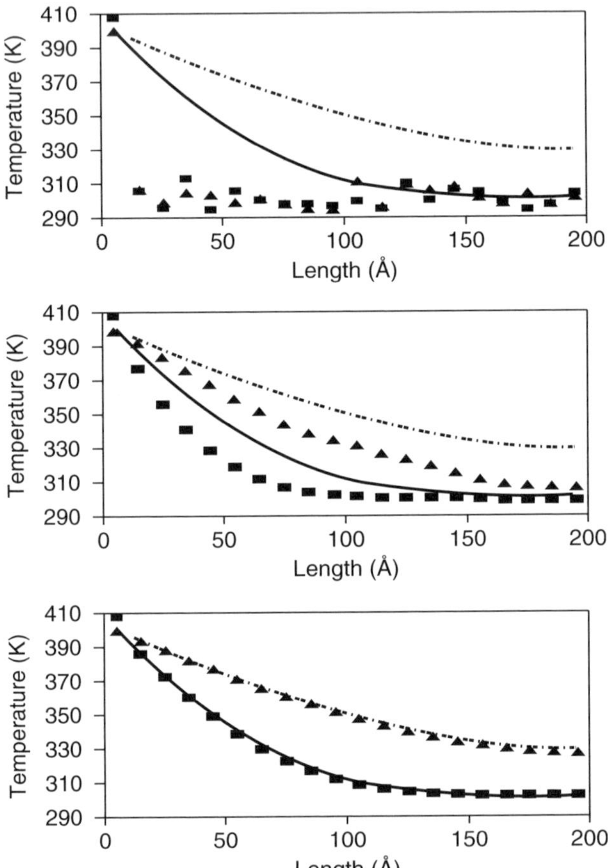

Figure 2 Kinetic energy (reported as a temperature) and temperature profiles along the direction of heat flow for single-crystal silver. The symbols are from the atomistic simulations; the solid lines are solutions to the continuum heat flow equation. Top: Heat flow from a silver (111) surface without the atomistic–continuum thermostat. The squares and solid line are for 0.1 ps and the triangles and dashed line are for 0.5 ps. The solid lines are a numerical solution to the heat flow equation using the D value determined experimentally. Middle: Simulation using velocity scaling only. The squares and solid line are for 0.1 ps; the triangles and the dashed line are for 0.5 ps. Bottom: Results using the Hoover thermostat. The symbols and lines are the same as for the middle panel.

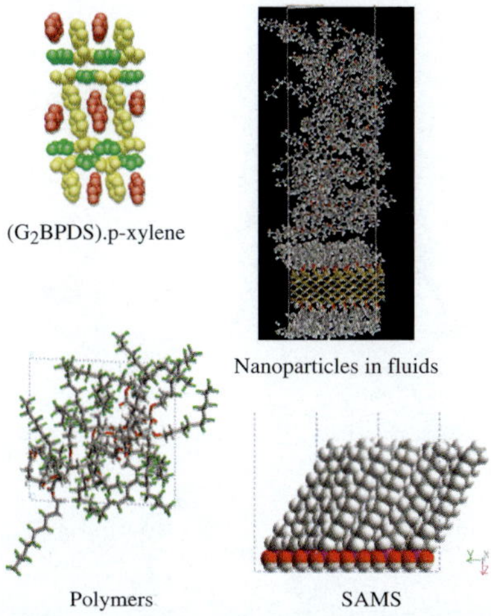

Chapter 1, Figure 1 Mesoscale phenomena.

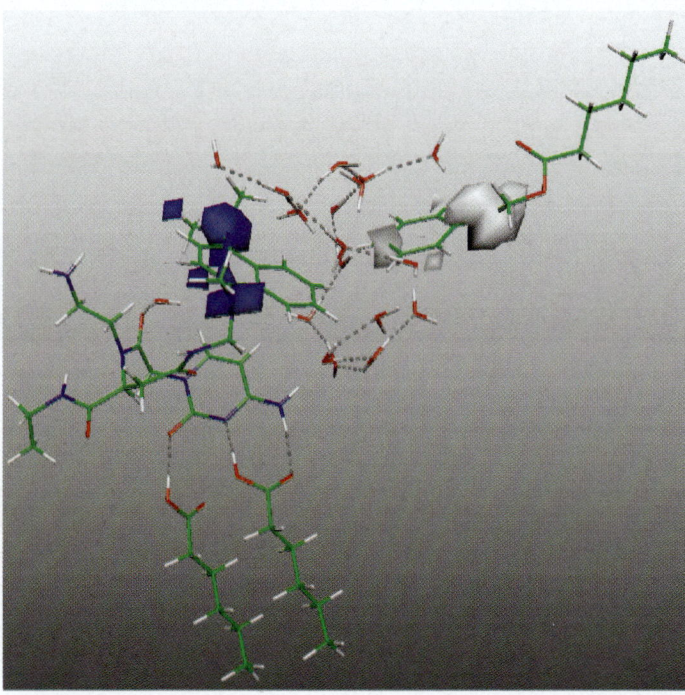

Chapter 2, Figure 7 Visualization of the electron charge transfer associated with the second excited state. The tunneling is from the sensitizer 1,4-bis(*N,N*-dimethylamino)naphthalene molecule to the pFA molecule. The electron cloud hole is indicated in blue, and the transferred electron cloud location is shown in gray.

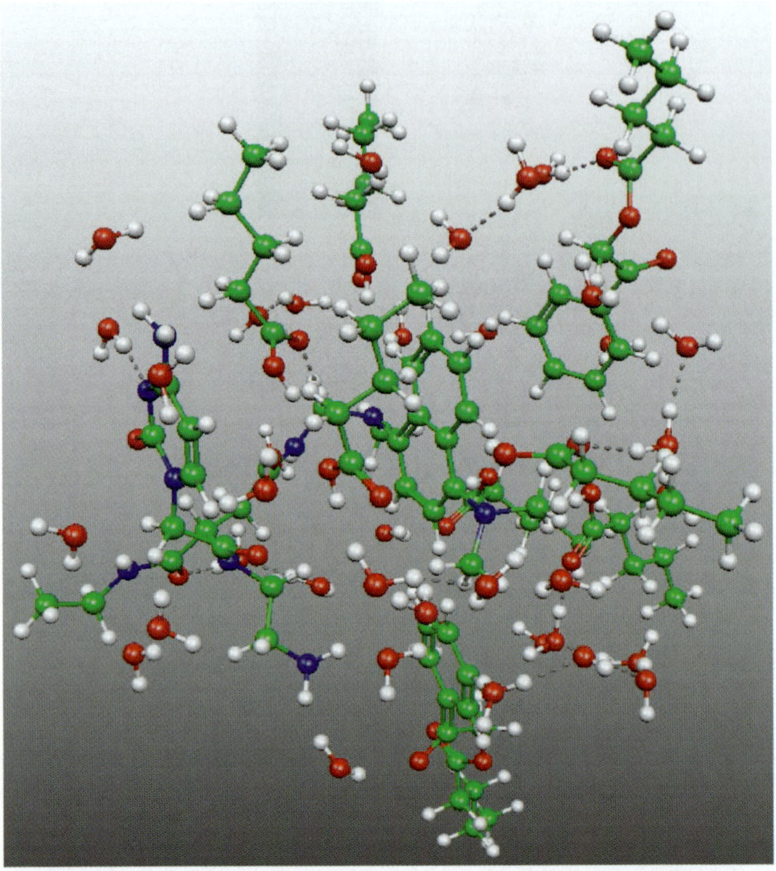

Chapter 2, Figure 8 Image of the geometric and electronic structure of a system consisting of a cytosine–PNA fragment covalently bonded to a 1,4-bis(*N*,*N*-dimethylamino)naphthalene sensitizer molecule (in the center), two pFA molecules (bottom and top right), six FA molecules, and water molecules that are optimized using the PM3 method. The water molecules organize into nano icelike substructures. Carbon atoms and their associated covalent bonds are shown as green spheres and sticks, hydrogens are shown in white/gray, oxygens are red, nitrogens and blue. Hydrogen bonds are depicted by dashed lines.

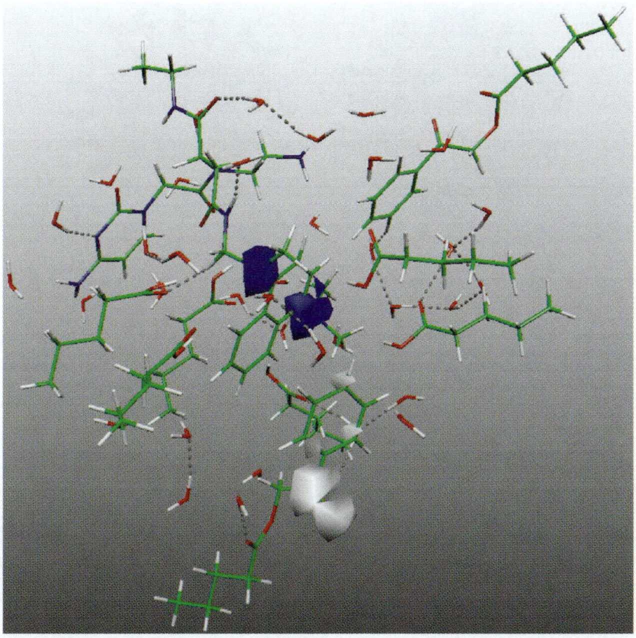

Chapter 2, Figure 9 Visualization of the electron charge tunneling associated with the ninth excited state. The transition is from the sensitizer 1,4-bis(*N,N*-dimethylamino)naphthalene molecule (in the center) to the first pFA molecule (bottom). The electron cloud hole is indicated in blue, and the transferred electron cloud location is visualized in white/gray.

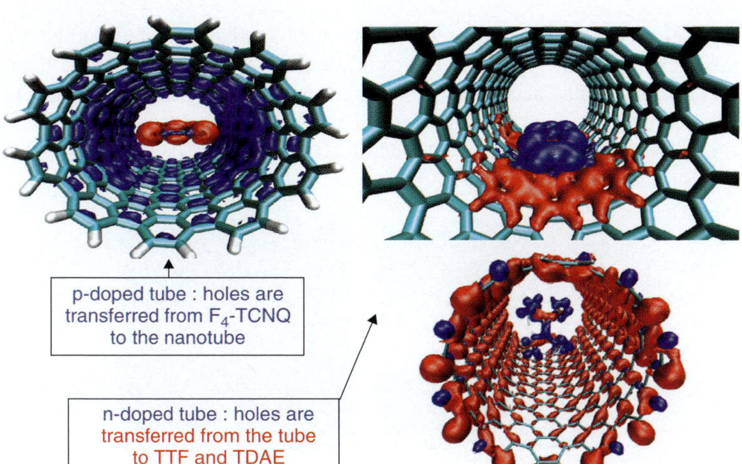

Chapter 3, Figure 4 Charge density isosurface resulting from charge transfer as computed from DFT(LDA)/6-31G* for one molecule of F_4TCNQ (top left panel), one molecule of TTF (top right panel), and one molecule of TDAE (bottom right panel).

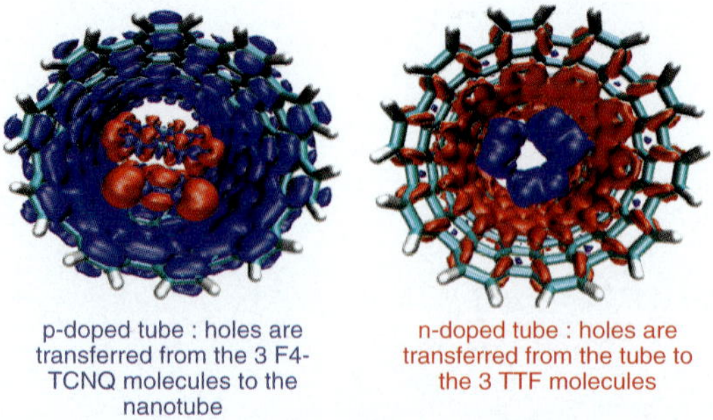

Chapter 3, Figure 5 Charge density isosurface resulting from charge transfer as computed from DFT(LDA)/6-31G* for three molecules of F_4TCNQ (left panel) and TTF (right panel).

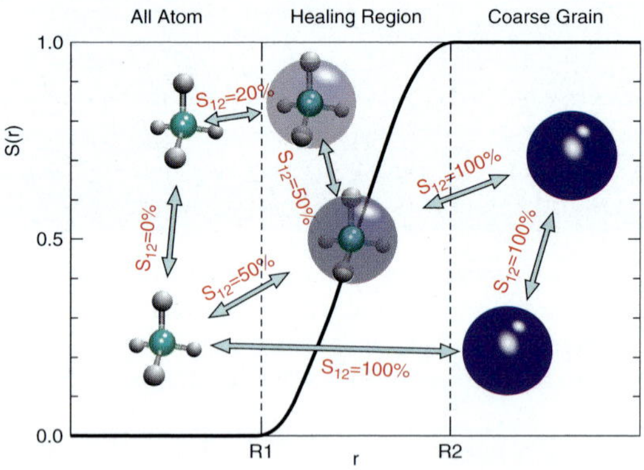

Chapter 5, Figure 1 Schematic representation of methane in the atomistic region (at the left), the coarse grain region (at the right), and the connecting healing region (in the middle). The solid black line shows the value of the switching function for a CG site as a function of its position, $S(r)$. The fractional CG character of each pair interaction equals the maximum $S(r)$ among the two interacting CG sites (illustrated in red).

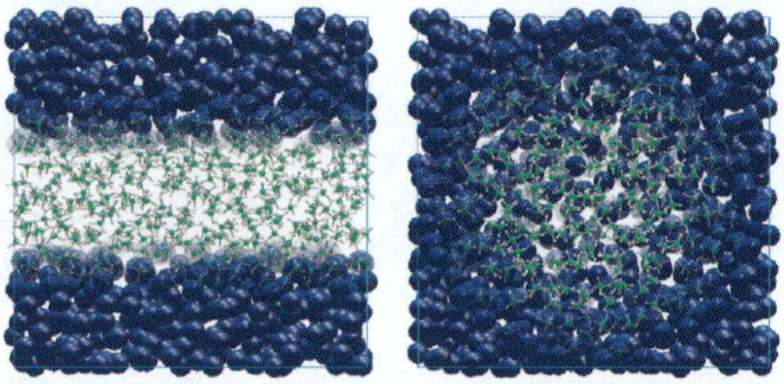

Chapter 5, Figure 2 Two examples of the spatial partitioning of a periodic unit cell of 1000 methane molecules into an atomistic region, a coarse-grained region, and an intermediate healing region. The transparency of the dark blue CG spheres reflects the values of the switching function $S(r)$. On the left is a rectangular partitioning, and on the right is a spherical partitioning.

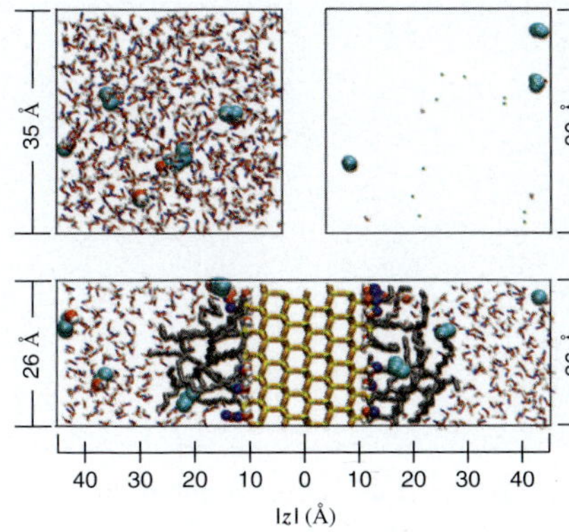

Chapter 11, Figure 1 The three boxes present in the Gibbs ensemble simulation of RPLC. The cubic vapor and bulk solvent boxes are shown in the upper right and upper left corners, respectively, with their approximate box lengths indicated. The box containing the stationary phase is shown on the bottom in its z–y plane. This box has fixed edge lengths of $x = 20$ Å, $y = 26$ Å and $z = 90$ Å. Solute molecules are shown as large spheres, with oxygen in red, hydrogen in white, and CH_3 in groups in cyan. Solvent molecules are shown in the stick representation, with oxygen in red, hydrogen in white, and CH_3 in groups in blue. Those solvent molecules that are invloved in hydrogen bonding with the silica substrate are shown as medium spheres for emphasis. The silicon and oxygen atoms in the substrate are shown as yellow and orange tubes, respectively, while the grafted alkyl chains are shown as dark gray tubes. Helium atoms in the vapor phase are shown as small green spheres.

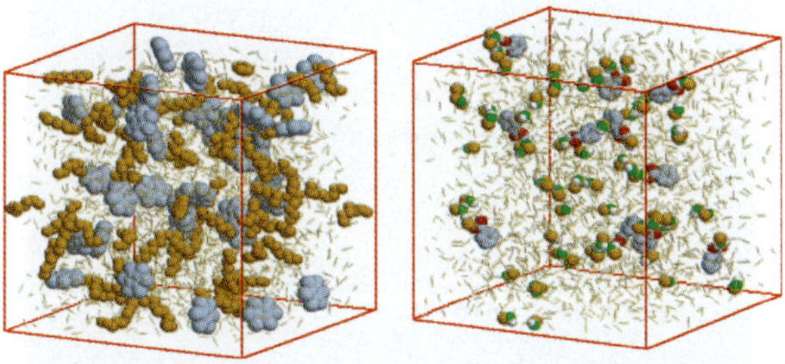

Chapter 11, Figure 4 Supercritical-CO_2/entrainer/solute systems. A stick representation is used to show the CO_2 molecules in both cases. The configuration on the left shows an *n*-octane/hexamethyl benzene system, with the pseudoatoms in *n*-octane shown as orange spheres and those in hexamethyl benzene shown as light blue spheres. Shown are 35 solute molecules in a box with side length 58.05 Å; the average number of solutes in the supercritical phase for this system is 34 ± 2. The configuration to the right shows a methanol/benzoic acid system. The coloring scheme for the spheres is as follows: methanol methyl pseudoatoms (orange), methanol oxygen atoms (green), methanol and carboxyl hydrogen atoms (white), carboxyl oxygen atoms (red), and benzene and carboxyl carbon pseudo atoms (light blue). Shown are 11 benzoic acid molecules in a box of side length 57.15 Å; the average number of solutes in the supercritical phase for this system is 9 ± 2. The red lines depict the edges of the periodic cell.

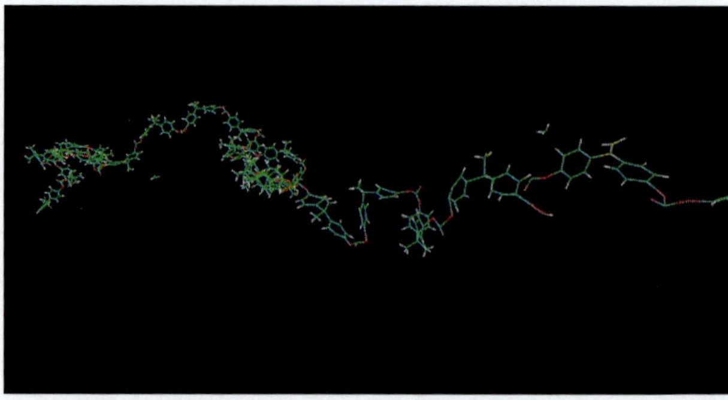

Chapter 13, Figure 2 Still frame from RMD simulations showing the formation of an OH-terminated fragment (oxygen atoms are colored red and hydrogens silver) and free methyl groups resulting from side-chain scission reactions.

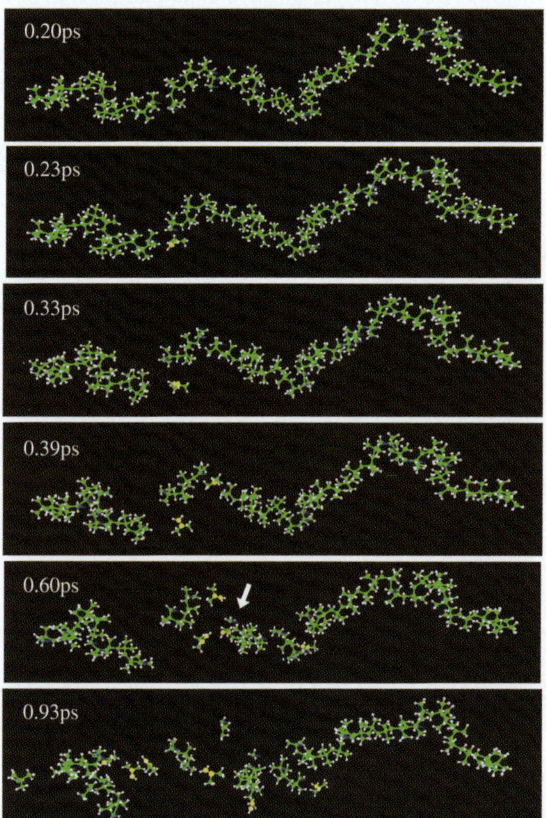

Chapter 13, Figure 4 Still frames from a PIB simulation at 2000 K showing chain scission followed by propagation and termination (gray atoms are hydrogen, green indicates single-bonded carbon bonds, yellow indicate's double-bonded carbon bonds, and blue indicates radical sites).

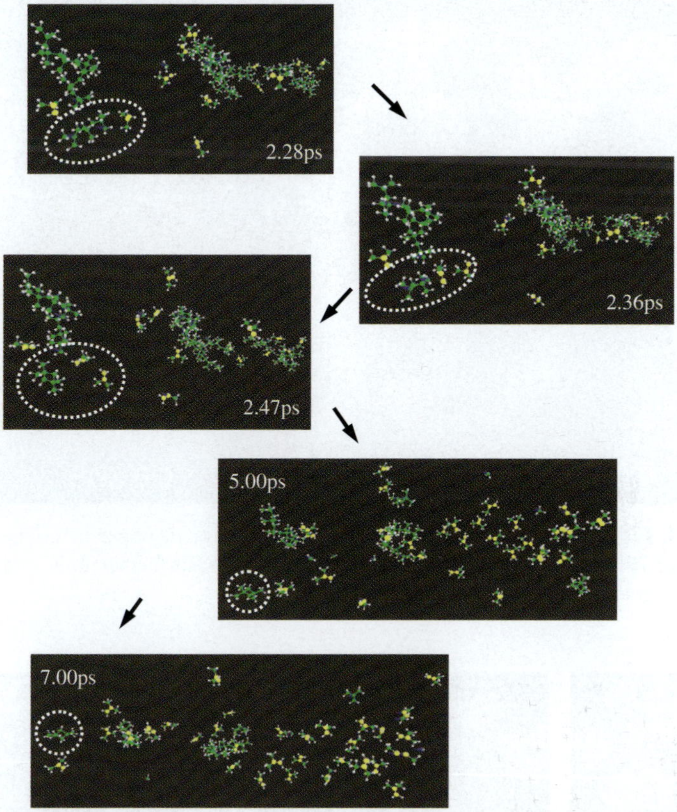

Chapter 13, Figure 5 Influence of added NaC1 concentration on the sphere to rod transition probability for SDS micelles. The line includes the values predicted and the points show experimental data. (Replotted using data from ref. 12).

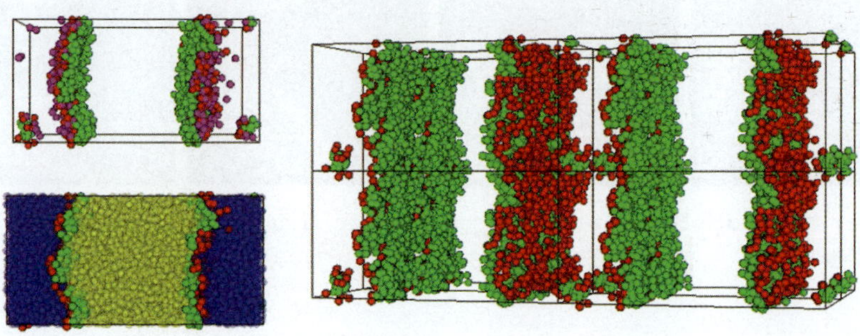

Chapter 14, Figure 13 Three different images of SDS molecules at their maximum packing density at an oil–water interface. The image at the top left shows the SDS surfactant molecules, with the sodium counterions (magenta) displayed. The image at the bottom left shows the oil(yellow), the water(blue), and the SDS molecules. The image on the right shows $2 \times 2 \times 2$ simulation cells. Only the SDS molecules are displayed (heads are red, tails are green).

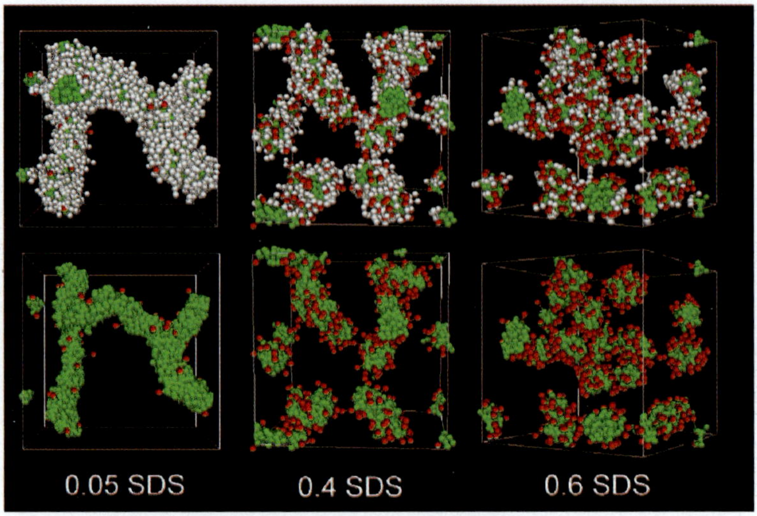

Chapter 14, Figure 18 Mesoscale structure predicted for mixtures of $C_{12}E_6$ ethoxylate and SDS for different mole fractions of SDS. Only the surfactant tails and SDS head groups are shown in the second row of images.

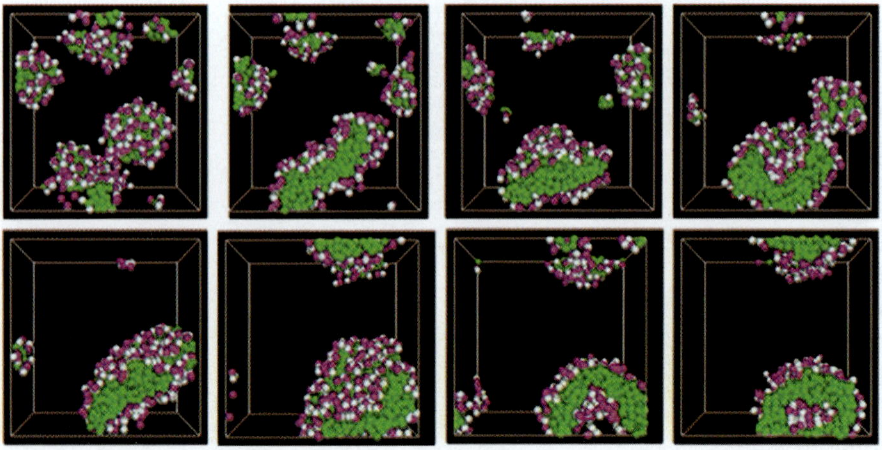

Chapter 14, Figure 20 Self-assembly of a mixture of anionic and cationic surfactants to form a vesicle.

A second issue was discovered for systems in which there is a strong variation in density from region to region within the grid. As described above, the temperature in the grid is calculated from the average kinetic energy in the simulation, which is a reasonable approximation if the system is not too far from equilibrium and the density is fairly uniform. If there is a large density variation between grid regions (i.e., at a free surface), movement of atoms across the grid boundaries can result in nonphysical heating (or cooling) of the system. For example, if a hot atom evaporates from a dense solid region to an empty grid region near a surface or void, the formerly empty region becomes hot. This additional heat then flows back into the surrounding high-density regions, resulting in nonphysical heating. To account for this and to conserve energy, the thermal diffusivity is scaled by the density factor ρ_i/ρ_0, where ρ_i is the density of a neighboring grid box and ρ_0 is the density of the current grid box. This results in a variable diffusion coefficient between a grid point and its neighboring grid points that enforces conservation of energy. Similarly, to incorporate temperature-dependent thermal properties (such as corrections to the heat capacity when below the Debye temperature) or defect-dependent thermal properties, the value used for D in Eq. (1) can be adjusted depending on the temperature or defect density of a given grid region, or an explicit functional dependence of thermal properties on temperature can be included in heat transport equation (1).

To demonstrate the effect on measured physical properties, the frictional force for a tip moving along a gold surface was calculated with and without the continuum–atomistic thermostat applied. Above sliding rates of approximately 10 Å/ps, the choice of thermostat had a very strong effect on the average normal and friction forces, with both the normal and frictional forces being higher with the thermostat. For example, at a sliding rate of 25 Å/ps, the frictional force is 50% larger for the case with the thermostat applied compared to the case without the thermostat applied.[15] At the higher sliding rate there is appreciable heating of the surface without the thermostat, whereas with the thermostat turned on, the heat is removed from the surface rapidly and the surface temperature remains constant. As the temperature increases, the bonds are softened and the atoms are more easily pushed out of the way by the moving tip. This results in lower normal and frictional forces. When the thermostat is applied, the surface temperature does not increase as rapidly, which results in a higher but more realistic normal and friction force.

Joule Heating

Joule heating results from resistance to the current flow induced by an applied potential. Standard methods for modeling the influence of electric currents on atomic dynamics range from purely classical to fully quantum mechanical. In the classical models, one-body forces are based on energy and momentum conservation model collisions between moving electrons and ions (an *electron wind*). Generally speaking, these approaches either require simple geometries or some method to specify current densities. Quantum mechanical approaches to modeling Joule heating include detailed multiple scattering models,[17,18] forces derived from first principles and tight-binding electronic structure calculations, perturbation models, and quantum-classical Hamiltonians.[19–26] Although more rigorous than purely classical modeling, quantum models are too computationally demanding to be used in large-scale simulations.

To include Joule heating in large-scale atomic simulations without adding explicit electronic degrees of freedom, an ad hoc coupling scheme was developed that builds on the thermostat described earlier. Current flow and resulting Joule heating were incorporated through a virtual network of resistors that connects the centers of the adjacent regions of the grid (solid lines in Figure 1). The resistivity of each grid region ρ_{grid} is first defined as

$$\rho_{\text{grid}} = \frac{\rho_{\text{metal}}}{d_{\text{grid}}/d_{\text{metal}} + C_{\text{air}}} \tag{5}$$

where d_{grid} and d_{metal} are the density of the current grid and bulk metal, respectively, ρ_{metal} is the bulk metal resistivity, and C_{air} is a small constant chosen to give ρ_{grid} the resistivity of air when d_{grid} is zero. The value of the resistor connecting two grid regions is taken as the average resistance of the two grid regions. With this virtual resistive network and an applied voltage across the system, the potential at every grid region is calculated using a sparse matrix solver to iteratively solve Kirchhoff's law [Eq. (5)] using the Maxwell loop method for circuit equations:[27]

$$\begin{aligned} \sum_{\alpha} i_{\alpha} &= 0 \\ \sum_{\alpha} V_{\alpha} &= \sum_{\alpha} i_{\alpha} R_{\alpha} \end{aligned} \tag{6}$$

The potential difference and the resistance between connected grid regions is used to calculate the current between the two points using Ohm's law, $V = IR$, and the heat resulting from the current flow is calculated using the power equation, $P = VI$. The heat generated is added back to the atomic velocities in the appropriate grid region via the Q term in Eq. (1). The atomic–continuum simulation is repeated stepwise, with the grid of resistors updated as needed. Test simulations of current flow and resulting Joule heating were carried out with voltage applied across single-crystal nanowires composed of silver. Two cases were examined: a resistivity independent of temperature, and a resistivity given by the equation

$$\rho_r = \rho_{r0}[1 + \alpha_r(T - T_0)] \tag{7}$$

where ρ_{r0} is the resistivity of the metal at temperature T_0 and α is the thermal coefficient of the resistivity. For both cases, the temperature at the ends of the nanowire were maintained at 300 K, and temperature profiles along the wire were calculated from the coupled continuum–atomic simulations as well as a numerical solution of the continuum equations without coupling to an MD simulation. The MD simulation was carried out with an embedded atom method (EAM) potential for silver. The silver nanowire had a potential applied in the range 0.01 to 0.10 V. The experimentally determined thermal diffusivity, electrical resistivity, and thermal coefficient of resistivity were used. As expected, the coupled continuum–atomic thermal profile matched the numerical solutions for both cases and for the various applied potentials. Since the temperature dependence of the resistivity is large, variable resistivity is needed to reproduce experimental results accurately. To show the effects of Joule heating on an MD simulation, a pinched silver nanowire subjected to a constant tensile strain was simulated both with and without Joule heating from current flow. In the simulations without current flow, the system deforms through six partial dislocations, three on each side of the pinch, starting at the side of the pinch and moving to the center of the wire [see Figure 3; atoms are shaded by centrosymmetry, with lighter shades having larger centrosymmetry values (centrosymmetry is a measure of an atom's local environmental deviation from a perfect face-centered-cubic lattice)]. The deformation mechanism consists of the motion of partial dislocations along close-packed planes followed by formation of short single-atom chains before failure, consistent with prior reports.[28–31] With Joule heating, an approximately 40-Å-long region at the center of the system melts. As the system is strained, the wire forms from the melt without the initial formation of dislocations in the regions of the system that remain crystalline (see Figure 3).

In addition to Joule heating, magnetic forces due to the current flow can be calculated and added to the forces from the interatomic potential. The force is

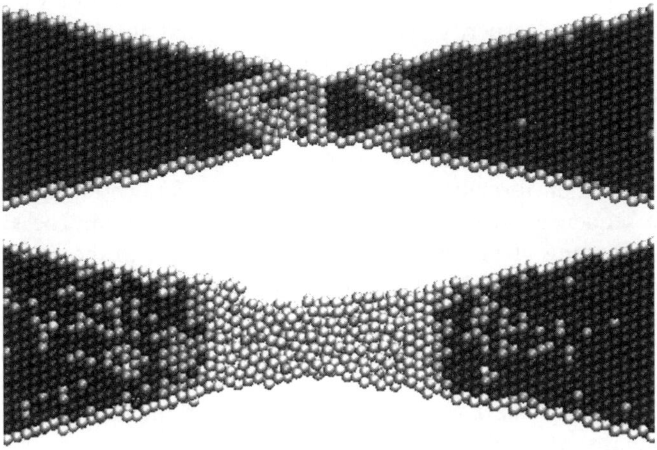

Figure 3 Top: Pinched nanowire strained for 10 ps at a rate of 1 Å/ps with an applied voltage of 0.0 V. Atoms are shaded by a centrosymmetry parameter (lighter shades have larger centrosymmetry values). Stacking faults show up as lines. Bottom: Pinched nanowire strained for 10 ps at a rate of 1 Å/ps with an applied voltage of 0.4 V. Atoms are shaded by centrosymmetry parameter.

obtained using the equation

$$F_{mag,mn} = \frac{\mu_0}{4\pi}\frac{i_m i_n}{r_{nm}} l^2 \tag{8}$$

where μ_0 is the magnetic permeability, l the grid box length, i the current vector flowing through the mth or nth boxes, and r_{nm} the distance between the nth and mth boxes. The relationship between the atomic dynamics, which alters the density distribution in the grid, the values of the resistor network, and calculation of the magnetic field from the current flow through the virtual resistors, provides an ad hoc feedback loop that couples the electromagnetic forces to the forces from the interatomic potential, again without explicit electrons in the calculation.

2. SIMULATION OF A HOT AND COLD ASPERITY CONTACT WITH AND WITHOUT SLIDING

The simulation of metal electrical contacts has many applications, from MEMS to electromagnetic launchers. Although these systems have generated a lot of attention lately, little modeling can be done at the atomistic level, due primarily to the lack of electrons in the atomistic potentials used. This fact makes these systems ideal candidates for simulation with the ad hoc coupling methods laid out above.

A series of metal–metal sliding simulations were carried out using single and multiple aluminum contacts sliding on a flat metal surface through which a variable electric current flows. Figure 4, top, shows a sequence of snapshots from a single spherical Al asperity (with radius 40 Å) sliding to the right on a flat Al surface at 500 m/s. The atoms are shaded by their respective centrosymmetry parameters so that defects are apparent (lighter shades have higher centrosymmetry values). The system has a normal load of 1.6 nN and an applied voltage of 0.0 V. As the bottom layer slides, the area of the contacting asperity deforms, followed by dislocation formation in the bottom layer. Shown in Figure 4, bottom, is the same sequence except that a voltage of 0.3 V is used. The system heats and melts beginning at the contact, and atoms are lost to the vacuum through evaporation at the interface. Dislocations form in parts of the bottom surface that remain crystalline, including a dislocation that forms starting at the bottom of the slab that goes in a direction opposite that of the others. Due to the thermal expansion of the system and the periodic boundaries in the two directions perpendicular to sliding, the internal stress generated in the substrate nucleates the dislocation that starts at the bottom of the slab.

To model the effect of applied voltage on dislocation formation in an asperity, a system was built consisting of a gold contact with a single, faceted asperity. The contact surface was brought down on the asperity tip at a constant velocity of 0.25 Å/ps. Two simulations were run, one with zero voltage applied, and another with 0.1 V applied. Figure 5 provides a central cross section of one system. The atoms

Figure 4 Sliding simulation of an Al asperity on a flat Al surface. Atoms are shaded by their centrosymmetry parameter. Top: Asperity sliding to the left at 500 m/s with 0.0 V across the system at 0, 10, and 16 ps. Bottom: Asperity sliding to the left at 500 m/s with 0.3 V across the system at 0, 10, and 20 ps.

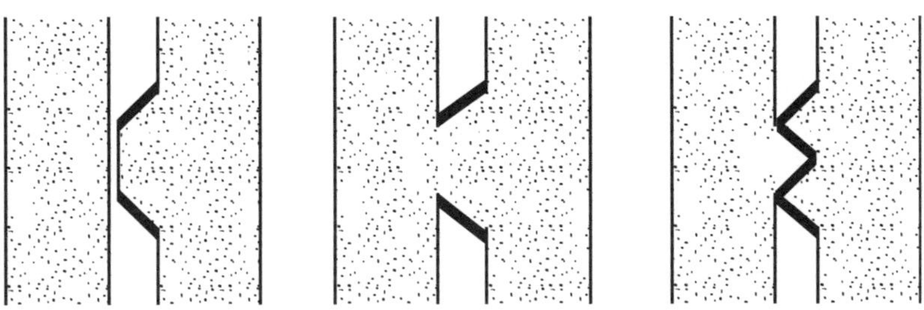

Figure 5 Sequential views of an Au contact closing on a single asperity. Atoms shaded by centrosymmetry; only atoms associated with defects and free surfaces are shown.

are shaded by centrosymmetry, and only atoms with the highest centrosymmetry values are depicted; therefore, only atoms associated with defects and free surfaces are presented. It can be seen that as the left contact moves continually, the force produces dislocations in the asperity. The dislocations form with and without voltage, but occur sooner with the voltage present; this is due to increased energy for dislocation nucleation and larger thermal expansion forces, both of which correspond to higher temperatures produced by the presence of a voltage.

3. OTHER AD HOC COUPLING APPLICATIONS

The schemes described above are only two examples of a potentially wide variety of multiscale procedures using an ad hoc coupling of numerical atomistic and continuum simulations. Next, we speculate on two of these possible coupling schemes.

Solid Diffusion and Discrete Dynamics

At the continuum level mass transport can be described with differential equations that are equivalent to Eq. (1), where concentration plays the role of temperature and a diffusion coefficient is equivalent to the thermal diffusivity. In an ad hoc coupling scheme, a continuum diffusion equation would be solved stepwise using the concentrations, temperature, and stress profiles given by a coarse-grained molecular dynamics simulation. As the second part of the feedback, the composition of the atomic simulation would be modified such that it matches the solution of the continuum equation. After one (or more) steps in the molecular simulation, the concentrations would be recalculated through one (or more) steps of the numerical solution to the continuum mass diffusion equation, and the procedure would be repeated using this ad hoc concentration coupling.

An interesting application would be in the simulation of diffusion creep in metals. Plastic deformation in metals occurs via motion of dislocations, a process called *slip*. In a crystalline solid, dislocations move easiest along specific close-packed directions and planes. Dislocation motion can be blocked, and hence materials strengthened, by introduction of defects such as grain boundaries, second phases, substitutional and interstitial impurity atoms, and other dislocations. At relatively high temperatures, blocked dislocations can begin to move via a process called *creep*. In this process vacancies diffuse to the edge of a dislocation, which allows the dislocation to "climb" away from the defect. When the dislocation climbs sufficiently far from the stress field associated with the defect, it can resume movement and the material continues to deform. This is an activated process, with activation energy typically close to that for vacancy diffusion. The motion of vacancies can be driven in this case not only by a concentration gradient, but also by the applied stress and the stress field associated with the dislocation itself.

The motion of a dislocation induced by an applied shear stress occurs on a time scale that can be captured in a molecular dynamics simulation, and indeed this has been modeled at the atomic level.[32] However, the diffusion associated with dislocation climb and the resulting creep is in general too slow to be modeled with molecular dynamics simulations except perhaps at elevated temperatures where other interfering processes (e.g., melting) may become active. In a continuum–atomistic coupling scheme, the continuum mass transport equation for motion of vacancies would be solved stepwise using the vacancy concentration and stress profiles in the simulation. The vacancy motion would produce dislocation climb on a time scale appropriate for diffusion until the dislocation was sufficiently far from the defect that its motion could be modeled with a molecular dynamics simulation. Hence a fast dynamic process, in this case dislocation motion, would be initiated by a longer time process, vacancy diffusion. A simulation of this type would therefore couple two time scales, one appropriate for diffusion and another appropriate for dislocation motion. In addition, the relationship between the diffusion coefficient and the stress and temperature profiles on the atomic simulation can be based on experimental data, producing an effective correction to the interatomic potential used in the atomistic simulation.

Chemical Kinetics and Mesoscale Dynamics

Several schemes have been developed in which a subset of atomic degrees of freedom are coarse-grained from a simulation so that longer times or larger systems can be modeled. A well-established variation on this approach is to freeze out high-frequency vibrations and leave only the low-frequency bending and torsional modes that are important for producing correct molecular shapes in large molecules. In dissipative particle dynamics, molecules or molecular aggregates (plus a solvent) are replaced with quasiparticles that interact via an effective soft potential.[33] This mesoscale approach allows both larger systems and longer time steps to be accessed compared to a conventional molecular simulation while in principle using information such as interparticle forces that can be directly related to the atomic scale. In a related approach, Zhigilei and co-workers have incorporated a breathing sphere model into simulations of laser ablation and sputtering.[34] In this mesoscale approach, quasiparticles can be created and combined to reflect the collisions conditions in the simulation.

In principle, continuum–atomistic coupling can be used to incorporate chemical reactivity into mesoparticle simulations. In this case the continuum expression would be a series of rate equations that can be numerically solved stepwise in time. The initial concentrations, as well as conditions such as temperature and pressure profiles and/or collision rates coarse-grained from the mesoscale simulation, would constitute the initial conditions for solving the kinetic rate equations. From the rate equations would come concentration profiles, which would in turn be used to define the creation or combination of mesoparticles (as well as temperature changes associated with these reactions) in the discrete dynamics simulations. Experimental data could enter the simulations through the rate constants, which would help to refine the approximations made in defining the quasiparticles relative to a full atomic-scale dynamics. Whether the simultaneous numerical solutions to the continuum-rate and discrete particle dynamics simulations would involve comparable or disparate time scales would depend on the values of the rate constants and the appropriate time step size in the mesoscale simulations.

4. CONCLUSIONS

A method for coupling continuum equations into a classical MD simulation has been presented in which continuum equations correct for deficiencies in the atomic potentials used during the MD simulation, as well as serving as a multiscale technique. Several examples were presented showing the usefulness of the ad hoc coupling method. These examples include thermal transport, heat generation from current flow, and the magnetic force resulting from that current flow. All were numerically solved simultaneously with a MD simulation using the EAM potential for metals, with an ad hoc feedback between the two simulations. With the EAM potential, the coupling scheme corrects for the lack of explicit electrons, which dominate thermal properties in metals and are required for current flow. The ad hoc scheme allows for these properties and properties dependent on them to be simulated accurately without explicitly treating the electronic degrees of freedom, a critical feature for large-scale systems. Other applications of this method, both as a multiscale method and a potential correction scheme are currently being explored.

Acknowledgments

This work was funded by a Georgia Tech MURI, which is supported by the Office of Naval Research. Helpful discussions with George C. Jordan, Doug Irving, and Yanhong Hu are acknowledged.

REFERENCES

1. Tully, J. C.; Gilmer, G. H.; Shigard, M. Molecular dynamics of surface diffusion: I. The motion of adatoms and clusters, *J. Chem. Phys.*, 1979, 71: 1630.
2. Hoover, W. G. Molecular dynamics, in *Lectures Notes in Physics*, Vol. 258, Springer-Verlag, Berlin, 1986.
3. Nosé, S. A unified formulation of the constant temperature molecular dynamics methods, *J. Chem. Phys.*, 1984, 81: 511.
4. Kopelevich, D. I.; Panagiotopoulos, A. Z.; Kevrekidis, I. G. *J. Chem. Phys.*, 2005, 122: 044907.
5. Abraham, F. F., et al. Spanning the length scales in dynamic simulation, *Comput. Phys.*, 1998, 12: 538.
6. Broughton, J. Q. Concurrent coupling of length scales: methodology and application, *Phys. Rev. B*, 1999, 60: 2391.
7. Tadmor, E.; Ortiz, M.; and Phillips, R. Quasicontinuum analysis of defects in solids, *Philos. Mag. A*, 1996, 73: 1529.
8. Wagner, G. J.; Liu, W. K. Coupling of atomistic and continuum simulations using a bridging scale decomposition approach, *J. Comput. Phys.*, 2003, 190: 249.
9. Voter, A. F.; Montalenti, F.; Germann, T. C. Extending the time scale in atomistic simulation of materials, *Annu. Rev. Mater. Res.*, 2002, 32: 321.
10. Laio, A.; Parrinello, M. Escaping free-energy minima, *Proc. Natl. Acad. Sci. USA*, 2002, 99: 12562.
11. Daw, M. S. Embedded-atom Method: derivation and application to impurities, surface and other defects in metals, *Phys. Rev. B*, 1984, 29: 6443.
12. Foiles, S. M., Embedded-atom and related methods for modeling metallic systems, *Matter. Res. Soc. Bull.* 1996, 21: 24.
13. Ivanov, D. S.; Zhigilei, L. V. Combined atomistic–continuum modeling of short-pulse laser melting and disintegration of metal films, *Phys. Rev. B*, 2003, 68: 64114.

14. Ivanov, D. S.; Zhigilei, L. V. Effect of pressure relaxation on the mechanisms of short-pulse laser melting, *Phys. Rev. Lett.*, 2003, 91: 105701.
15. Schall, J. D.; Padgett, C. W.; Brenner, D. W. Ad hoc continuum-atomistic thermostat for modeling heat flow in molecular dynamics simulations, *Mol. Sim.*, 2005, 31: 283.
16. Padgett, C. W. Brenner, D. W. A continuum–atomistic method for incorporating joule heating into classical molecular dynamics simulations, *Mol. Sim.*, 2005, 31: 749.
17. Rous, P. J. Driving force for adatom electromigration within mixed CuAl overlayers on Al(111), *J. Appl. Phys.*, 2001, 89: 4809.
18. Rous, P. J. Electromigration wind force at stepped Al surfaces, *Phys. Rev. B*, 1999, 59: 7719.
19. Horsfield, A. P., et al. Power dissipation in nanoscale conductors: classical, semi-classical and quantum dynamics, *J. Phys. Condens. Matter*, 2004, 16: 3609.
20. Todorov, T. N. Local heating in ballistic atomic-scale contacts, *Phil of. Mag. B*, 1998, 77: 965.
21. Hoekstra, J., et al. Electromigration of vacancies in copper, *Phys. Rev. B*, 2000, 62: 8568.
22. Todorov, T. N. Time dependent tight binding theory, *J. Phys. Condens. Matter*, 2001, 13: 10125.
23. Montgomery, M. J.; Todorov, T. N.; Sutton, A. P. Power dissipation in nanoscale conductors, *J. Phys. Condens. Matter*, 2002, 14: 5377.
24. Brandbyge, M., et al. Origin of current-induced forces in an atomic gold wire: a first-principles study, *Phys. Rev. B*, 2003, 67: 193104.
25. Chen, Y. C.; Zwolak, M.; Di Ventra, M. Local heating in nanoscale conductors, *Nano Lett.*, 2003, 3: 1691.
26. Horsfield, A. P.; Bowler, D. R.; and Fisher, A. J. Open-boundary Ehrenfest molecular dynamics: towards a model of current induced heating in nanowires, *J. Phys. Condens. Matter*, 2004, 16: L65.
27. Kip, A. F. *Fundamentals of Electricity and Magnetism*, McGraw-Hill, New York, 1969.
28. Landman, U., et al. Atomistic mechanisms and dynamics of adhesion, nanoindentation and fracture, *Science*, 1990, 248: 454.
29. daSilva, E. Z.; da Silva, A. J. R.; Fazzio, A. How do gold nanowires break? *Phys. Rev. Lett.*, 2001, 87: 256102.
30. Diao, J. K.; Gall, K.; Dunn, M. L. Yield strength in asymmetry in metal nanowires, *Nano Lett.* 2004, 4: 1863.
31. Hakkinen, H., et al. Nanowire gold chains: formation mechanisms and conductance, *J. Phys. Chem. B*, 2000, 104: 9063.
32. Chang, J. P.; Cai, W.; Bulatov, V. V.; Yip, S. Molecular dynamics simulations of motion of edge and screw dislocations in a metal, *Comp. Mater. Sci.*, 2002, 23: 111.
33. Groot, R. D.; Warren, P. B. Bridging the gap between atomistic and mesoscopic simulation, *J. Chem. Phys.*, 1997, 107: 4423.
34. Zhigilei, L.V.; Kodali, P. B. S.; Garrison, B. J. Molecular dynamics model for laser ablation and desorption of organic solids, *J. Phys. Chem. B*, 1997, 101: 2028.

9

FULLY ANALYTIC IMPLEMENTATION OF DENSITY FUNCTIONAL THEORY FOR EFFICIENT CALCULATIONS ON LARGE MOLECULES

RAJENDRA R. ZOPE
Department of Chemistry, George Washington University, Washington, DC, and Department of Physics, University of Texas at El Paso, El Paso, Texas

BRETT I. DUNLAP
Theoretical Chemistry Section, U.S. Naval Research Laboratory, Washington, DC

Computer simulations are playing an increasingly important role in our understanding of materials. Generally, the choice of computational models that are employed in studying the properties of materials depends on the property of interest and the length scale or size of the system.[1] The latter is the most important factor in the selection of an appropriate level of theory. Our interest is in the electronic and structural properties of large carbon fullerenes and fullerenelike cages of aluminum and boron nitride containing a few hundred atoms. At these sizes, the current toolbox of methods that are available include semiempirical quantum mechanical models such as ZINDO,[2] PM3[3] methods, or tight-binding approaches.[4] More accurate descriptions of electronic properties require use of more involved methods, such as density functional (DF)–based models.[5,6] The traditional quantum chemical methods beyond Hartree–Fock or quantum Monte Carlo are, in general, more accurate than DF models. However, they are suitable for systems with a few tens of atoms. At present, the applicability of DF

Multiscale Simulation Methods for Nanomaterials, Edited by Richard B. Ross and Sanat Mohanty

models is restricted to 200 to 300 atoms, depending on the schemes used to approximate kinetic and exchange energy functionals, the basis sets used to expand Kohn–Sham orbitals, the treatment of core electrons (use and quality of pseudopotentials), and the types of atoms in the system. We are working toward development of a fully analytic implementation of density functional theory (ADFT).[7,8] The computationally efficient ADFT and efficient use of the available point-group symmetry of molecules allow us to optimize large inorganic and carbon fullerenes containing more than 2000 atoms.[9,10]

ADFT uses analytic atom-centered localized Gaussian basis sets. These basis sets are used to expand the molecular (Kohn–Sham) orbitals and the one-body effective (Kohn–Sham) potential using variational and robust fitting methodology.[11,12] The exchange-correlation part of the Kohn–Sham potential is obtained using a functional form based on Slater's exchange functional.[13] For this reason, the analytic implementation is also called the Slater–Roothaan (SR) method,[7] a method that allows an arbitrary scaling of the exchange potential around each type of atom in heteroatomic systems. These scaling factors can be used to parametrize the SR method. Using a suitable choice of these scaling parameters, accurate total and atomization energies that are comparable to some of the most sophisticated density functional models can be obtained.[8,14,15] In the following section we describe the analytic implementation of the density functional model and details of the SR method.

1. ANALYTICAL METHODS

Analytic Formulation of the Gàspàr–Kohn–Sham–Slater Density Functional Model

In the Hohenberg–Kohn–Sham formulation of the density functional theory[5,6] the total electronic energy of system containing N electrons and M nuclei is given by

$$E^{\mathrm{HKS}}[\rho] = \sum_i^N \langle \phi_i | f_1 | \phi_i \rangle + E_{ee} + E_{xc}[\rho_\uparrow, \rho_\downarrow] \tag{1}$$

where the first term contains the kinetic energy operator and the nuclear attractive potential due to the M nuclei,

$$f_1 = -\frac{\nabla^2}{2} - \sum_A^M \frac{Z_A}{|\vec{r} - \vec{R}_A|} \tag{2}$$

The second term in Eq. (1) represents the classical Coulomb interaction energy of electrons, and the last term is the exchange-correlation energy, which represents contributions that are quantal in origin. Equation (1) is an exact expression for the total energy, but practical application requires approximation to E_{xc}. Over the years, numerous parameterizations of different accuracy and complexity have been devised to model E_{xc} and are available in the literature. Most, however, have complex functional form that makes use of numerical grids necessary in implementations of DF models. A number of groups have developed numerical integration methods for computation of integrals over the exchange-correlation contributions.[16] Today, practically all implementations of DF models use numerical grids to compute the contribution to the total energy and matrix elements from the exchange-correlation terms. This is true even if analytic basis sets such as Gaussians are used to express KS orbitals. However, it turns out that if we model the E_{xc} according to Gàspàr–Kohn–Sham–Slater, the contribution to total energy from this term can also be calculated analytically using Gaussian basis sets and variational methodology.[12,17,18]

The Gàspàr–Kohn–Sham–Slater (GKS) exchange energy functional is given by

$$E_{xc}[\rho_\uparrow, \rho_\downarrow] = -\frac{9}{8}\alpha\left(\frac{6}{\pi}\right)^{1/3} \int d^3r \left[\rho_\uparrow^{4/3}(\vec{r}) + \rho_\downarrow^{4/3}(\vec{r})\right] \tag{3}$$

where $\alpha = \frac{2}{3}$ is the Gàspàr–Kohn–Sham value and $\alpha = 1$ is Slater's original value. To calculate E_{xc} analytically, the one-third and two-thirds powers of the electron density are expanded in Gaussian basis sets:

$$\rho^{1/3}(\vec{r}) \approx \overline{\rho}^{1/3} = \sum_i e_i E_i \tag{4}$$

$$\rho^{2/3}(\vec{r}) \approx \overline{\rho}^{2/3} = \sum_i f_i F_i \tag{5}$$

Here $\{E_i\}$ and $\{F_i\}$ are independent Gaussian basis functions, and e_i and f_i are expansion coefficients. The exchange energy is then given by[12,17,18]

$$E_{xc} = C_\alpha \left[\frac{4}{3}\langle\rho\overline{\rho}^{1/3}\rangle - \frac{2}{3}\langle\overline{\rho}^{1/3}\overline{\rho}^{1/3}\overline{\rho}^{2/3}\rangle + \frac{1}{3}\langle\overline{\rho}^{2/3}\overline{\rho}^{2/3}\rangle\right] \tag{6}$$

where $C_\alpha = -\frac{9}{8}\alpha\left(\frac{6}{\pi}\right)^{1/3}$. Thus, using four LCGO basis sets (one for orbital expansion and three for fitting the Kohn–Sham potential), the total energy is calculated analytically.

Similarly, to compute the Coulomb energy,

$$E_{ee} = \langle\rho||\rho\rangle = \frac{1}{2}\int\int \frac{\rho(\vec{r})\rho(\vec{r'})}{|\vec{r} - \vec{r'}|} d^3r\, d^3r' \tag{7}$$

we use the first robust and variational fitting methodology and express the charge density as a fit to a set of Gaussian functions,

$$\rho(\vec{r}) \approx \overline{\rho}(\vec{r}) = \sum_i d_i G_i(\vec{r}) \tag{8}$$

Here $\overline{\rho}(\vec{r})$ is the fitted density and d_i the expansion coefficient of the charge density Gaussian basis function G_i. The elimination of the first-order error in total energy due to the fit leads to the unique robust expression for the self-Coulomb energy.[11] The LCAO (linear combination of atomic orbitals) coefficients and the vectors **d**, **e**, and **f** are found by constrained variation. It is easy to obtain the contribution from the first term in Eq. (1) in analytic fashion. Thus, in ADFT four sets of Gaussian basis are required: one for KS orbitals and three for the KS effective potential. This methodology was implemented successfully by Werepentski and Cook, who demonstrated that noise-free forces and smooth potential energy can be obtained using a *fully analytic* (grid-free) implementation.[17,18]

Slater–Roothaan Method

Although the analytic implementation described above is computationally efficient, its performance for the atomization of molecules is limited, due to the limitation of the functional form adopted. We have tested its performance by computing atomization energies of a set of 56 molecules from the G2 data set. For $\alpha = \frac{2}{3}$, the mean absolute error in atomization of 56 molecules is about 16 kcal/mol. This can be improved to 12 kcal/mol by allowing the value of α to change.[19] Thus, the α in Eq. (6) can be viewed as a scaling parameter that scales the exchange potential. The model above can then be modified so that each type of atom in the heteroatomic system has its own value of scaling parameter. This led to development of the Slater–Roothaan (SR) method.[7] Apart from the advantage that the calculations can be performed in completely analytic fashion, it also allows molecules to dissociate correctly in their atomized limit.[20] The exchange energy in the SR method has the form[7,15]

$$\begin{aligned} E^{\mathrm{SR}} &= \sum_i \langle \phi_i | f_1 | \phi_i \rangle + 2\langle \rho || \overline{\rho} \rangle - \langle \overline{\rho} || \overline{\rho} \rangle \\ &\quad + \sum_{\sigma=\uparrow,\downarrow} C_x \left[\frac{4}{3} \langle g_\sigma \overline{g}_\sigma^{1/3} \rangle - \frac{2}{3} \langle \overline{g}_\sigma^{1/3} \overline{g}_\sigma^{1/3} \overline{g}_\sigma^{2/3} \rangle + \frac{1}{3} \langle \overline{g}_\sigma^{2/3} \overline{g}_\sigma^{2/3} \rangle \right] \end{aligned} \tag{9}$$

Here $C_x = C_\alpha/\alpha$; the partitioned $\frac{3}{4}$ power of the exchange energy density,

$$g_\sigma(\vec{r}) = \sum_{ij} \alpha(i)\alpha(j) D_{ij}^\sigma(\vec{r}) \tag{10}$$

where $D_{ij}^\sigma(\vec{r})$ is the diagonal part of the spin density matrix multiplied by the partitioning function,

$$\alpha(i) = \alpha_i^{3/8} \tag{11}$$

which contains α_i, the α in the Xα model for the atom on which the atomic orbital i is centered. The fits to powers of g_σ are obtained variationally from Eq. (9).

2. COMPUTATIONAL DETAILS

As noted earlier, the analytic SR method requires four Gaussian basis sets: one for the orbital expansion and three to fit diffierent powers of electron density, which we obtain from the literature. We choose Pople's triple-ζ (TZ) 6-311G** basis[21,22] and the DGauss[24] valence double-ζ (DZ) basis set[25] called DZVP2 for orbital basis sets. The *s*-type fitting bases are obtained by scaling and uncontracting the *s* part of the orbital basis. The scaling factors are 2 for the density, $\frac{2}{3}$ for $\overline{\rho}^{1/3}$, and $\frac{4}{3}$ for $\overline{\rho}^{2/3}$. These scaled bases are used for all *s*-type fitting bases. Ahlrichs' group has generated a RI-J basis for fitting the charge density of a valence triple-ζ orbital basis set used in the Turbomole program.[26] The non-*s* parts of Ahlrich's fitting bases are used in combination with 6-311G** orbital basis sets. We use this combination of basis sets (6-311G**/RIJ) for boron nitride cages and carbon fullerenes. In combination with the DZVP2 orbital basis, we use the *pd* part of the A2 charge density fitting basis. The combination DZVP2/A2 is used for studying aluminum nitride cages. The geometries of molecules were optimized using the Broyden–Fletcher–Goldfarb–Shanno (BFGS) algorithm.[23,27] The forces on atoms are rapidly computed nonrecursively using the 4-*j* generalized Gaunt coefficients.[9,28] The atomic energies are obtained in the highest symmetry for which the self-consistent solutions have integral occupation numbers. The atomization energy is computed from the total energy difference of optimized molecule and its constituent atoms.

3. BORON AND ALUMINUM NITRIDE CAGES

The discovery of carbon fullerene, C_{60}, followed by the discovery of higher fullerenes and carbon nanotubes, has led to an intense search for hollow cagelike and tubelike structures of other materials. In this search, boron nitride (BN) is probably the second most studied material, after carbon. A number of groups have reported observation of BN nanotubes as well as cagelike structures.[29–32] Particular relevant to this chapter are a series of experiments by Oku and co-workers in which they detected BN clusters in a mass spectrum.[29,32] These authors have proposed a number of cagelike structures for the BN clusters detected in the mass spectrum. Here we report the electronic structure of these cages and their aluminum nitride (AlN) analogs. We note that whereas BN cages have been reported, no cagelike structure of AlN has been observed, although observations of the AlN nanotubes have been noted.[33–36] The optimized cage structures of BN and the symmetries of these structures are given in Figure 1. All these structures have been found to be energetically stable with a binding energy of about 14 to 16 eV per pair of BN atoms. Notable among these is the octahedral $B_{24}N_{24}$ cage proposed by Oku and co-workers as a candidate structure for one of the most abundant clusters in the mass spectrum. This cage is perfectly round, as is C_{60} fullerene, where each carbon atom is equivalent to all other carbon atoms. A pair of BN atoms in this cluster is equivalent to all other pairs in the cluster. It is to be

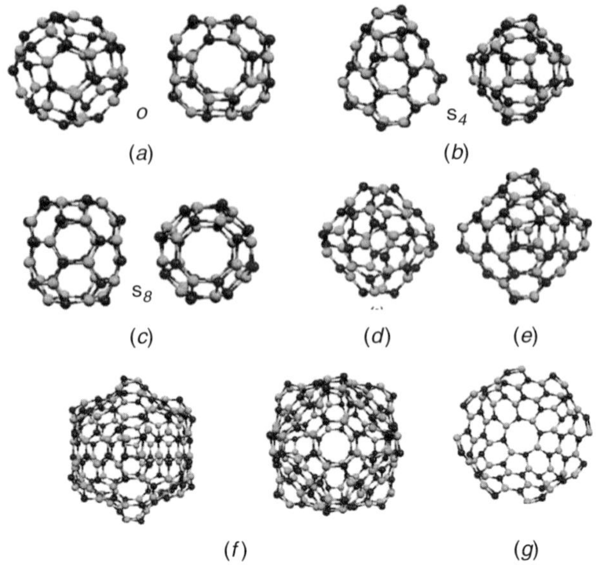

Figure 1 Optimized BN cages: (a) two views of an octahedral $B_{24}N_{24}$ cage; (b) two views of an S_4 $B_{24}N_{24}$ cage; (c) an S_8 $B_{24}N_{24}$ cage; (d) a $B_{28}N_{28}$ cage of T symmetry; (e) a $B_{36}N_{36}$ cage of T_d symmetry; (f) an octahedral $B_{96}N_{96}$ cage; (g) a hemispherical cap of an (8,8) BN nanotube based on half of $B_{96}N_{96}$.

noted that the exact analog of carbon fullerene is not possible for alternate boron nitride cages. The presence of pentagonal rings in carbon fullerenes does not permit full alternation of B and N atoms. Thus, even-membered rings are necessary to make alternate fullerenes close. The octahedral round $B_{24}N_{24}$ cage contains twelve squares and six octagons. This structure can be used to form caps for (4,4) BN nanotubes.[37] However, unlike C_{60} fullerenes, the round $B_{24}N_{24}$ octahedral cage is not energetically special. The alternate $B_{24}N_{24}$ cages with symmetry S_4 and S_8 are energetically nearly degenerate with octahedral cages,[38] so it is not clear which structure is likely to be observed in the experiment.

A C_{4h} $B_{28}N_{28}$ cage (Figure 2*c*) can be generated from a base $B_{24}N_{24}$ cage by cutting the latter into two halves after orienting them along the C_4 axis, then inserting a ring of eight alternate B and N atoms perpendicular to the axis (i.e., horizontally) and then rotating the top half by an eighth of a revolution. The resulting cage contains eight inequivalent atoms and has C_{4h} symmetry. If two rings of four alternate BN pairs are inserted instead of one, the resulting $B_{32}N_{32}$ cage is a tubular structure with S_8 symmetry. The binding energy increases systematically by going from $B_{24}N_{24}$ to the $B_{28}N_{28}$ cage (0.26 eV per BN pair) and from $B_{28}N_{28}$ to the $B_{32}N_{32}$ cage (0.06 eV/BN pair). The successive additions of alternate BN rings energetically stabilizes the BN tubular cages and results in (4,4) BN nanotubes with round caps that are based on octahedral $B_{24}N_{24}$ cages. Note that the same (4,4) tube can also be generated by starting with the S_8 $B_{24}N_{24}$

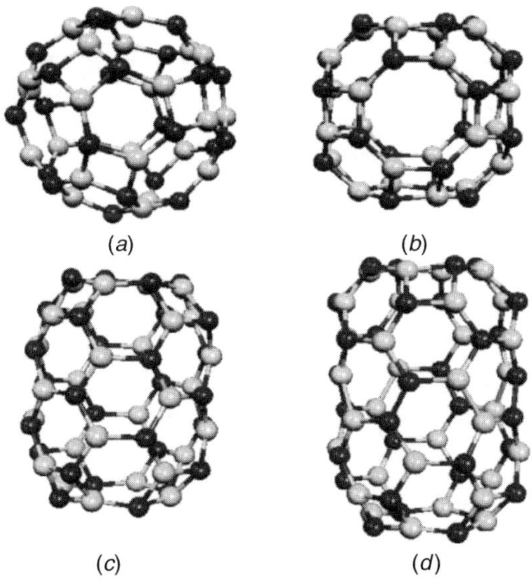

Figure 2 Optimized structures of capped BN nanotubes. Parts (a) and (b) are two different views of a $B_{24}N_{24}$ cage: (a) along the C_3 axis; (b) along the C_4 axis; (c) a $B_{28}N_{28}$ (C_{4h}) cage obtained by adding a ring of eight alternate B and N atoms; (d) a tubular $B_{32}N_{32}$ (S_8) cage obtained by inserting two rings of eight alternating B and N atoms (see the text for more details).

cage structure. The hemispherical caps of (4,4) tubes based on octahedral $B_{24}N_{24}$ that we have proposed have also been observed in a molecular dynamics study of the growth mechanism of BN nanotubes.[39]

The $B_{24}N_{24}$ can be enlarged by adding hexagons. This leads to another round cage $B_{96}N_{96}$ of octahedral symmetry. The optimized $B_{96}N_{96}$ cage is shown in Figure 1*f*. Energetically the $B_{96}N_{96}$ cage is more stable than the $B_{24}N_{24}$ cage. It is clearly different from $B_{24}N_{24}$ in that while being mostly round, its 12 squares stick out significantly, like the detonators of a marine mine. Its halves can form a round cap for the (8,8) BN nanotube (see Figure 1*g*).

We have reoptimized the BN cages by replacing boron with aluminum. We would like to point out that unlike BN cages, which have been observed experimentally, the AlN cages studied here are predictions. The optimized cage structures in AlN are similar to those of BN except that they are larger than that of BN, due to the larger AlN bond distance. The exception to this trend is the $Al_{96}N_{96}$ cluster (see Figure 3). The optimization of $Al_{96}N_{96}$ structure starting from $B_{96}N_{96}$ cluster results in the formation of a double-shell onionlike structure. This onion cluster has $Al_{24}N_{24}$ cage at its core.[40] On the other hand, if we scale the $B_{36}N_{36}$ to account for greater AlN bond length and then replaces B by Al and optimize we do get a fullerenelike hollow cage with squares sticking out. This cage is referred to here as $Al_{96}N_{96}$–II. We find that all AlN cages are energetically stable with a binding energy of about 10 to 11 eV per pair of AlN.

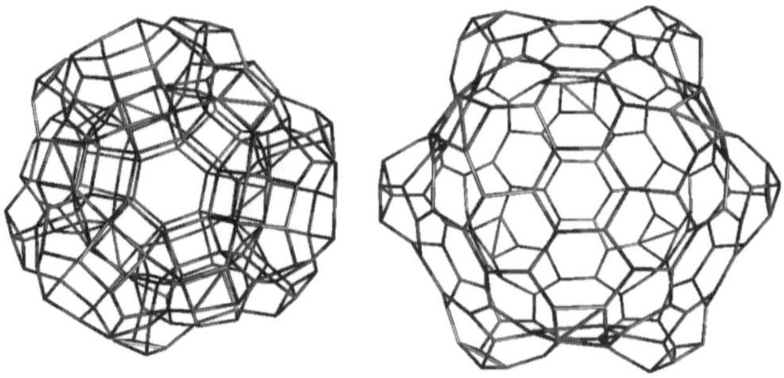

(a) (b)

Figure 3 Optimized octahedral $Al_{96}N_{96}$ cages: (a) two shell onionlike octahedral cages with an $Al_{24}N_{24}$ cage at its interior ($Al_{96}N_{96}$–I); (b) fullerenelike cage ($Al_{96}N_{96}$–II).

However, the binding energy of AlN cages is less than that of BN cages, which have a binding energy of 14 to 16 eV per BN pair. Similarly, the ionization potential (IP) and the energy gap between the energy eigenvalues of the highest occupied molecular orbital (HOMO) and the lowest unoccupied molecular orbital (LUMO) is smaller in AlN cages. The vertical IP of BN cages is in the range 7 to 9 eV, whereas it is about 6 to 7 eV in AlN cages. Due to the large HOMO–LUMO gap in the BN cages, the BN cages bind an extra electron quite weakly. Consequently, the electron affinity of boron nitride cages is practically zero within our model. The AlN cages have an electron affinity of about 1 to 2 eV. We have summarized the electronic structure data of AlN cages in Table 1. The last row of the table contains the range of values for the BN

TABLE 1. Electronic Structure Data for AlN Cages[a]

	Symmetry	BE	Gap	VIP	VEA	ΔSCE
$Al_{24}N_{24}$	O	10.24	2.97	7.05	1.46	5.59
$Al_{24}N_{24}$	S_4	10.34	2.47	6.84	1.72	5.12
$Al_{24}N_{24}$	S_8	10.34	2.63	6.79	1.58	5.21
$Al_{28}N_{28}$	C_{4h}	10.42	2.74	6.81	1.59	5.22
$Al_{28}N_{28}$	T	10.45	2.67	6.84	1.69	5.15
$Al_{32}N_{32}$	S_8	10.49	2.79	6.77	1.61	5.16
$Al_{36}N_{36}$	T_d	10.54	2.70	6.73	1.76	4.95
$Al_{48}N_{48}$	S_d	11.09	2.81	6.56	1.76	4.8
$Al_{96}N_{96}$	O	10.48	2.18	6.15	2.34	3.8
BN range		15	4–5	7–9	0	7–9

[a]Calculated values for the binding energy (BE) pair, the energy gap between the highest occupied molecular orbital and the lowest unoccupied molecular orbital, the vertical ionization potential (VIP), the electron affinity (VEA), and the energy gap obtained from the ΔSCF calculation for the optimized AlN cages. Last row gives a range of values for the same set of BN clusters. All energies are in eV.

cages. The HOMO–LUMO gap calculated by the ΔSCF method, in which the first ionization potential is subtracted from the first electron affinity, is given in the last row ofthe table.

4. CARBON FULLERENES

Carbon fullerene structures larger than 100 atoms have been studied by several groups.[41–43] Most of these studies have used semiempirical models or tight-binding methods or the Hartee–Fock theory plus minimal basis sets. Except for very recent calculations[44] on C_{240}, fullerenes have not been studied using reasonable-quality basis sets.[45] This is principally because of the high computational cost. We have used the computationally efficient ADFT described above to optimize the geometries of several carbon fullerenes, from C_{60} to C_{2160}, using large polarized Gaussian basis sets of triple-ζ quality (6-311G**).[10] The ADFT code developed in our group exploits the icosahedral symmetry of these fullerenes in an efficient manner. Therefore, very large calculations on C_{2160} with about 39,000 orbital basis functions is still doable with modest computation resources. To get accurate geometries of larger fullerenes, we parametrize the ADFT to get the exact geometry of C_{60}. This is accomplished by minimizing the mean-square deviation between the experimental and predicted bond lengths of C_{60}. This is possible without much difficulty, as optimization of C_{60} using ADFT takes less than 5 minutes on a single-processor Linux box (Intel XEON, CPU 2.20 GHz with 2 gigabytes of random access memory). The exact geometry of C_{60} can be obtained for $\alpha = 0.684667$. We use this value of α for optimizing larger carbon fullerenes and hope that this will give accurate estimates of their geometries. The C_{960} fullerene can be optimized on a single processor in 5 days. The C_{2160} optimization was performed on a Linux cluster using 48 processors and took about 5 days. The median bond distance of optimized carbon fullerenes is given in Table 2, and optimized structures are shown in Figure 4. We also made an attempt to get accurate atomization energies using the optimized geometries of fullerenes. For this purpose we reparametrize ADFT to get

TABLE 2. Data for Optimized Carbon Fullerenes[a]

Fullerene	Median Bond Distance	Mean Radius	AE (eV)
C_{60}	1.4244	3.5481	−7.140
C_{240}	1.4306	7.0728	−7.373
C_{540}	1.4264	10.5528	−7.431
C_{960}	1.4249	14.0342	−7.459
C_{1500}	1.4244	17.5225	−7.474
C_{2160}	1.4241	21.0137	−7.484

[a]Median nearest-neighbor bond distance, mean radius for the fullerenes of this work computed using $\alpha = 0.684667$. The rightmost column gives the atomization energy per atom (AE) that we compute using $\alpha = 0.64190$.

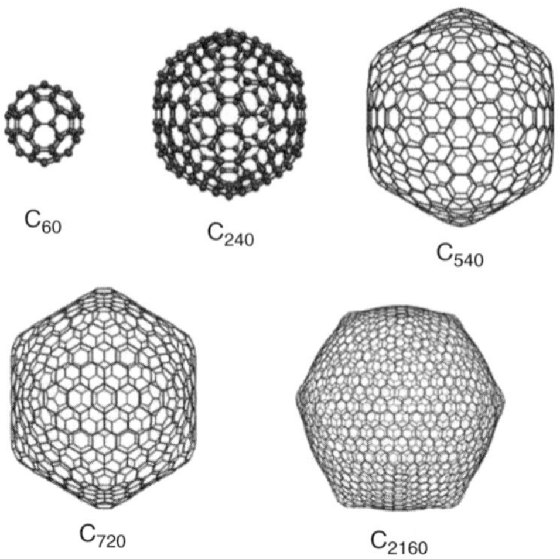

Figure 4 Fully optimized structures of carbon fullerenes (basis set: 6-311G**/Ahlrichs) (see the text for more details).

the exact binding energy of C_{60} fullerene and use the α value thus determined to compute the atomization energies of larger fullerenes. These values are also given in Table 2. However, such a procedure fails in that the atomization energy of C_{240} is already lower than that of graphite. Thus, to get an accurate estimate of atomization energy we need to go beyond the functional form that we have chosen in parameterizing the ADFT. Work is progress in our laboratory in this direction.

5. CONCLUSIONS

We have presented fully analytic implementation of density functional theory. It uses analytic Gaussian basis sets variationally to express the Kohn–Sham molecular orbitals, electron density, and the one-body effective potential of density functional theory. The resulting formulation is computationally very efficient and allows for calulations on relatively large systems. It permits use of atomic number–dependent potential by means of Slater's exchange parameters. Using the ADFT code, which uses the full-point-group symmetry of the molecule efficiently, we have optimized large inorganic fullerenelike cages and carbon fullerenes containing more than 2000 atoms.

REFERENCES

1. See, for example, other contributions from this volume.
2. Bacon, A. D.; Zerner, M. C. *Theochim. Acta*, 1979, 53: 21. Anderson, W. P.; Edwards, W. D.; Zerner, M. C. *Inorg. Chem*., 1986, 25: 2728.

3. Stewart, J. J. P. *J. Comput. Chem*. 1989, 10: 209, 221.
4. Andersen, O. K.; Jepsen, O. *Phys. Rev. Lett*., 1984, 53: 2571. Cohen, R. E.; Mehl, M. J.; Papaconstantopoulos, D. A. *Phys. Rev. B* 1994, 50: 14694. Menon, M.; Connolly, J. *Phys. Rev. B* 1994, 50: 8903.
5. Hohenberg, P.; Kohn, W. *Phys. Rev. B*, 1964, 136: 864.
6. Kohn, W.; Sham, L. J. *Phys. Rev. A*, 1965, 140: A1133.
7. Dunlap, B. I. *J. Phys. Chem. A*, 2003, 107: 10082.
8. Zope, R. R.; Dunlap, B. I. *J. Chem. Theory Comput*., 2005, 1: 1193.
9. Dunlap, B. I. *Comput. Phys. Commun*., 2005, 165: 18.
10. Dunlap, B. I; Zope, R. R. *Chem. Phys. Lett*., 2006, 422: 451; cond-mat/0603225.
11. Dunlap, B. I.; Connolly, J. W. D.; Sabin, J. R. *J. Chem. Phys*. 1979, 71: 3396, 4993.
12. Dunlap, B. I.; Rösch, N. *J. Chim. Phys*., 1989, 86: 671.
13. Slater, J. C. *Phys. Rev*., 1951, 81: 385.
14. Zope, R. R.; Dunlap, B. I. *Phys. Rev. B*, 2005, 71: 193104.
15. Zope, R. R.; Dunlap, B. I. *J. Chem. Phys*., 2006, 124: 044107.
16. Pederson, M. R.; Jackson, K. A. *Phys. Rev. B*, 1990, 41: 7453. Jackson, K. A.; Pederson, M. R. *Phys. Rev. B*, 1990, 42: 32763281. Gill, P. M. W.; Johnson, B. G.; Pople, J. A. *Chem. Phys. Lett*., 1993, 209: 506. Treutter, O.; Ahlrichs, R. *J. Chem. Phys*., 1995, 102: 346.
17. Werpetinski, K. S.; Cook, M. *Phys. Rev. A*, 1995, 52: 3397.
18. Werpetinski, K. S.; Cook, M. *J. Chem. Phys*., 1997, 106: 7124.
19. Zope, R. R.; Dunlap, B. I. *Chem. Phys. Lett*., 2004, 399: 417.
20. Slater, J. C. *J. Chem. Phys*., 1972, 57: 2389. Slater, J. C.; Johnson, K. H. *Phys. Rev. B*, 1972, 5: 844.
21. Krishnan, R.; Binkley, J. S.; Seeger, R.; Pople, J. A. *J. Chem. Phys*., 1980, 72: 650.
22. McLean, A. D.; Chandler, G. S. J. *J. Chem. Phys*., 1980, 72: 5639.
23. Broyden, G. *J. Inst. Math. Appl*. 1970, 6: 222.
24. Andzelm, J.; Wimmer, E. *J. Phys. B*, 1991, 172: 307; *J. Chem. Phys*., 1992, 96: 1280.
25. Godbout, N.; Salahub, D. R.; Andzelm, J.; Wimmer, E. *Can. J. Chem*., 1992, 70: 560.
26. Eichkorn, K.; Weigend, F.; Treutler, O.; Ahlrichs, R. *Theor. Chem. Acc*., 1997, 97: 119.
27. Press, W. H.; Flannery, B. P.; Teukolsky, S. A.; Vetterling, W. T. *Numerical Recipes: The Art of Scientific Computing*, Cambridge University Press, Cambridge, 1986, p. 309.
28. Dunlap, B. I. *Phys. Rev. A*, 2002, 66: 032502.
29. Oku, T.; Nishiwaki, A.; Narita, I.; Gonda, M. *Chem. Phys. Lett*., 2003, 380: 620.
30. Stephan, O.; Bando, Y.; Loiseau, A.; Willamie, F.; Shramchenko, N.; Tamiya, T.
31. Goldberg, D.; Bando, Y.; Ste'pahan, O.; Kurashima, K. *Appl. Phys. Lett*., 1998, 73: 2441.
32. Oku, T.; Hirano, T.; Kuno, M.; Kusunose, T.; Nihara, K; Suanuma, K. *Mater. Sci. Eng*. B, 2000, 74: 206.
33. Tondare, V. N.; Balasubramanian, C.; Shende, S. V.; Joag, D. S.; Godbole, V. P.; Bhoraskar, S. V. *Appl. Phys. Lett*., 2002, 80: 4813.

34. Balasubramanian, C.; Bellucci, S.; Castrucci, P.; De Crescenzi, M.; Bhoraskar, S. V. *Chem. Phys. Lett*., 2004, 383: 188.
35. Balasubramanian, C.; Godbole, V. P.; Rohatgi, V. K.; Das A. K.; Bhoraskar, S. V. *Nanotechnology*, 2004, 15: 370.
36. Wu, Q.; Hu, Z.; Wang, X.; Lu, Y.; Chen, X.; Xu, H.; Chen, Y. *J. Am. Chem. Soc*., 2003, 125: 10176.
37. Zope, R. R.; Dunlap, B. I. *Chem. Phys. Lett*., 2004, 386: 403.
38. Zope, R. R.; Baruah, T.; Pederson, M. R.; Dunlap, B. I. *Chem. Phys. Lett*., 2004, 393: 300.
39. Blase, X.; De Vita, A.; Charlier, J.-C.; Car, R. *Phys. Rev. Lett*., 1998, 80: 1666.
40. Zope, R. R.; Dunlap, B. I. *Phys. Rev. B*, 2005, 72: 045439.
41. Dunlap, B. I.; Brenner, D. W.; Mintmire, J. W.; Mowery, R. C.; White, C. T. *J. Phys. Chem*., 1991, 95: 8737.
42. Scuseria, G. E. *Science*, 1996, 271: 942.
43. Itoh, S.; Ordejn, P.; Drabold, D. A.; Martin, R. M. *Phys. Rev. B*, 1996, 53: 2132.
44. Geudtner, B.; Janetzko, F; Koster, A. M.; Vela, A.; Calaminici, P. *J. Comput. Chem*., 2006, 27: 483.
45. Scuseria, G. E. *Chem. Phys. Lett*., 1995, 243: 193.

10

ALUMINUM NANOPARTICLES: ACCURATE POTENTIAL ENERGY FUNCTIONS AND PHYSICAL PROPERTIES

NATHAN E. SCHULTZ, AHREN W. JASPER, DIVESH BHATT, J. ILJA SIEPMANN, AND DONALD G. TRUHLAR

Department of Chemistry and Supercomputing Institute, University of Minnesota, Minneapolis, Minnesota

Aluminum, which has a very negative enthalpy of combustion, is a widely used ingredient in solid rocket propellants. Conventional formulations of solid rocket fuel use micrometer-sized aluminum particles; however, a number of studies have shown that the incorporation of nanometer-sized aluminum particles can greatly increase the energetic properties of the fuel.[1–4] There are two main ways in which nanoparticle formulations are beneficial for propellants and other energetic materials: (1) Smaller particles have a higher extent of conversion, which means that a higher percentage of fuel can be oxidized under practical combustion conditions as the particles decrease in size; (2) the burn rate increases greatly as the particles decrease in size.

The issues noted above and the fundamental need to understand the properties of nanoparticles have motivated the study of Al nanoparticles. One of the key goals in nanoparticle modeling is the ability to model particle properties as a function of size. A classic example of a size-dependent property is the melting point.[5,6] A number of studies have been carried out on the size-dependent melting points of Al clusters and nanoparticles,[7–9] and all indicate a melting-point that increases with particle size. The melting-point depression of small particles is important in understanding the physical characteristics of nanoparticles,

Multiscale Simulation Methods for Nanomaterials, Edited by Richard B. Ross and Sanat Mohanty

and it also has technological implications, affecting, for example, the ignition temperature. Micrometer-sized Al particles are typically coated by a thick oxide layer, which cracks during combustion because the pure Al interior of the particle melts and significantly increases the internal pressure.[2] It is therefore possible that an earlier phase change during the heating process accompanying combustion affects the ignition temperature, but quantitative estimates of the effect are uncertain because the burn mechanism for nano-Al may be different from that for micrometer-sized Al particles. Nanoparticle simulations can provide useful information about this mechanism and about other aspects of technologically important systems composed of metal nanoparticles.

In typical materials simulations, the Born–Oppenheimer separation of nuclear and electronic motions[10] is assumed, and nuclear motion is governed by a single ground-electronic-state potential energy function. This assumption is tenuous for systems involving bulk metals, and one suspects that it is only approximately valid for metal nanoparticles as well. For small clusters, we may compute electronic energy gaps accurately. For example, the first excited electronic state of Al_2, $^3\Sigma_g^-$, is only 0.03 eV higher in energy than the ground state, $^3\Pi_u$,[11] and the first excited electronic state of Al_3, $^2A''_2$, is only 0.23 eV higher in energy than the ground electronic state, $^2A'_2$.[11] The size dependence of the highest occupied molecular orbital/lowest unoccupied molecular orbital gap for larger Al clusters and nanoparticles is uncertain, but it is reasonable to expect that low-lying electronic states may be involved in controlling the size dependence of certain properties of nanoparticles, such as enhanced reactivities. We note that nonadiabatic effects have been observed in the reactions of Al atoms with hydrogen.[12] Nevertheless, the Born–Oppenheimer ground-electronic-state potential energy surface is a practical starting point for simulations involving Al particles and their reactions with hydrogen or other heteroatoms.

When we assume that the Born–Oppenheimer approximation is valid, the initial step in any simulation is the development and validation of a potential energy function. Reliable electronic structure calculations are quite affordable for systems with up to $\sim 10^2$ electrons[13], but the computations become less and less reliable as the particle size increases, due to the approximations that must be made. Therefore, it is common[14,15] in nanoparticle simulation to use computationally inexpensive, atomistic, analytic potential energy functions parameterized to reproduce experimental or computed bulk properties. One of the most important characteristics of nanoparticle systems, however, is the dependence of their physical properties on particle size. Therefore, one of the key goals of nanoparticle simulations is to model and predict trends in this size dependence, but using potentials fit to bulk properties may introduce systematic size-dependent errors.

The approach that we have taken in our nanoparticle work is to adjust analytic potential energy functions to fit electronic structure calculations for small Al molecules, Al clusters, Al nanoparticles, and various bulk crystal phases as well as experimental data for diatomic molecules and the observable bulk crystal phase. The primary property that we use for fitting is the geometry-dependent atomization energy (both absolute and relative), and we include a large number of geometries both near and far from local minima. By doing this we implicitly include equilibrium bond lengths and forces. For Al, prior to our work, accurate data were unavailable for particles larger than a few atoms. The development of an Al database to fill this void is discussed in Section 1. We then discuss analytic potential energy functions and provide an example involving the size-dependent cohesive energy of Al in Section 2. In Section 3 we discuss in detail some recent simulations on the size-dependent properties of Al nanodroplets.

The phrase *particle size* is used to refer to the diameter or volume of a particle with a fixed number of atoms. (The phrase can also refer to the number of atoms in a particle.) There are many ways to compute the diameter of a particle. The approach that we take here is to compute the diameter d_{max} as the maximum Al–Al distance plus twice the van der Waals radius of Al, denoted r_{vdW}. For Al, r_{vdW} is 2.436 Å (discussed in Section 3). We can use this simple method to calculate how particle size depends on the number of Al atoms. We consider roughly spherical clusters and nanoparticles in which the atomic positions correspond to face-centered-cubic lattice sites. (The lattice constant used is 4.022 Å.[16]) The diameters for Al_n, with $n = 13$, 19, 55, 177, and 381, are then 1.04, 1.27, 1.61, 2.18, and 2.74 nm, respectively. To make a distinction between nanoparticles and clusters, we arbitrarily consider systems with fewer than 20 atoms to be clusters and those with 20 or more atoms to be nanoparticles.

1. NANO-AL DATABASE AND EFFECTIVE CORE SCHEME

The first step in developing the nano-Al database was to identify an affordable level of electronic structure theory that would provide accurate results for a wide range of clusters. There are two general classes of electronic structure theory: wave function theory[17,18] (WFT) and density functional theory[19,20] (DFT). State-of-the-art WFT methods are generally accurate to better than 0.04 eV per bond (1 kcal/mol per bond) for bond energies, and the most accurate WFT methods are accurate enough that they probably do not need to be specifically verified for Al clusters. The computational cost of these methods limits the feasibility of reliable ab initio WFT methods to approximately 10 Al atoms. DFT offers a more computationally affordable approach to calculating atomization energies, but due to the empirical nature of the best density functionals, the DFT methods themselves usually have to be specifically validated.

The first phase of the analytic potential energy function development was to determine how accurately DFT methods can treat small Al clusters. We used multicoefficient Gaussian-3/version 3[21] (MCG3/3) computations to develop[22] a small database of accurate bond energies for Al_n ($n = 2$ to 7). The MCG3/3 method[21] is a WFT method that is accurate to within 0.02 eV/atom when tested against the Database/3[21] collection of main group atomization energies. Using this database, we were able to assess the error of several DFT methods and identify the PBE0 functional[23,24] as a promising functional with an accuracy of 0.01 eV/atom when tested against the database of MCG3/3 calculated Al_n

atomization energies.[22] Because the general accuracy of MCG3/3 is 0.02 eV/atom, we can assign an error of 0.02 eV/atom to PBE0.

The PBE0 functional was assessed using the modified G3Large (MG3) basis set,[25,26] which is an all-electron basis set. [The MG3 basis set is equivalent for Al to the 6-311+G(3 d2f)[27–31] basis set.] The computational time required for a DFT calculation with the MG3 basis set becomes intractable as the size of the system grows, and the largest system we addressed at this level is Al_{13}, which is a 1.0-nm particle. This system requires 110 hours of computer time for a single-point energy calculation on a single processor of a Hewlett-Packard (HP) Itanium-2 computer. The affordability of these calculations can be increased greatly by replacing the all-electron basis set with a valence-electron basis set and replacing the core electrons by an effective core potential (ECP).[32] The combination of valence-electron basis set and ECP that we have developed (specifically for use with DFT methods) is labeled MEC (Minnesota effective core potential).[33] When tested against PBE0/MG3 calculations, the accuracy of the PBE0/MEC method is 0.01 eV/atom for atomization energies per atom (also called *cohesive energies*) and 0.005 Å for bond lengths. The database that we used for development of the analytic potential energy functions was created using the PBE0 functional with the MG3 basis set for systems with $n \leq 13$ and using the PBE0 functional with the MEC scheme for systems with $n = 14$ to 177. The computer time required for a single-point energy calculation with $n = 177$ is ~8000 hours (31 hours on 256 processors of a HP Itanium-2 computer).

We enforced the correct bulk limit on the analytic potential energy functions by including experimental values[16] for the cohesive energies, lattice constants, and bulk moduli of the face-centered-cubic (FCC) crystal phase. Accurate cohesive energies for the hexagonal-close-packed (HCP) and body-centered-cubic (BCC) crystal phases are included by adjusting calculated values[34] by a procedure that is described elsewhere.[35] Details of the database are also given elsewhere.[22,35] Some additional information about the database is provided in Section 3. We note briefly that we have included multiple points on the potential energy surface for each size cluster. By doing this, we fit to regions of the potential energy surface that would be visited during a finite-temperature simulation.

2. ANALYTIC POTENTIAL ENERGY FUNCTIONS

In previous work we tested and developed several analytic potential energy functions for Al. Here we present results for six analytic potential energy functions, and additional results can be found elsewhere.[35,36] We note that new and accurate potential energy functions[37] are being developed for condensed phase Al, but we do not survey those methods exhaustively.

The embedded atom model[38] has been widely used to study metal systems. For the embedded atom model, the potential energy, E, of the system is written as

$$E = \sum_{i>j} U_2(r_{ij}) + \sum_i F_i(\rho_i) \tag{1}$$

where r_{ij} is the distance between atomic centers i and j, U_2 is a pairwise interaction between atoms i and j, and F_i is a functional of the local electron density at the nucleus of atom i due to the other atoms; this density is called ρ_i. In many embedded-atom models, the embedding functional F is the square root of ρ_i, and ρ_i is approximated as a sum of pairwise additive terms. Note that the overall potential is not pairwise additive because F is nonlinear; nevertheless, U_2 and ρ_i are functions of single-pair distances, so the cost for evaluating the potential is just as manageable as that for a pairwise additive potential.

Several embedded-atom models that differ in their prescriptions for F_i, U_2, and ρ_i have been proposed[35,38–44] for Al. In this chapter we discuss five embedded-atom models: Ercolessi–Adams,[41] Mei–Davenport,[42,43] Sutton–Chen,[39] Streitz–Mintmire,[44] and NP-B.[35] The Mei–Davenport and NP-B models were chosen for detailed study here because in previous work[35] we reoptimized seven embedded-atom models against our database and found that the reparameterized Mei–Davenport model (which, as just explained, is called NP-B) gave the most accurate results. We also consider the Sutton–Chen, Streitz–Mintmire, and Ercolessi–Adams potentials because they have been used previously to simulate Al nanoparticles.[9,45–51]

The mean unsigned error per atom for NP-B is 0.05 eV/atom, whereas the Mei–Davenport fit has a mean unsigned error per atom of 0.18 eV/atom. This improvement in accuracy shows that the physical form of the Mei–Davenport potential energy function is flexible enough to describe the bonding of Al atoms in different bonding situations, but that the data used to obtain the original parameters (which included only bulk data) were not diverse enough to provide an accurate potential energy function for Al clusters and nanoparticles. This comparison shows that it is important to have a robust data set in addition to having an appropriate physical form for the potential energy function.

We also consider the NP-A analytic potential energy function,[35] which was also developed using the Al database discussed above. This function has the form

$$E = \sum_{i>j} V_2(r_{ij}) f_{ij}^{\mathrm{MB}} \tag{2}$$

where V_2 is the two-body interaction fitted to the extended-Rydberg[52,53] functional form, and f_{ij}^{MB} is a many-body function that deviates from unity when atoms i and j interact with other atoms. Several prescriptions for f_{ij}^{MB} were tested[35] using the database discussed above, and an accurate fit was obtained with $f_{ij}^{\mathrm{MB}} = f_{ij}^{\mathrm{S}} f_{ij}^{\mathrm{CN}}$, where f_{ij}^{S} is a screening function that weakens the bond between atoms i and j if other atoms are in between atoms i and j, and f_{ij}^{CN} incorporates the dependence of the bond order on the coordination numbers of the participating atoms.

To illustrate the screening term, we consider a system of three Al atoms (see Figure 1). In the figure the coordinates of atoms 1 and 2 are held fixed and atom 3 moves along coordinate R. Physically speaking, the interaction between atoms 1 and 2 is screened by the presence of atom 3 as atom 3 moves along a R. The

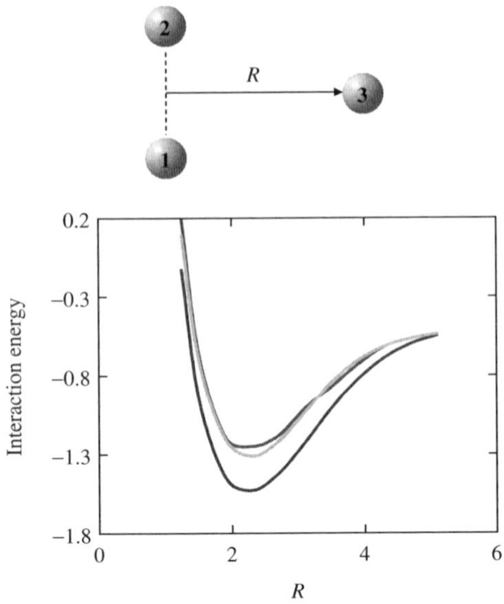

Figure 1 In this model system, atoms 1 and 2 are held fixed and atom 3 moves along coordinate R (Å). Atoms 1 and 2 are separated by 2.8635 Å. The interaction energy (in eV/atom) for this system is plotted with a well depth of <1.3 for PBE0/MG3, with a well depth of 1.55 for an accurate two-body interatomic potential [Eq. (3)], and with a well depth of ~ 1.3 for a two-body interatomic potential that is modified by a screening term [Eq. (4)] with parameters of the ER2+ES potential.

interaction energy is plotted in Figure 1 as a function of R for the following two potentials:

$$E = \sum_{i>j} V_2(r_{ij}) \tag{3}$$

and

$$E = \sum_{i>j} V_2(r_{ij}) f_{ij}^{\mathrm{S}} \tag{4}$$

where Eq. (3) is simply the two-body interaction without many-body effects and Eq. (4) is the two-body interaction modified only by the screening term. [In previous work,[36] Eq. (4) was denoted ER2 + ES.] We can see from Figure 1 that the two-body interaction alones significantly overestimates the three-body interaction energy and the screening function allows for a more accurate description of the three-body interaction. In addition to predicting a more accurate binding energy for Al_3, Eq. (4) also predicts a more physical repulsive wall.

In addition to screening, which is a three-body effect, we also consider the effect of coordination number, which in bulk Al is 12. We note that screening and coordination number effects are related; that is, the presence of the third atom in Figure 1 raises the coordination number of atoms 1 and 2, and a highly coordinated atom involves pairs of atoms being screened by nearby atoms. However, we found it useful to treat these effects separately and to include coordination number effects explicitly. To illustrate the effectiveness of the coordination number term, we consider a potential energy function of the form

$$E = \sum_{i>j} V_2(r_{ij}) f_{ij}^{\mathrm{CN}} \tag{5}$$

In previous work,[36] Eq. (5) was denoted ER2 + ECN. The cohesive energy of the ground state of icosahedral Al_{13} (which consists of a central atom coordinated to 12 atoms and 12 surface atoms each coordinated to six atoms) computed with PBE0/MEC is 2.5 eV/atom. The cohesive energies of the same structure computed with Eqs. (3) and (5) are 5.5 and 2.5 eV/atom, respectively. The pairwise additive potential energy function in Eq. (3) overestimates the interaction energy of Al_{13} by 3 eV/atom, and the coordination number factor corrects this error.

As mentioned above, the screening and coordination number factors are related, and in fact Eqs. (4) and (5) have similar overall errors when tested against the full database. However, the cohesive energy of Al_{13} computed with Eq. (4) is 3.4 eV/atom and is less accurate than Eq. (5) for this property. We find in general that the effect of including the coordination number term is more significant for bigger clusters (which have the largest contribution to their total energies from coordination effects), whereas the reduction in the error due to the incorporation of the screening term is more evenly distributed.

Physically, one expects that there is some cutoff distance at which the interaction between two atoms may be set to zero. We have therefore built cutoffs into the functional forms of NP-A and NP-B. When using a cutoff distance, the cost to evaluate the analytic potential energy function scales as n in the large-n limit. Such linear scaling is achieved for the NP-A potential by multiplying the terms in Eq. (2) by a cutoff function,[54] which goes smoothly to zero at $r_{ij} = 6.5$ Å. Without a cutoff, the computational cost of the screening and coordination number factors both scale as n^3. The range parameter of the cutoff function was optimized during the fitting procedure to avoid numerical and convergence problems that can arise when applying cutoffs during simulations. The cutoff function for the NP-B analytic potential energy function goes to zero at 5.38 Å. Without the cutoff function, the computational cost of evaluating the embedding term scales as n^2, where n is the number of atoms in the system. Both potentials (NP-A and NP-B) begin to scale linearly at ~10,000 atoms. However, the cutoff functions give significant cost reductions for smaller clusters. For example, the average CPU cost of an energy evaluation of Al_{1055} with NP-A and NP-B on an IBM Power4 computer is reduced by factors 5 and 2, respectively.

In Figure 2 we plot the mean unsigned error[35] (in eV/atom) for the five potentials for groups of various particle sizes. The groups contain particle sizes with $n = 2, 3, 4, 7, 9$ to 13, 14 to 19, 20 to 43, 50 to 55, 56 to 79, 80 to 87, and 89 to 177, respectively, and are labeled by the average number of atoms in the particles of that bin, which are 2, 3, 4, 7, 13, 18, 33, 53, 71, 86, and 124, respectively.

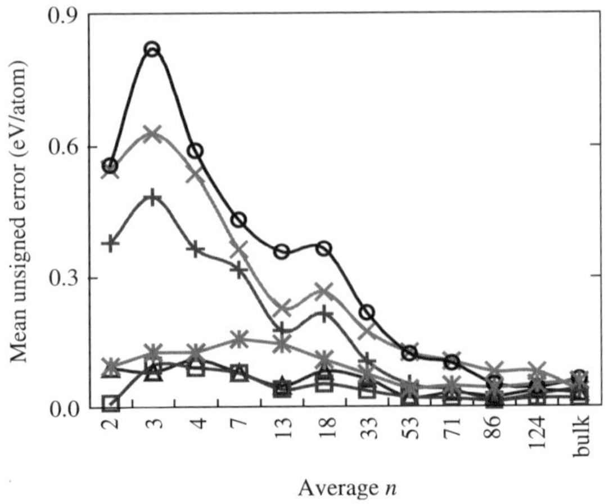

Figure 2 Mean unsigned error (in eV/atom) grouped by particle size for the Ercolessi–Adams (*), Mei–Davenport (×), NP-A (□), NP-B (Δ), Streitz–Mintmire (+), and Sutton–Chen (O) potential energy functions of the average particle size in a bin.

The most accurate potential for clusters ($n = 2$ to 20), nanoparticles (21 to 177), and the bulk crystal phase is NP-A. The Mei–Davenport, Streitz–Mintmire, and Sutton–Chen PEFs were not fit to nanoparticle or cluster data and have a more size-dependent error. On the one hand, it might be argued that it is unfair to test the bulk- fitted potentials against nanoparticles and clusters, but on the other hand, it can be argued that these studies are very important because these potentials are sometimes used in nanoparticle simulations without validation.[9,45–49]

The Ercolessi and Adams analytic potential energy function[41] was fit (by the original authors[41]) to Al cluster and surface data and to bulk crystal data. The mean unsigned error (MUE) per atom for this analytic potential energy function (when evaluated with our database) is less dependent on the number of atoms than the error for the Mei–Davenport, Streitz–Mintmire, and Sutton–Chen analytic potential energy functions; however, it has a larger mean unsigned error (MUE) per atom than either the NP-A or NP-B PEF. The total MUE for the Ercolessi and Adams analytic potential energy function is 0.11 eV/atom, whereas the NP-A and NP-B analytic potential energy functions have MUE values of 0.03 and 0.05 eV/atom, respectively. The fitting data used by Ercolessi–Adams was obtained (by the original authors) from the local density approximation to DFT,[55,56] which is not usually quantitatively accurate for metals,[57] whereas our data was obtained from a validated[22] hybrid DFT method (PBE0). The improved accuracy of our analytic potential energy functions is due, then, to the quality of fitting data, which again highlights the need to have not only physical functional forms but also accurate fitting data.

An interesting example of how the errors depend on the number of atoms is to look at the cohesive energies of nanocrystals, which are nanometer-sized objects with a structure cut from a bulk crystal. In this chapter we discuss FCC nanocrystals, which are nanoparticles that have the same local arrangement of atoms that is found in FCC crystals. An FCC crystal is generated around a central atom using a lattice parameter. The distance from the central atom in an FCC cluster to another atom i in the cluster is denoted R_i, and due to the periodic nature of the crystal, there is a unique set, S_m, of values for R_i, where m is an index. Nanocrystal m is defined as a nanocrystal containing all of the atoms with $R_i < S_m$. For an FCC crystal, the nanocrystals studied here have $n = 13$, 19, 43, 55, 79, 87, 135, and 177, where n is the number of Al atoms. Thus, this sequence of values defines a unique set of FCC nanocrystals that have geometric magic numbers. It also possible to define nonunique FCC nanocrystals for $n = 14$ to 18 and 20 to 42, which are also studied in this chapter. The strategy that we employ for determining the coordinates of these nonunique nanocrystals is based on our determination of the lowest-energy geometry. For Al_{14}, an atom is placed at one of the available and equivalent FCC lattice sites between nanocrystal Al_{13} and Al_{19}. There are now four nonequivalent unoccupied lattice sites in which an atom can be located to form Al_{15}, and the energy with each of these lattice sites occupied is evaluated with PBE0/MEC to determine which isomer of Al_{15} is the lowest in energy. The same procedure is followed for $n = 16$ to 18 and $n = 20$ to 42.

In Figure 3 we plot the cohesive energies computed with the PBE0/MEC DFT method and by the MeiD, NP-A, NP-B, StrM, and SutC analytic potential energy functions for FCC nanocrystals with $n = 13$ to 43, 55, 79, 87, 135, and 177. The lattice constant is optimized for each nanocrystal with the same method, PBE0/MEC or an analytic potential energy function, that is used to calculate the cohesive energy of that nanocrystal. For example, the cohesive energies calculated with NP-A also use lattice constants that are calculated with NP-A. The only potentials that are accurate across this entire size range are NP-A and NP-B, with NP-A being more accurate. The StrM potential is accurate for $n \gtrsim 20$ and is less accurate for $n \lesssim 20$. This behavior in the StrM potential can also be seen in Figure 2. The other two potentials, MeiD and SutC, have errors that are approximately 0.1 eV/atom for nanoparticles larger than $n = 55$, and the errors grow to 0.3 to 0.4 eV/atom for smaller nanocyrstals.

All of the analytic potential energy functions presented in this chapter break down to some extent for small n, where n is the number of atoms. NP-A is built on an accurate two-body interaction, so the dimer is quantitatively accurate for NP-A. The NP-B analytic potential energy function does reasonably well for the dimer (see Figure 2), but both NP-A and NP-B cannot predict the correct geometries for Al_4 or Al_5. Al_4 and Al_5 are known to be planar,[11,58,59] but the analytic potential energy functions predict Al_4 and Al_5 to be three-dimensional. It is possible to develop analytic potential energy functions to predict planar geometries for Al_4 or Al_5,[58] but these analytic potential energy functions are inaccurate for larger clusters. For example, the analytic potential energy function of Pettersson et al.[58] that predicts Al_4 and Al_5 to be planar also predicts Al_{13} to be planar and has a 0.7-eV/atom error for the bulk cohesive energy.[58] One way to understand this problem is by comparing the total coordination numbers of

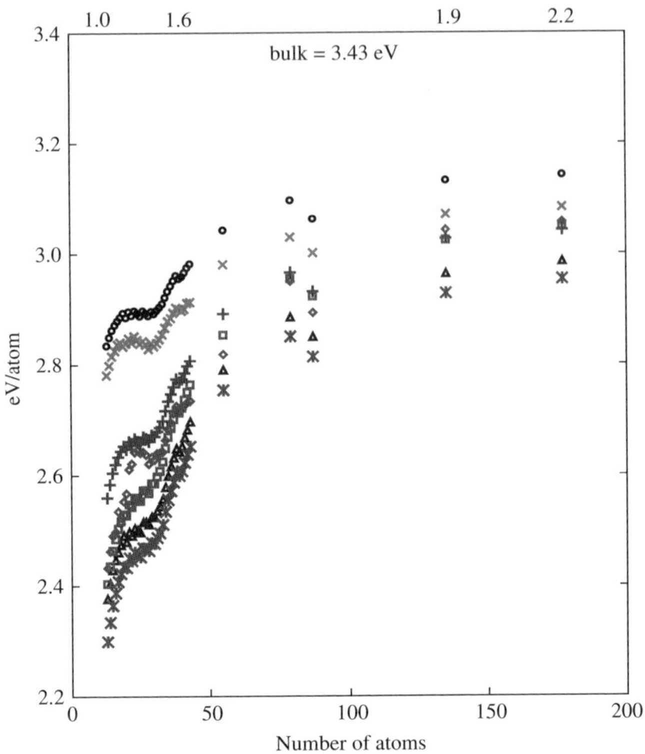

Figure 3 Cohesive energy (in eV/atom) for FCC nanocrystals computed with PBE0/MEC (◇) and the Ercolessi–Adams (⌈), Mei–Davenport (×), NP-A (□), NP-B (⌈), Streitz–Mintmire (+), and Sutton–Chen (O) potential energy functions as functions of the number of atoms (along the bottom) and particle diameter (along the top).

the Al atoms in planar and nonplanar clusters. The total coordination number is defined as the sum of coordination numbers for all of the atoms in a cluster. For example, Al_3 (equilateral triangle) has a total coordination number of 6 because each of the atoms is bonded to two other Al atoms. For Al_4, the planar structure (D_{2h} symmetry) has four atoms with a coordination number of 2 for a total coordination number of 8, whereas the nonplanar structure (T_d symmetry) has four atoms with a coordination number of 3, for a total coordination number of 12. Thus, a structure with a total coordination number of 8 is lower in energy than a structure with a total coordination number of 12. For Al_5, the planar structure (C_{2v} symmetry) has a total coordination number of 16, whereas the nonplanar structure (T_d symmetry) has a total coordination number of 20. However, the ground state of Al_6 is three-dimensional (O_h symmetry) and has a total coordination number of 24, and the lowest-energy planar structure of Al_6 (C_{2h} symmetry) has a total coordination number of 20. These considerations show why it is very difficult to develop many-body functional forms that fit all these clusters; such functions must favor low coordination numbers for Al_4 and Al_5 but higher coordination numbers for Al_6 and larger. We have only examined structural isomers where the coordination number differs between isomers and not clusters, and that have the same coordination number but different structural isomers.[60]

3. NANOPARTICLE SIMULATIONS

Simulation Procedure

The potential used for the simulations is the NP-B embedded-atom model. The nanoparticles are simulated via Metropolis Monte Carlo[61] in a canonical ensemble where the number of atoms, box size, and temperature are fixed. Al_n nanoparticles with $n = 55$, 400, and 1000 were simulated with periodic boundary conditions with cubic box lengths of 35, 45, and 60 Å, respectively. For the 55-atom system, the starting structure is an energy (NP-B potential) minimized icosahedral nanoparticle. The starting structure for 400- and 1000-atom systems is a FCC nanocrystal. Accordingly, the size of the box in each case is larger enough that the periodic images of a nanoparticle do not interact. Consequently, the particles are treated essentially as isolated nanodroplets in each case.

Nanoparticle Diameters

There are many ways that one can compute the diameters of nanoparticles. We compute the particle diameters as the maximum distance between two atoms plus twice the van der Waals radius of Al. The van der Waals radius for Al is 2.346 Å (see below). The particles discussed in this section were optimized with the NP-B potential unless otherwise specified. The diameters for all the particles in this section are given in Table 1.

The first particle that we discuss is Al_{13}. Al_{13} is a special cluster because it is the first cluster that can have an atom with a coordination number of 12. An Al atom in a periodic FCC lattice also has a coordination number of 12; therefore, Al_{13} is the smallest cluster to have an interior "bulklike" atom and surface atoms. The global minimum[33] of Al_{13} (icosahedron) found with NP-B at 0 K has $d_{max} = 1.09$ nm. The FCC-nanocrystal for Al_{13} has $d_{max} = 1.02$ nm. The global minimum of Al_{19} with the NP-B potential (double-icosahedron) has $d_{max} = 1.26$ nm, and the FCC-nanocrystal has $d_{max} = 1.29$ nm. The ground-state structure of Al_{55} with the NP-B potential is icosahedral with $d_{max} = 1.55$ nm, whereas the FCC-nanocrystal has $d_{max} = 1.58$ nm. From these results we can see that the diameters of the particles for a given number of atoms are not very sensitive to the crystal structure, as the FCC-nanocrystals and icosahedral nanoparticles for a given number of atoms differ by an average of 0.04 nm. We conclude that the particle diameter is, to a first approximation, independent of crystal structure.

The structures for Al_{400} and Al_{1000} were optimized with the NP-B potential. The starting geometries were the globally optimized geometries for the

TABLE 1. Maximum Al–Al Distance d_{max} (nm) for Several Al Particles Computed Using the NP-B Embedded-Atom Model

Number of Atoms	Structure	D_{max}
	$T = 0\ K$	
13	FCC-nanocrystal	1.09
	global minimum[a]	1.02
19	FCC-nanocrystal	1.26
	global minimum[a]	1.29
55	FCC-nanocrystal	1.58
	global minimum[a]	1.55
400	global minimum[b]	2.72
1000	global minimum[b]	3.77
	$T = 1000\ K$	
55	ensemble average	1.71
400	ensemble average	3.00
1000	ensemble average	3.92
	$T = 1500\ K$	
55	ensemble average	1.80
400	ensemble average	3.11
1000	ensemble average	4.10
	$T = 2000\ K$	
55	ensemble average	1.88
400	ensemble average	3.24
1000	ensemble average	4.21
	$T = 2500\ K$	
55	ensemble average	1.97
400	ensemble average	3.42
1000	ensemble average	4.38

[a]Icosahedral.
[b]The geometry was optimized by starting with the global minimum for a Lennard-Jones particle (see the text).

Lennard-Jones system,[62] where the initial coordinates were scaled by 3.00. The coordinates for the Lennard-Jones system were obtained from the Cambridge Cluster Database.[62,63] For optimized Al_{400} and Al_{1000}, $d_{max} = 2.72$ and 3.77 nm, respectively. It is reasonable to expect that the diameters would change by less 0.1 nm if a more exhaustive search for the global minimum were conducted.

The average $d_{\max}$ values for nanodroplets (Al_n with $n = 55$, 400, and 1000 with $T = 1000$ to 2500 K) are also given in Table 1. We note initially that for all droplet sizes $d_{\max}$ with $T \geq 1000$ K is greater than $d_{\max}$ with $T = 0$ K, which is an expected result. The interesting aspect of these $d_{\max}$ values is that the diameters increase at different rates, depending on the number of atoms in the droplet. To explore this, we have fit ($d_{\max}$,T) to a linear equation,

$$d_{\max} = \beta T + b \tag{6}$$

where the slope of the line, β, indicates how rapidly $d_{\max}$ increases with T. The intercept, b, would in principle be the value of $d_{\max}$ at $T = 0$, but we do not expect $d_{\max}$ to remain linear with T as the particles undergo a phase change from liquid to solid. The values of β for Al_n with $n = 55$, 400, and 1000 are 3.0×10^{-3}, 2.8×10^{-3}, and 1.7×10^{-3} K^{-1}, respectively. We can see that β increases with an increasing number of atoms and that the response of $d_{\max}$ to T is a size-dependent property.

Density and Thermal Expansion

A fundamental property of any material is the density. The density is unambiguous for bulk materials, but for nanoparticles it requires a definition of the volume of the nanoparticle. Here, we calculate the nanoparticle volume by using the method of overlapping van der Waals spheres.[64] We denote the density computed from the number of particles and the volume of overlapping van der Waals spheres as ρ_{vdW}. In this method, the only input that is required is the van der Waals radius for Al. The van der Waals radius, r_{vdW}, that we use for Al is 2.346 Å. This value was obtained from the bond length of AlNe and Ne_2 by the relationship

$$r_{vdw}(\mathrm{Al}) = r_e(\mathrm{AlNe}) - \frac{1}{2} r_e(\mathrm{Ne_2}) \tag{7}$$

where r_e(AlNe) and $r_e(Ne_2)$ are the bond lengths of AlNe and Ne_2, respectively. We computed r_e(AlNe) and $r_e(Ne_2)$ to be 3.894 and 3.099 Å, respectively, using WFT. The electron correlation method used was coupled cluster theory with single and double substitutions and quasiperturbative triples, CCSD(T),[65–67] and the one-electron basis set used was the aug-cc-pV5Z[68] basis set.

We can also compute the covalent radii, r_{cov}, of Al by first computing the bond length of Al_n particles that, by symmetry, have one unique Al–Al bond length. We have computed the bond length using the PBE0 density functional with the MEC scheme. The Al_n clusters that we examined are Al_2, Al_3 with D_{3h} symmetry, and Al_{13}, Al_{55}, and Al_{177} FCC-nanocrystals. The small clusters, Al_2 and Al_3, have $r_{cov} = 1.365$ and 1.253 Å, respectively. The FCC-nanocrystals have $r_{cov} = 1.923$, 1.938, and 1.985 Å, respectively. The zero-point-exclusive experimental lattice constant[16] at 0 K implies an equilibrium internuclear distance of nearest neighbors of 4.022 Å, which corresponds to $r_{cov} = 2.011$ Å.

As stated earlier, the densities are computed from overlapping van der Waals spheres, which use r_{vdW}. We have seen that r_{cov} can depend on the number of Al atoms. It is also likely that r_{vdW} depends on the number of atoms in the clusters. This is one unsatisfactory part of our method for computing volumes. An additional unsatisfactory aspect of this method for computing volumes is that voids are excluded from the total volume, and the total volume associated with voids may be nonneglible as the temperature is increased. The major drawback to excluding voids is that it does not allow a meaningful comparison to experiment or a bulk simulation. In a bulk simulation, the density of a liquid is computed from the number of atoms within the simulation box and volume of the box. In this manner, voids are included in the total volume of the liquid. In bulk experiments, the density of liquid aluminum is determined by melting a millimeter-sized piece of Al and measuring the diameter and mass of the drop.[69] In this manner, voids are once again included in the total volume. An alternative way[70] to obtaining the nanoparticle volumes would be to roll a probe sphere over the surface of the nanoparticle. This would eliminate the contribution of internal voids, but it will introduce parameter-dependent oscillations in the surface area.

In Table 2 we give the computed nanoparticle densities, ρ_{vdW}. We also include experimental values of the density of bulk liquid.[69] We denote the experimental value of the density of the bulk liquid as ρ_{exp}. We can see that the ρ_{vdW} values are always lower than the experimental values for the bulk liquid. This is not entirely unexpected, as the nanoparticle densities should be lower due to surface effects, but it is not clear what significance should be attached to this

TABLE 2. Coefficient of Thermal Expansion, α (10^{-5} K^{-1}) for Al_{55}, Al_{400}, and Al_{1000}; Density, ρ (g/mL); and the Sphericality Parameter, L (Unitless)

	ρ_{vdW}			Bulk Liquid	
Property	Al_{55}	Al_{400}	Al_{1000}	ρ_{bulk}[a]	ρ_{exp}[b]
V					
1000 K	1.67	1.96	2.06	2.30	2.36
1500 K	1.61	1.88	1.97	2.19	2.25
2000 K	1.55	1.81	1.89	2.09	2.13
2500 K	1.50	1.73	1.82	2.00	2.01
α	6.77	7.79	7.93	8.86[c]	9.91[d]
L					
1000 K	0.88	0.94	0.95		
1500 K	0.84	0.93	0.94		
2000 K	0.82	0.91	0.94		
2500 K	0.79	0.88	0.94		

[a]Calculated by extrapolating the nanodroplets volumes using $\rho_{vdW} = aN^{-1/3} + \rho_{bulk}$, where N is the number of atoms.
[b]Experimental density.
[c]Calculated from ρ_{bulk}.
[d]Calculated from ρ_{exp}.

finding because of the volume of nanoparticle voids. For example, if we fit the density of Al_{55}, Al_{400}, and Al_{1000} with $T = 1000$ K to $\rho_{vdW} = \alpha N^{-1/3} + \rho_{bulk}$, we find that $\rho_{bulk} = 2.30$ g/mL. The value of ρ_{bulk} corresponds to the bulk density, and it differs from the experimental value of 2.36 g/mL by 0.06 g/mL. The discrepancy between the calculated and expected values cannot be attributed only to the presence of voids because correcting for voids would increase the calculated density. We have also found that we can change the values of ρ_{bulk} and ρ_{vdW} by changing the value of r_{vdW}. For example, decreasing the van der Waals radius by 23% to 1.9 Å increases the density of Al_{1000} at 1000 K to 2.31 g/mL (+12%) and increases ρ_{bulk} at 1000 K by 2.41 g/mL (+5%). Due to the differences in ρ_{bulk} and ρ_{exp}, it might be more appropriate to compare ρ_{vdW} to ρ_{bulk} than to ρ_{exp}. Values of ρ_{bulk} and ρ_{exp} for all of the temperatures are given in Table 2.

By comparing the nanoparticle densities (ρ_{vdW}) to the extrapolated bulk densities (ρ_{bulk}), we can see that the nanodroplets expand with temperature at a different rate than does the bulk liquid. To quantify this, we calculate the coefficient of thermal expansion. For the nanodroplets the coefficient of thermal expansion, α_{vdW}, is calculated as

$$\alpha_{vdW} = \frac{1}{V}\frac{dV_{vdW}}{dT} \tag{8}$$

where V_{vdW} is the volume computed from overlapping van der Waals spheres. For the bulk liquid, we compute α_{bulk} as

$$\alpha_{bulk} = \rho_{bulk}\frac{d(1/\rho_{bulk})}{dT} \tag{9}$$

where ρ_{bulk} is defined in the preceding paragraph. We have computed ρ_{bulk} at $T = 1000$, 1500, 2000, and 2500 K and then fitted ($1/\rho_{ex}$, T) to a linear line to obtain α_{bulk} for the bulk liquid. The computed values of α_{bulk} are given in Table 2 along with the experimental value, α_{exp}, for this quantity. We compute α_{exp} by replacing ρ_{bulk} with ρ_{exp} in Eq. (9).

We first talk about α_{bulk} and α_{exp}. The values of α_{bulk} and α_{exp} are 8.9×10^{-5} and 9.9×10^{-4} K^{-1}, respectively. The agreement between α_{bulk} and α_{exp} is very good, despite the approximations that are involved in calculating α_{bulk}. Turning now to the nanodroplets, we can see from Table 2 that α_{vdW} is a size-dependent property that decreases with decreasing particle size. For the smallest droplet, Al_{55}, the computed α_{vdW} is 6.8×10^{-5} K^{-1}, and that is 69% of α_{bulk}; however, for the largest particle, Al_{1000}, the computed α_{vdW} is 7.9×10^{-5} K^{-1}, 89% of α_{bulk}.

The size dependence of the coefficient of thermal expansion has previously been studied by Pathak and Shenoy for systems below the melting point.[51] The coefficient of thermal expansion that Pathak and Shenoy[51] calculated is denoted as α_{stress}, because it is computed from the temperature-dependent stress tensor, whereas we calculate α_{vdW} from volume changes. Also, α_{stress} is computed for nanometer-thick slabs with two-dimensional periodicity, and α_{vdW} is for liquid droplets. The values of α_{stress} for 2.0-, 3.2-, and 4.0-nm-thick slabs are 4.6×10^{-5}, 5.3×10^{-5}, and 5.6×10^{-5} K^{-1}, respectively.

In this chapter we chose to focus on the similar behavior of α_{stress} and α_{vdW} rather than the differences between the two quantities. The results of Pathak and Shenoy[51] showed that α_{stress} increases with decreasing slab thickness for a Lennard-Jones system, whereas they showed that α_{stress} decreases with decreasing slab thickness for an Al system. (Pathak and Shenoy[51] modeled the Al slab with the embedded atom model of Ercolessi and Adams.[41]) Whether α_{stress} increases or decreases with decreasing particle size depends on the type of system being studied. The results of Pathak and Shenoy[51] agree qualitatively with our results, as we find that α_{vdW} decreases with decreasing system size.

Earlier we discussed the dependence of particle diameter, d_{max}, on T, where this relationship was quantified through β in Eq. (6). We saw that β increases with increasing particle size as α_{vdW} does. It is clear that these two quantities, β and α_{vdW}, are related, as they both pertain to changes in particle size with temperature.

Shapes

Another fundamental property of a nanoparticle is its shape. It is sometimes assumed,[4] due to lack of better information, that Al nanoparticles are spherical. We are able to quantify the shape of a nanodroplet by using the sphericality parameter, L, of Mingos et al., which is defined as

$$L = \frac{3I_{unique}}{\sum_{i=1}^{3} I_i} \tag{10}$$

where I_i is the principal moment of interia i and I_{unique} is the unique principal moment of inertia. I_{unique} is defined as the principal moment of interia that deviates most from the average principal moment of inertia. Using this definition, $L = 1$ for a sphere, $0 \leq L < 1$ for an prolate spheroid, and $1 < L \leq 1.5$ for an oblate spheroid. The sphericality parameter for a cylinder that has the length and width of a football is 0.51, whereas L for a hockey puck is 1.40.

The sphericality parameters for the nanodroplets are reported in Table 2. All of the droplets are prolate spheroids, with the smaller droplets having smaller L values than those of the larger particles. The shape of the largest droplet, Al_{1000}, is relatively independent of temperature, where $L = 0.95$ for $T = 1000$ K and $L = 0.94$ for $T = 1500$, 2000, and 2500 K. The shape of the smallest droplet, Al_{55}, has a stronger dependence on temperature, in particular $L = 0.88$, 0.84, 0.82, and 0.79 for $T = 1000$, 1500, 2000, and 2500 K, respectively. Al_{400} is intermediate between Al_{55} and Al_{1000}, with $L = 0.94$, 0.93, 0.91, and 0.89 for $T = 1000$, 1500, 2000, and 2500 K. We can infer from these results that particles larger than Al_{1000} are essentially spherical, and the shape is almost temperature independent; however, the shapes of smaller particles are prolate and temperature dependent.

4. CONCLUSIONS

In this chapter we summarized the development of analytic potential energy functions for simulating Al nanoparticles. One of the key steps in the development of the analytic potential energy functions was the development of a diverse data set of geometry-dependent atomization energies for Al_2 to Al_{177} that were calculated with validated density functional theory. We have developed two potentials, NP-A[35] and NP-B,[35] that are accurate for clusters, nanoparticles, and bulk crystal properties.

The NP-B[35] analytic potential energy function is an embedded atom model; in particular, it is a reparameterized version of the embedded atom model of Mei and Davenport.[42,43] The original parameterization by Mei and Davenport is less accurate for modeling aluminum clusters and nanoparticles; however, this inaccuracy does not mean that the physical form is not flexible enough to model clusters and nanoparticles. Our results show that the embedded atom functional form of Mei and Davenport is promising when the parameters are adjusted against our cluster and nano-Al data in addition to data for the bulk crystal phases.

The development of the NP-A potential began with an accurate description of diatomic Al. The many-body effects are incorporated through explicit many-body terms. The many-body terms used in the NP-A potential involve screening function and a dependence on coordination number. The physical nature of the screening function is that it weakens the bond between atoms i and j in the presence of other atoms. The coordination number term incorporates the dependence of the bond strength on the coordination numbers of the participating atoms. This bond strength dependence allows for weakening of the bond between atoms i and j as the number of neighboring atoms is increased. We note that the philosophy of NP-A is quite different from that of NP-B. The NP-B analytic potential energy function incorporates the many-body effects through an embedding term, whereas the NP-A potential energy function begins with an accurate description of diatomic Al and uses explicit many-body effects to correct the two-body interaction in the presence of various atomic environments.

We have used the NP-B potential to simulate Al nanodroplets to study the size dependence of densities, thermal expansion, and particle shapes. We have proposed computing the nanoparticle densities by first computing the volumes using overlapping van der Waals spheres. By computing the densities in this way, we obtain densities for the nanodroplets that can be used for comparing the bulk values. We have been able to show that nano-Al droplets have a decreasing coefficient of thermal expansion with decreasing particle size. We have also shown that particle shape is size dependent, with smaller particles being prolate spheroids. The shape of the smallest drop studied, Al_{55}, is more dependent on temperature than is the largest drop, Al_{1000}.

Acknowledgments

We thank Grażyna Staszewska, Przemyslaw Staszewski, and Zhen Hua Li for helpful discussions. This work was supported in part by the Defense-University

Initiative in Nanotechnology (DURINT) of the U.S. Army Research Laboratory and the U.S. Army Research Office under agreement DAAD190110503 and by the National Science Foundation under grant ITR-0428774.

REFERENCES

1. Armstrong, R. W.; Baschung, B.; Booth, D. W.; Samirant, M. *Nano Lett.*, 2003, 3: 253.
2. Rai, A.; Lee, D.; Park, K.; Zachariah, M. R. *J. Phys. Chem. B*, 2004, 108: 14793.
3. Meda, L.; Marra, G.; Galfetti, L.; Inchingalo, S.; Severini, F.; De Luca, L. *Compos. Sci. Technol.*, 2005, 65: 769.
4. Park, K.; Lee, D.; Rai, A.; Mukherjee, D.; Zachariah, M. R. *J. Phys. Chem. B*, 2005, 109: 7290.
5. Borel, J.-P. *Surf. Sci.*, 1981, 106: 1.
6. Wales, D. J.; Berry, S. R. *J. Chem. Phys.*, 1990, 92: 4283.
7. Mitev, P.; Papageorgiou, D. G.; Lekka, C. E.; Evangelakis, G. A. *Surf. Sci.*, 2004, 566–568: 937.
8. Werner, R.*Eur. Phys. J. B*, 2005, 43: 47.
9. Alavi, S.; Thompson, D. L. *J. Phys. Chem. A*, 2005, 110: 1518.
10. Born, M.; Huang, K. *The Dynamical Theory of Crystal Lattices*, Oxford University Press, London, 1954.
11. Zhan, C.-G.; Zheng, F.; Dixon, D. A. *J. Am. Chem. Soc.*, 2002, 124: 14795.
12. Chaben, G.; Gordon, M. S.; Yarkony, D. R. *J. Phys. Chem. A*, 1997, 101: 7953.
13. Frisch, M. J.; Trucks, G. W.; Schlegel, H. B.; Scuseria, G. E.; Robb, M. A.; Cheeseman, J. R.; Montgomery, J.; et al. *Gaussian 03*, Gaussian, Inc., Pittsburgh, PA, 2003.
14. Shim, J.-H.; Lee, B.-J.; Cho, Y. W. *Surf. Sci.*, 2002, 512: 262.
15. Delogu, F. *Phys. Rev. B*, 2005, 72: 205418.
16. Gaudoin, R.; Foulkes, W. M. C. *Phys. Rev. B*, 2002, 66: 52104.
17. Hehre, W. J.; Radom, L.; Schleyer, P. v. R.; Pople, J. A. *Ab Initio Molecular Orbital Theory*, Wiley, New York, 1986.
18. Pople, J. A.*Rev. Mod. Phys.*, 1999, 71: 1267.
19. Kohn, W. *Rev. Mod. Phys.*, 1998, 71: 1253.
20. Kohn, W.; Becke, A. D.; Parr, R. G. *J. Phys. Chem.*, 1996, 100: 12974.
21. Lynch, B. J.; Truhlar, D. G. *J. Phys. Chem. A*, 2003, 107: 3898.
22. Schultz, N. E.; Staszewska, G.; Staszewski, P. *J. Phys. Chem. B*, 2004, 108: 4850.
23. Perdew, J. P.; Burke, K.; Ernzerhof, M. *Phys. Rev. Lett.*, 1996, 77: 3865.
24. Adamo, C.; Cossi, M.; Barone, V. *Theochem*, 1999, 493: 145.
25. Fast, P. L.; Sánchez, M. L.; Truhlar, D. G. *Chem. Phys. Lett.*, 1999, 306: 407.
26. Curtiss, L. A.; Redfern, P. C.; Raghavachari, K.; Rassolov, V.; Pople, J. A. *J. Chem. Phys.*, 1999, 110: 4703.
27. McLean, A. D.; Chandler, G. S. *J. Chem. Phys.*, 1980, 72: 5639.
28. Gordon, M. S.; Binkley, J. S.; Pople, J. A.; Pietro, W. J.; Hehre, W. J. *J. Am. Chem. Soc.*, 1982, 104: 2797.

29. Francl, M. M.; Pietro, W. J.; Hehre, W. J.; Gordon, M. S.; DeFrees, D. J.; Pople, J. A. *J. Chem. Phys.*, 1982, 77: 3654.
30. Spitznagel, G. W.; Clark, T.; Schleyer, P. v. R. *J. Comput. Chem.*, 1987, 8: 1109.
31. Frisch, M. J.; Pople, J. A.; Binkley, J. S. *J. Chem. Phys.*, 1984, 80: 3265.
32. Krauss, M.; Stevens, W. J. *Annu. Rev. Phys. Chem.*, 1984, 35: 357.
33. Schultz, N. E.; Truhlar, D. G. *J. Chem. Theory Comput.*, 2005, 1: 41.
34. Jaffe, J. E.; Kurtz, R. J.; Gutowski, M. *Comput. Mater. Sci.*, 2000, 18: 199.
35. Jasper, A. W.; Schultz, N. E.; Truhlar, D. G. *J. Phys. Chem. B*, 2005, 109: 3915.
36. Jasper, A. W.; Staszewski, P.; Staszewska, G.; Schultz, N. E.; Truhlar, D. G. *J. Phys. Chem. B*, 2004, 108: 8996.
37. Zope, R. R.; Mishin, Y. *Phys. Rev. B*, 2003, 68: 24102.
38. Daw, M. S.; Baskes, M. I. *Phys. Rev. B*, 1984, 29: 6443.
39. Sutton, A. P.; Chen, J. *Philos. Mag. Lett.*, 1990, 61: 139.
40. Gollisch, H. *Surf. Sci.*, 1986, 166: 87.
41. Ercolessi, F.; Adams, J. *Europhys. Lett.*, 1993: 26.
42. Mei, J.; Davenport, J. W. *Phys. Rev. B*, 1992, 46: 21.
43. Mei, D.; Davenport, J. W.; Fernando, G. W. *Phys. Rev. B*, 1991, 43: 4653.
44. Streitz, F. H.; Mintmire, J. W. *Phys. Rev. B*, 1994, 50: 11996.
45. Joswig, J. O.; Springborg, M. *Phys. Rev. B*, 2003, 68: 85408.
46. Chui, Y. H.; Chan, K. Y. *Phys. Chem. Chem. Phys.*, 2003, 13: 2869.
47. Rodríguez-López, J. L.; Montejano-Carrizales, J. M.; Pal, U.; Sánchez-Ramirez, J. F.; Troiani, H. E.; García, D.; Miki-Yoshida, M.; José-Yacaman, M. *Phys. Rev. Lett.*, 2004, 92: 196102.
48. Campbell, T. J.; Aral, G.; Ogata, S.; Kalia, R. K.; Nakano, A.; Vashishta, P. *Phys. Rev. B*, 2005, 71: 205413.
49. Vashishta, P.; Kalia, R. K.; Nakano, A. *J. Nano. Res.*, 2003, 5: 119.
50. Sun, D. Y.; Gong, X. G. *Phys. Rev. B*, 1998, 57: 4730.
51. Pathak, S.; Shenoy, V. B. *Phys. Rev. B*, 2005, 72: 113404.
52. Rydberg, R. *Z. Phys.*, 1931, 73: 376.
53. Murrell, J. N.; Carter, S.; Farantos, S. C.; Huxley, P.; Varandas, A. J. C. *Molecular Potential Energy Functions,* Wiley, New York, 1984.
54. Stillinger, F. H.; Weber, T. A. *Phys. Rev. B*, 1985, 31: 5262.
55. Ceperley, D. M.; Alder, B. J. *Phys. Rev. Lett.*, 1980, 45: 566.
56. Perdew, J. P.; Zunger, A. *Phys. Rev. B*, 1981, 23: 5048.
57. Schultz, N. E.; Zhao, Y.; Truhlar, D. G. *J. Phys. Chem. A*, 2005, 109: 4388.
58. Pettersson, L. G. M.; Bauschlicher, C. W.; Halicioglu, T. *J. Chem. Phys.*, 1987, 87: 2205.
59. Geske, G.; Boldyrev, A. I. *J. Chem. Phys.*, 2000, 113: 5130.
60. Zope, R. R.; Baruah, T. *Phys. Rev. A*, 2001, 64: 53202.
61. Metropolis, N.; Rosenbluth, A. W.; Rosenbluth, M. N.; Teller, A. H.; Teller, E. *J. Chem. Phys.*, 1953, 21: 1087.
62. Xiang, Y.; Cheng, L.; Cai, W.; Shao, X. *J. Phys. Chem. A*, 2004, 108: 9516.

63. Wales, D. J.; Doye, J. P. K.; Dullweber, A.; Hodges, M. P.; Naumkin, F. Y.; Calvo, F.; Hernández-Rojas, J.; Middleton, T. F. *The Cambridge Cluster Database*, http://www-wales.ch.cam.ac.uk/CCD.html.
64. Silla, E.; Tuñón, I.; Pascual-Ahuir, J. L. *J. Comput. Chem.*, 1991, 12: 1077.
65. Čížek, J. *Adv. Chem. Phys.*, 1969, 14: 35.
66. Purvis, G. D.; Bartlett, R. J. *J. Chem. Phys.*, 1982, 76: 1910.
67. Raghavachari, K.; Trucks, G. W.; Pople, J. A.; Head-Gordon, M. *Chem. Phys. Lett.*, 1996, 157: 479.
68. Woon, D. E.; Dunning, T. H. J. *J. Chem. Phys.*, 1993, 98: 1358.
69. Sarou-Kanian, V.; Millot, F.; Rifflet, J. C. *Int. J. Thermo.*, 2003, 24: 277.
70. Connolly, M. L. *J. Am. Chem. Soc.*, 1985, 107: 1118.

11

LARGE-SCALE MONTE CARLO SIMULATIONS FOR AGGREGATION, SELF-ASSEMBLY, AND PHASE EQUILIBRIA

JAKE L. RAFFERTY, LING ZHANG, NIKOLAJ D. ZHURAVLEV, KELLY E. ANDERSON, BECKY L. EGGIMANN, MATTHEW J. MCGRATH, AND J. ILJA SIEPMANN

Departments of Chemistry and of Chemical Engineering and Material Science, University of Minnesota, Minneapolis, Minnesota

Over more than 50 years, particle-based computer simulations have evolved from a numerical tool, used mainly to validate theoretical predictions for simple model problems, to become "computer experiments" using sophisticated interaction potentials that allow the simulator to predict macroscopic properties and to provide microscopic understanding for "real" systems.[1,2] The trajectory of the "in silico" system through its statistical mechanical phase space can be explored via deterministic molecular dynamics algorithms or via stochastic Monte Carlo processes. This review article provides five examples of large-scale Monte Carlo simulations where being termed "large-scale" is due either to the complexity of the system (number of components and phases) or the complexity of the method used to compute the interactomic potentials.

1. STRUCTURE AND RETENTION IN REVERSED-PHASE LIQUID CHROMATOGRAPHY

Reversed-phase liquid chromatography (RPLC) is among the most widely used and versatile analytical techniques.[3] Due to this importance, there is a need

Multiscale Simulation Methods for Nanomaterials, Edited by Richard B. Ross and Sanat Mohanty

to understand the RPLC retention process at the molecular level, but despite numerous experimental and theoretical studies, there is no general agreement on whether analyte retention can be described by a partition or by an adsorption process.[4–6] By carrying out atom-based Monte Carlo simulations of a model RPLC system, we are able to obtain a much more detailed view of the separation process than what is accessible through experiment or theory.

In this study we made use of efficient configurational-bias Monte Carlo[7,8] simulations in the isobaric–isothermal Gibbs ensemble[9] to study the retention of *n*-butane and 1-propanol solutes in a model RPLC system. To this extent, three simulation boxes were used (see Figure 1): (1) an elongated stationary-phase box that contained an explicit silica substrate with its surface parallel to the $x-y$ plane on which dimethyl octadecyl silanols are grafted at a coverage of 2.9 μmol/m^2 (the residual silanol groups on the surface are unprotected), and the remainder of this box is filled by the mobile phase; (2) a separate mobile-phase box that contained the bulk hydroorganic solvent mixture (67% mole fraction methanol); and (3) a helium vapor phase used as transfer medium and reference state. The

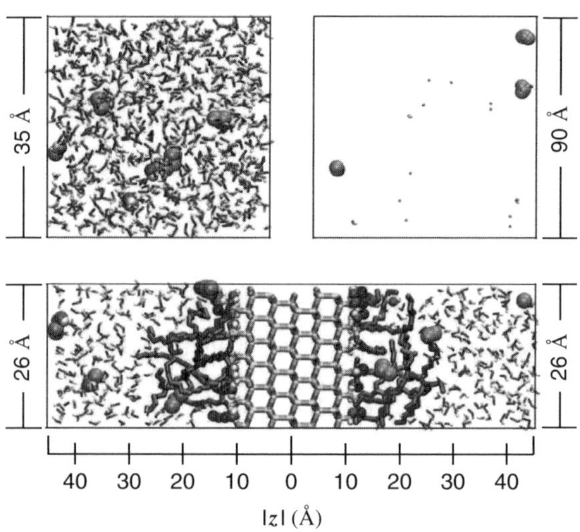

Figure 1 The three boxes present in the Gibbs ensemble simulation of RPLC. The cubic vapor and bulk solvent boxes are shown in the upper right and upper left corners, respectively, with their approximate box lengths indicated. The box containing the stationary phase is shown on the bottom in its $z-y$ plane. This box has fixed edge lengths of $x = 20$ Å, $y = 26$ Å and $z = 90$ Å. Solute molecules are shown as large spheres, with oxygen in red, hydrogen in white, and methyl in groups in cyan. Solvent molecules are shown in the stick representation, with oxygen in red, hydrogen in white, and methyl in groups in blue. Those solvent molecules that are involved in hydrogen bonding with the silica substrate are shown as medium spheres for emphasis. The silicon and oxygen atoms in the substrate are shown as yellow and orange tubes, respectively, while the grafted alkyl chains are shown as dark gray tubes. Helium atoms in the vapor phase are shown as small green spheres. (*See insert for color representation of figure.*)

TraPPE (transferable potentials for phase equilibria) force field[10–12] is used to represent the silanols, analyte molecules, and methanol, while water is represented by the TIP4P model.[13] The simulations were carried out a temperature of 323 K and a pressure of 1 atm.

Selected density profiles obtained for the stationary-phase box are presented in Figure 2. These density profiles indicate that the width of the bonded-phase/mobile-phase interface is about 10 Å and that some methanol solvent molecules are present in the bonded-phase region.[14] One can also observe a weak density depletion (partial dewetting) near the Gibbs dividing surface.[14] The analyte distribution coefficients (see Figure 2) show a striking dependence on z, the distance away from the center of the substrate. Whereas n-butane shows a strong bimodal distribution with one peak in the center of the bonded phase ($|z| \approx 16$ Å) and a second peak in the interfacial region ($|z| \approx 21$ Å) just below the Gibbs dividing surface, 1-propanol exhibits only a weak preference to reside in the interfacial region. Thus, our simulations demonstrate that the bonded phase does not participate in the retention process as a homogeneous phase, but that multiple analyte-specific "sorption regions" control retention.[15]

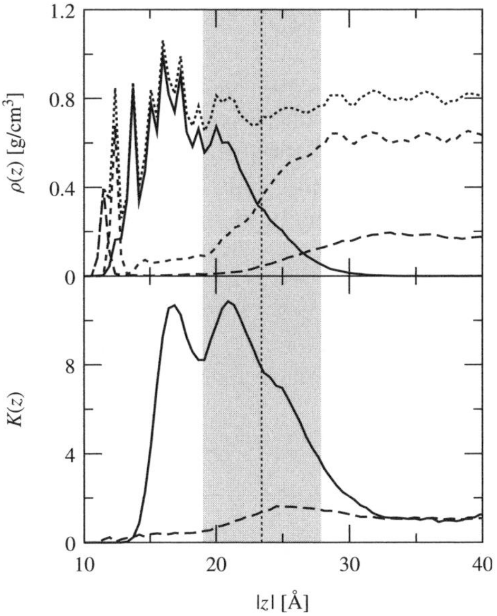

Figure 2 Density profiles (top) and solute distribution coefficients (bottom) in a model RPLC system. The C_{18}, water, and methanol densities are shown as solid, long-dashed, and short-dashed lines, respectively. The total system density, computed as the sum of bonded phase and solvent densities, is depicted as a dotted line. The z-dependent distribution coefficients for n-butane and 1-propanol are represented by solid and dashed lines, respectively. The interfacial region, defined by the position at which the total solvent density is at 10% of its bulk value to where it is 90%, is shaded in gray. The Gibbs dividing surface fitted to the total solvent density is shown by the dotted vertical line.

2. SOLUBILITY OF HELIUM IN n-HEXADECANE

Helium is the rare gas atom with the weakest dispersive interactions, and hence a helium vapor phase is often used for high-precision studies of vapor–liquid partitioning (to obtain Gibbs free energies of solvation) and as mobile phase in gas–liquid chromatography. Thus, an accurate force field for helium is needed for these applications. The parameters for the TraPPE force field are usually derived by fitting to vapor–liquid coexistence densities and, in particular, the critical point. Due to its very weak interactions, helium's critical temperature is only 5.2 K,[16] a temperature where nuclear quantum effects are very important. This poses the question whether one should include or ignore these quantum effects when deriving interaction parameters for helium. When quantum effects are included, one obtains a Lennard-Jones well depth of about 10 K (in thermal energy units) and a diameter of about 3.0 Å,[17] whereas adjusting the parameters to reproduce the classical critical point of Lennard-Jonesium yields a much smaller well depth of 4 K and a slightly larger diameter of 3.11 Å. To test which parameter set is more appropriate at elevated temperatures and pressures (encountered often in gas-liquid chromatography), we have carried out configurational-bias Monte Carlo simulations[7,8] in the constant-pressure Gibbs ensemble[9] using classical statistics to determine the solubility of helium in n-hexadecane. The simulated systems consisted of 64 n-hexadecane molecules and 96 helium atoms, which were both allowed to swap between the liquid and vapor phases.

Figure 3 shows the helium solubilities calculated using the Lennard-Jones parameters obtained without accounting for nuclear quantum effects. A comparison with the experimental data of Lin et al.[18] demonstrates that this parameter set can quantitatively reproduce the temperature and pressure

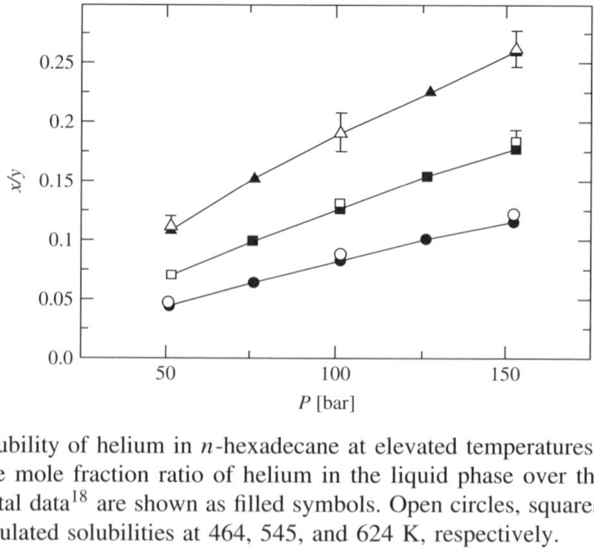

Figure 3 Solubility of helium in n-hexadecane at elevated temperatures and pressures. x/y denotes the mole fraction ratio of helium in the liquid phase over the vapor phase. The experimental data[18] are shown as filled symbols. Open circles, squares, and triangles denote the calculated solubilities at 464, 545, and 624 K, respectively.

dependence of the helium solubility. In contrast, the Lennard-Jones parameter set of de Boer and Michels[19] ($\varepsilon/k_{B)} = 10.2$ K and $\sigma = 2.56$ Å) yields much higher solubilities when classical statistics are used at these elevated temperatures (path integral simulations were not undertaken here to account for nuclear quantum effects because of the much greater computational expense). Thus, we can conclude that an effective helium force field for use at $T \gg T_c$ can be obtained by ignoring quantum effects when determining Lennard-Jones parameters from helium's critical point.

3. STRUCTURE AND SOLUBILITY IN SUPERCRITICAL CARBON DIOXIDE

Entrainers are low-molecular-weight compounds that are often added in small concentrations to supercritical carbon dioxide with the purpose of enhancing the solubility and selectivity for the solvation of low-volatility solutes in the supercritical phase.[20] Dobbs and co-workers[21,22] have studied extensively the effects of polar and nonpolar entrainers on the solvation of various solutes. The enhancement of polar solute solubility due to the presence of a polar and hydrogen-bonding entrainer was attributed to acid–base interactions, while the effects of the nonpolar entrainer were attributed to dispersion interactions. To provide molecular-level insight on these solvation processes, we performed constant-pressure Gibbs ensemble Monte Carlo simulations[9] for a polar and a nonpolar solute, benzoic acid and hexamethylbenzene, respectively.[23] Methanol and *n*-octane were chosen as the polar and nonpolar entrainers. Instead of explicitly modeling the crystalline phase of the solute molecules, these are placed in a vapor phase that is subject to an external pressure that equals the solute's sublimation pressure, including a Poynting correction.[23] The pressure for the supercritical solvent phase was set at 12 MPa and carbon dioxide and entrainer molecules were not allow to swap into the vapor phase. The simulations were carried out at 310 K for systems consisting of 2000 solvent molecules (1930 CO_2 and 70 entrainer molecules) and 60 to 100 solute molecules that are allowed to swap between the two phases. The TraPPE force field is used for carbon dioxide,[24] *n*-octane,[11] methanol,[12] hexamethylbenzene,[25] and benzoic acid.[25,26]

The simulation results qualitatively follow the experimental solubility data.[21,22] The presence of a nonpolar entrainer leads to a slight enhancement of the solubility of the nonpolar solute with an equal effect on the solubility of the polar solute. This was determined from the ratio of the mole fraction of solute in the carbon dioxide–entrainer phase relative to the pure carbon dioxide phase. For the nonpolar solute this enhancement is 1.3, whereas for the polar solute it is 1.2. The polar entrainer greatly enhances the solubility of the polar solute (2.3) while having little effect on the nonpolar solute (0.9).

The *n*-octane/hexamethylbenzene and methanol/benzoic acid systems are shown in Figure 4, and the four solute–entrainer number integrals (i.e., the average number of entrainer molecules that are found within a given center-of-mass separation around a solute) are given in Figure 5. Examining these figures and

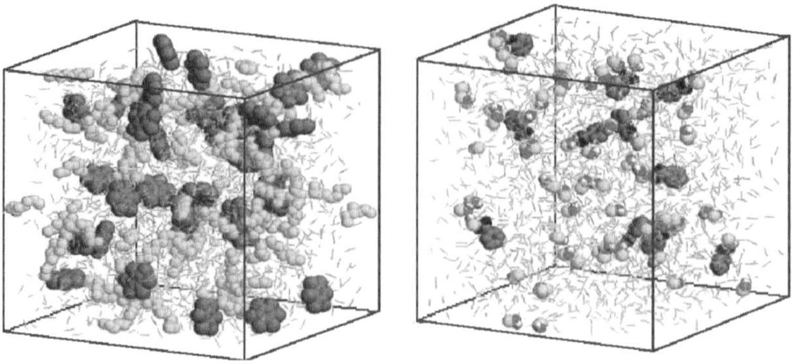

Figure 4 Supercritical-CO_2/entrainer/solute systems. A stick representation is used to show the CO_2 molecules in both cases. The configuration on the left shows an n-octane/hexamethylbenzene system, with the pseudoatoms in n-octane shown as orange spheres and those in hexamethylbenzene shown as light blue spheres. Shown are 35 solute molecules in a box with side length 58.05 Å; the average number of solutes in the supercritical phase for this system is 34 ± 2. The configuration to the right shows a methanol/benzoic acid system. The coloring scheme for the spheres is as follows: methanol methyl pseudoatoms (orange), methanol oxygen atoms (green), methanol and carboxyl hydrogen atoms (white), carboxyl oxygen atoms (red), and benzene and carboxyl carbon pseudo atoms (light blue). Shown are 11 benzoic acid molecules in a box of side length 57.15 Å; the average number of solutes in the supercritical phase for this system is 9 ± 2. The red lines depict the edges of the periodic cell. (*See insert for color representation of figure.*)

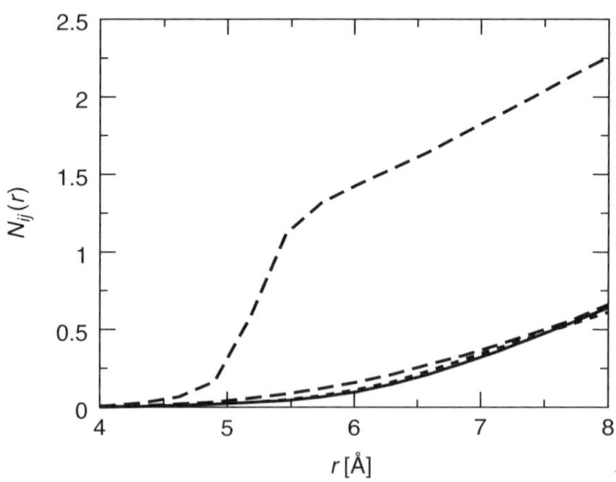

Figure 5 Number integrals showing the average number of entrainer molecules around a given solute molecule in supercritical carbon dioxide. The entrainer concentration is about 3.5 mol %. The solid and short-, long-, and medium-dashed lines show the number integrals for hexamethylbenzene/n-octane, hexamethylbenzene/methanol, benzoic acid/n-octane, and benzoic acid/methanol, respectively.

number integrals, it is evident that the methanol entrainer molecules preferentially aggregate around the polar benzoic acid molecules with about one entrainer per solute. None of the other systems exhibit this behavior. This entrainer aggregation creates a more favorable environment for the solute within the supercritical phase leading to the enhanced solubility in the presence of the entrainer. The nonpolar entrainer–nonpolar solute system shows a smaller enhancement of the solubility of the solutes, but without any specific aggregation for the hexamethylbenzene/*n*-octane and benzoic acid/*n*-octane pairs.

4. INTERFACIAL PROPERTIES OF AN AQUEOUS SOLUTION CONTAINING IONS WITH A RANGE OF SIZES

Driven by the desire for a better understanding of the reactivity of atmospheric aerosols, the aqueous liquid–vapor interface has become the topic of many recent investigations. Of particular interest is the surface composition of aqueous salt solutions. That common inorganic ions raise the surface tension of water has been known for nearly a century.[27] The traditional explanation that this phenomenon is caused by a surface depletion of these ions[28] is based on the Gibbs adsorption isotherm, which correlates an increase in surface tension with a negative surface excess of solute ions. While this surface excess is inherently a macroscopic thermodynamic property, it is often used to infer the microscopic properties of the solute and solvent molecules. Recently, however, both experimental and simulation studies capable of atomic resolution have suggested that some halide anions may show enrichment at the interface.[29–31] Although multiple factors are believed to contribute to the propensity of larger halide anions for the water surface, the polarizability of water and the anions has received the main attention.[32]

We have carried out Gibbs ensemble Monte Carlo simulations[9] to explore whether ion size can lead to surface enrichment of larger anions in a solution containing anions of various sizes. To separate the size effect from the polarizability, we used fixed-charge models for both water and monovalent ions.[33] The solutions contain nine different types of anions, with the size being increased by 20% in volume from one anion type to the next, but the same Lennard-Jones well depth is used for all anions. The size of the smallest anion type corresponds approximately to F^- and the largest to I^-. A single type of cation is used with a size corresponding to Na^+. Water is modeled with the rigid four-site TIP4P model.[13] The simulated liquid slab (with two liquid–vapor interfaces) consists of 1528 water molecules and a total of 72 ions. Sampling of the spatial distribution of the anions is greatly increased through special Monte Carlo moves that perform identity switches involving two ions of similar size.[33]

In Figure 6, the number density profiles near the liquid–vapor interface are compared for the largest and smallest anions and the cation. It is clear that the largest anion is found preferentially near the interface, while the smallest anion is, by comparison, depleted near the interface. Thus, anion size alone can result

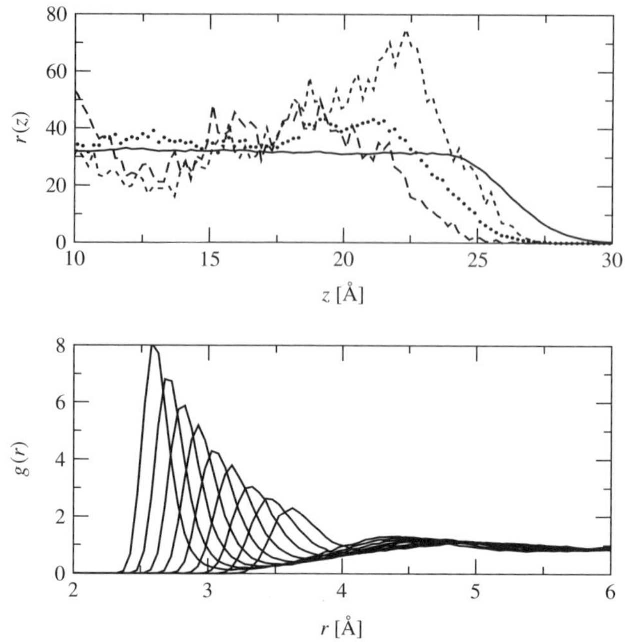

Figure 6 Number density profiles (top) and hydrogen–anion radial distribution functions computed for the interior region of the slab (bottom) in an aqueous solution containing nine types of anions that differ in size. The number densities for water, the smallest and largest anion, and the counterion are depicted as solid, long-dashed, short-dashed, and dotted lines, respectively. The radial distribution functions for the nine anion types follow a regular pattern, with the smallest anion possessing the highest peak at the shortest separation.

in large interfacial concentration imbalances. The anion–water hydrogen radial distribution functions averaged only for the interior region of the liquid slab (see Figure 6) show a 40% increase in hydrogen-bond length with anion size, whereas the solvation number only increases from six to seven water molecules (i.e., the larger anions are less well solvated than the smaller anions and hence are driven to the surface).

5. FIRST-PRINCIPLES MONTE CARLO SIMULATIONS OF THE VAPOR–LIQUID COEXISTENCE CURVE OF WATER

The prevalence of water on Earth, its important role in all biological systems and many technological processes, and its many unique properties (e.g., extensive polymorphism and density maximum for the liquid phase at atmospheric pressure) have led to numerous attempts to find models that while computationally tractable, provide an accurate representation of water's inter- and intramolecular interactions. With the goal in mind of finding a model that can describe water's physical and chemical properties in a consistent way, we have recently embarked on first-principles (employing Kohn–Sham density functional theory, KS-DFT[34]) Monte Carlo simulations of the vapor–liquid phase equilibria of water.[35] These simulations show that the Becke–Lee–Yang–Parr (BLYP) exchange/correlation functional combination,[36,37] together with the norm-conserving pseudopotentials of Goedecker and co-workers (GTH)[38,39] and a triple-ζ basis set with two sets of p- or d-type polarization functions, yields saturated liquid densities and a critical temperature that fall significantly below the experimental data. To provide a more complete view of the accuracy of density functional theory for predicting phase equilibria, additional Gibbs ensemble simulations were carried out to explore the influence of the density functional [by performing simulations for the Perdew–Burke–Ernzerhof (PBE) representation[40] with the TZV2P basis set] and of the basis set (by performing simulations for the BLYP functional using a smaller basis set, double-ζ with a single set of polarization functions).[41] All simulations were carried out for systems containing a total of 64 molecules, which is a system size shown previously to be sufficient not to show significant finite-size effects.[35]

Figure 7 shows that the PBE-GTH-TZV2P representation for water yields saturated liquid densities that are in much better agreement with experiment than found for BLYP-GTH-TZV2P, but in turn the saturated vapor densities are significantly too low and hence its critical temperature is overestimated.[41] Similarly, the use of a smaller basis set for the BLYP functional results in increased saturated liquid density and heat of vaporization and decreased saturated vapor

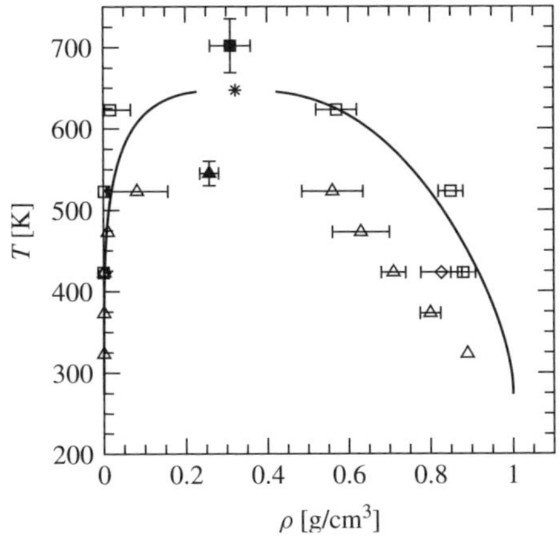

Figure 7 Vapor–liquid coexistence curve of water. The solid lines and asterisk depict the experimental saturated vapor and liquid densities and the critical point. Triangles, squares, and diamonds show the data calculated for the BLYP-GTH-TZV2P, PBE-GTH-TZV2P, and BLYP-GTH-DZVP representations of water, respectively. Open and filled symbols are used for the coexistence densities and critical points, respectively.

density.[41] In contrast, negligible basis set effects were found in previous simulations for liquid water in the microcanonical ensemble.[42] Thus, the sensitivity of vapor–liquid coexistence curves allows for a better evaluation of the accuracy of a given model, and future research will concentrate on finding an improved KS-DFT representation for water.

6. CONCLUSIONS

The five examples presented in this short review illustrate how large-scale Monte Carlo simulations can be used to provide molecular-level insight for multicomponent and/or multiphase systems and to assess the accuracy of empirical potentials and first-principles representations for the model system. These Monte Carlo simulations benefit from the use of open ensembles with fluctuating particles numbers and special Monte Carlo moves to sample spatial distributions where mass transport is slow on the time scale of molecular dynamics simulations.

Acknowledgments

Financial support from the National Science Foundation and through Graduate School Fellowships (K.E.A., B.L.E., M.J.M., and J.L.R.), 3M Foundation Graduate Fellowships (K.E.A. and M.J.M.), and a DOE Computational Science Graduate Fellowship (M.J.M.) is gratefully acknowledged. Part of the computer resources were provided by the Minnesota Supercomputing Institute and Lawrence Livermore National Laboratory.

REFERENCES

1. Allen, M. P.; Tildesley, D. J. *Computer Simulation of Liquids*, Oxford University Press, Oxford, 1987.
2. Frenkel, D.; Smit, B. *Understanding Molecular Simulation*, 2nd ed., Academic Press, San Diego, CA, 2002.
3. Braithwate, A.; Smith, F. J. *Chromatographic Methods*, 5th ed., Blackie Academic, London, 1996.
4. Vailaya, A.; Horvath, C. *J. Chromatogr. A*, 1998, 829: 1.
5. Ratunga, R. P. J; Carr, P. W. *Anal. Chem.*, 2000, 72: 5679.
6. Devido, D. R., Dorsey, J. G.; Chan, H. S.; Dill, K. A. *J. Phys. Chem. B*, 1998, 102: 7272.
7. Siepmann, J. I.; Frenkel, D. *Mol. Phys.*, 1992, 75: 59.
8. Martin, M. G.; Siepmann, J. I. *J. Phys. Chem. B*, 1999, 103: 4508.
9. Panagiotopoulos, A. Z; Quirke, N.; Stapleton, M.; Tildesley, D. *J. Mol. Phys.*, 1988, 63: 527.
10. See http://www.chem.umn.edu/groups/siepmann/trappe/intro.php for information on the TraPPE force field.

11. Martin, M. G.; Siepmann, J. I. *J. Phys. Chem. B*, 1998, 102: 2569.
12. Chen, B.; Potoff, J. J.; Siepmann, J. I. *J. Phys. Chem. B*, 2001, 105: 3093.
13. Jorgensen, W. L.; Chandrasekhar, J.; Madura, J. D.; Impey, R. W.; Klein, M. L. *J. Chem. Phys.*, 1983, 79: 926.
14. Zhang, L.; Rafferty, J. L.; Chen, B.; Siepmann, J. I.; Schure, M. R. *J. Chromatogr. A*, 2006, 1126: 219.
15. Rafferty, J. L.; Zhang, L.; Siepmann, J. I.; Schure, M. R. *Anal. Chem.*, 2007, 79: 6551.
16. Lemmon, E. W.; McLinden M. O; Friend, D. G. NIST *Chemistry WebBook: NIST Standard Reference Database 69*, National Institute of Standards and Technology, Gaithersburg, MD, 2005; http://webbook.nist.gov.
17. Aziz, R. A; Nain, V. P. S; Carley, J. S.; Taylor, W. L.; McConville, G. T. *J. Chem. Phys.*, 1979, 70: 4330.
18. Lin, H. M.; Lee, R. J.; Lee, M. J. *Fluid Phase Equil.*, 1995, 111:89.
19. deBoer, J.; Michels, A. *Physica*, 1938, 5: 945.
20. McHugh, M. A.; Krukonis, V. J. *Supercritical Fluid Extraction: Principles and Practice*, 2nd ed., Butterworth, Boston, 1994.
21. Dobbs, J. M.; Wong, J. M.; Johnston, K. P. *J. Chem. Eng. Data*, 1986, 31: 303.
22. Dobbs, J. M.; Wong, J. M.; Lahiere, R. J.; Johnston, K. P. *Ind. Eng. Chem. Res.*, 1987, 26: 56.
23. Anderson, K. E.; Siepmann, J. I. Submitted for publication.
24. Potoff, J. J.; Siepmann, J. I. *AIChE J.*, 2001, 47: 1676.
25. Wick, C. D.; Martin, M. G.; Siepmann, J. I. *J. Phys. Chem. B*, 2000, 104: 8008.
26. Kamath, G.; Cao, F.; Potoff, J. J. *J. Phys. Chem. B*, 2004, 108: 14130.
27. Heydweiler, A. *Ann. Phys.*, 1910, 33: 145.
28. Onsager, L.; Samaras, N. N. T. *J. Chem. Phys.*, 1934, 2: 528.
29. Knipping, E. M.; Lakin, M. J.; Foster, K. L.; Jungwirth, P.; Tobias, D. J.; Gerber, R. B., Dabdub, D.; Finlayson-Pitts, B. J. *Science*, 2000, 288: 301.
30. Jungwirth, P.; Tobias, D. J. *J. Phys. Chem. B*, 2001, 105: 10468.
31. Jungwirth, P.; Finlayson-Pitts, B. J.; Tobias, D. J. *Chem. Rev.*, 2006, 106: 1137.
32. Petersen, P. B.; Saykally, R. J. *Annu. Rev. Phys. Chem.*, 2006, 57: 333.
33. Eggimann, B. L.; Siepmann, J. I. *J. Phys. Chem. C*, in press.
34. Kohn, W.; Sham, L. J. *Phys. Rev.*, 1965, 140: A1133.
35. McGrath, M. J.; Siepmann, J. I.; Kuo, I. W.; Mundy, C.J.; Van de Vondele, J.; Hutter, J.; Mohamed, F.; Krack, M. *J. Phys. Chem. A*, 2006, 110: 640.
36. Becke, A. D. *Phys. Rev. A*, 1988, 38: 3098.
37. Lee, C.; Yang, W.; Parr, R. G. *Phys. Rev. B*, 1988, 37: 785.
38. Goedecker, S.; Teter, M.; Hutter, *J. Phys. Rev. B*, 1996, 54: 1703.
39. Hartwigsen, C.; Goedecker, S.; Hutter, *J. Phys. Rev. B*, 1998, 58: 3641.
40. Perdew, J. P.; Burke, K.; Ernzerhof, M. *Phys. Rev. Lett.*, 1996, 77: 3865.
41. McGrath, M. J.; Siepmann, J. I.; Kuo, I. W.; Mundy, C. J. *Mol. Phys.*, 2006, 104: 3619.
42. Van deVondele, J.; Mohamed, F.; Krack, M.; Hutter, J.; Sprik, M.; Parrinello, M. J. *Chem. Phys.*, 2005, 122: 014515.

12

NEW QM/MM MODELS FOR MULTISCALE SIMULATION OF PHOSPHORYL TRANSFER REACTIONS IN SOLUTION

KWANGHO NAM, JIALI GAO, AND DARRIN M. YORK

Department of Chemistry, Supercomputing Institute, and Digital Technology Center, University of Minnesota, Minneapolis, Minnesota

The study of chemical reactions in complex condensed phase environments presents considerable challenges to simulation. The chemical nature of bond rearrangement requires an accurate description of the electronic structure, whereas the reactive event itself may involve the coordinated participation of many thousands of atoms and molecules in the environment and require very long time scales to observe. From this perspective, biocatalysis is inherently a problem that requires design of new methods for multiscale simulation: highly accurate quantum methods for the reaction under study that are sufficiently fast to allow long-time simulation in order to adequately sample the relevant phase space of rare events.

Of particular relevance to the present work is the study of phosphoryl transfer reactions. In cellular systems, the transfer of phosphoryl groups and hydrolysis of phosphate chemical bonds plays a central role in signaling pathways,[1] in the storage and interconversion of energy in metabolic pathways,[2,3] and in processes of transcription and translation,[4,5] RNA synthesis and degradation,[6,7] and ribozyme catalysis.[8–10]

Multiscale Simulation Methods for Nanomaterials, Edited by Richard B. Ross and Sanat Mohanty

From a quantum mechanical viewpoint, phosphoryl transfer reactions involve the making and breaking of bonds with phosphorus, and frequently require changes in the valence state of phosphorus along the reaction path. For example, a dissociative reaction proceeds from a tetravalent phosphate reactant to a trivalent metaphosphate intermediate before it is hydrolyzed to form the final phosphate product. An associative mechanism, on the other hand, typically proceeds through a pentavalent phosphorane intermediate and/or transition state. These changes in valency of phosphorus require an accurate d-orbital representation to capture proper quantum mechanical behavior.[11,12]

From a simulation perspective, phosphoryl transfer reactions are typically highly anionic reactions that interact strongly with solvent and metal ions. Reactions often involve the association (or dissociation) of anionic species. In the gas phase, these interactions are highly energetic, particularly for dianionic (or even more negatively charged) species. The solvation effect, to within a linear-response approximation, varies as the square of the ionic charge, and hence preferentially stabilizes the dianionic species relative to the separated anionic species. Hence, the balance of solvation effects and ionic interactions are critical in the determination of chemical reactivity. Rigorous treatment of long-range electrostatic interactions is thus critical in simulations of phosphoryl transfer reactions in solution.[13,14]

In the present work, we discuss recent advances in the combined quantum mechanical/molecular mechanical (QM/MM) modeling of phosphoryl transfer reactions in solution. Very recently, a new semiempirical quantum model been introduced that accurately models phosphoryl transfer reactions of a variety of biological phosphates.[15] The model, designated AM1/d-PhoT, has been parameterized to reproduce high-level density-functional results from a recently constructed database of quantum calculations for RNA catalysis.[16,17] The AM1/d-PhoT model has been integrated with a new linear-scaling Ewald method[18] to calculate long-range electrostatic interactions efficiently in combined QM/MM simulations and implemented into the CHARMM molecular simulation package. Together, these methods allow accurate models of phosphoryl transfer reactions in complex condensed phase environments. In the present work, these methods are applied together to the study of the relative hydrolysis rates of cyclic and acyclic phosphates for which experimental values are available: dimethyl phosphate, ethylene phosphate, and trimethylene phosphate. The results illustrate the robustness of the new methods and establish an important set of benchmarks for phosphate hydrolysis reactions in solution.

1. METHODS

The semiempirical AM1, MNDO/d, and AM1/d-PhoT models used in the present work have been discussed in detail elsewhere[15] and are only summarized here. For more extensive reviews of semiempirical methods in general, see refs. 19–22. The QM/MM-Ewald method has been presented elsewhere.[18]

Semiempirical AM1, MNDO/d, and AM1/d-PhoT Models

The AM1 and MNDO/d semiempirical models discussed here are all based on the neglect of diatomic differential overlap (NDDO) approximation, as is the new AM1/d-PhoT model.[15] The MNDO/d and AM1 Hamiltonians differ in the way that core–core repulsion interactions are treated and in the inclusion of d-orbitals on third-row atoms in the MNDO/d method. In the MNDO/d method, the repulsion between two nuclear cores (A and B) is determined as

$$E_N^{\mathrm{MNDO}}(A,B) = Z_A Z_B \langle s_A s_B | s_A s_B \rangle (1 + e^{-\alpha_A R_{AB}} + e^{-\alpha_B R_{AB}}) \quad (1)$$

where Z_A and Z_B are the effective core charges, $\langle s_A s_B | s_A s_B \rangle$ is a Coulomb repulsion integral between two s-orbitals centered on atoms A and B, and α_A and α_B are parameters that decrease screening of the nuclear charge by the electrons at small interatomic distances. For the O–H and N–H bonds, a slightly different screening form is used, a detailed description of which is provided in the original paper[23] and elsewhere.[24]

In an AM1 Hamiltonian, the core–core repulsion includes an additional set of Gaussian terms that take the form

$$E_N(A, B) = E_N^{\mathrm{MNDO}}(A, B) + \frac{Z_A Z_B}{R_{AB}} \left[\sum_k a_{Ak} e^{-b_{Ak}(R_{AB} - c_{Ak})^2} + \sum_k a_{Bk} e^{-b_{Bk}(R_{AB} - c_{Bk})^2} \right] \quad (2)$$

The addition of Gaussian core–core terms leads to significant improvements in hydrogen bonding[25] where the MNDO/d method is known to be highly underbound due to the excessive repulsion just outside bonding distances.

The AM1/d-PhoT model combines the d-orbital description of MNDO/d with the Gaussian core–core terms of AM1.[15] In this way a balanced model may be achieved that accurately models reactive intermediates in transphosphorylation with, at the same time, improved treatment of hydrogen bonding. The Gaussian core–core terms in the present AM1/d-PhoT model are given by

$$E_N(A, B) = E_N^{\mathrm{MNDO}}(A, B) + \frac{Z_A Z_B}{R_{AB}} G_{\mathrm{scale}}^A G_{\mathrm{scale}}^B \times \left[\sum_k a_{Ak} e^{-b_{Ak}(R_{AB} - c_{Ak})^2} + \sum_k a_{Bk} e^{-b_{Bk}(R_{AB} - c_{Bk})^2} \right] \quad (3)$$

where G_{scale}^A and G_{scale}^B are scaling parameters for atoms A and B, and in the present work vary from zero to 1 (values of zero recover the conventional MNDO/d core–core model, whereas values of 1 recover the AM1 core–core model). The G_{scale} scaling parameters provide the flexibility to attenuate (or even shut off) Gaussian core–core interactions between certain atoms and offer a simple mechanism for interconverting between AM1-like and MNDO/d-like models.

The G_{scale} terms are constants (i.e., they are set for each atom and do not depend on the molecular environment). Proper adjustment of the G_{scale} parameter for P, along with the conventional AM1/d atomic parameters, was needed to obtain a robust, transferable AM1/d-PhoT model to achieve high accuracy for both ground- and transition-state energies and geometries in biological phosphoryl transfer reactions.[15] Implementation of the AM1/d-PhoT model poses no specific problems relative to AM1 and MNDO/d that might limit its use in QM/MM modeling, and the present work represents an encouraging first validation step.

AM1/d-PhoT Parameter Optimization

Although the main focus of the present work is to validate the AM1/d-PhoT model developed in the solution phase, in this section we describe the AM1/d-PhoT parametrization procedure briefly; the detailed procedure has been presented elsewhere.[15] It is the hope that the AM1/d-PhoT model developed affords greater accuracy and transferability to reactions involving biological phosphorus compounds in the solution, ribozyme, and enzyme environment. In the development of a model that reproduce the structures, energetics, and other properties for the phosphoryl transfer reactions accurately, the parameterization has been carried out based on a high-level density functional QCRNA database.[16,17]

The optimized AM1/d-PhoT parameters are determined by minimizing a χ^2 merit function that is a sum of weighted square errors of properties predicted by the trial parameter set, λ, against the target data. In the nonlinear optimization of the χ^2 function, a quadratically convergent direction set optimization method[26,27] with narrow parameter bands to avoid large changes in atomic parameters from their starting values and a genetic algorithm[27] have been employed. The form of χ^2 merit function used is given by

$$\chi^2(\lambda) = \sum_{i}^{\text{mol}} \sum_{\alpha}^{\text{prop}(i)} w_{i\alpha} \left[Y_{i\alpha}^{\text{AM1/d}}(\lambda) - Y_{i\alpha}^{\text{Ref}} \right]^2 \tag{4}$$

where the first summation with index i runs over molecules, complexes, or reactions ("mol"), and the second summation with index α runs over properties associated with the ith "molecule" ["prop(i)"], $Y_{i\alpha}^{\text{AM1/d}}(\lambda)$ and $Y_{i\alpha}^{\text{Ref}}$ are the values calculated for property α for molecule i and the corresponding reference value from experiment or from density-functional calculations, respectively, and the term $w_{i\alpha}$ is the associated weight, defined as the inverse square $\sigma_{i\alpha}$ values

$$w_{i\alpha} = (\sigma_{i\alpha}^2)^{-1} \tag{5}$$

where the $\sigma_{i\alpha}$ values have the same units as the molecular property $Y_{i\alpha}$ and specify the significance or importance of this particular property. The properties considered in the parameterization include heats of formation, gas-phase proton affinities, dipole moments, geometries of molecules and molecular complexes, intermolecular interaction energies, and relative reaction energies and

TABLE 1. Optimized AM1/d-PhoT Parameters for Hydrogen, Oxygen, and Phosphorus Atoms

Parameter	H	O	P
U_{ss} (eV)	−10.934610	−96.760676	−46.250810
U_{pp} (eV)	—	−78.776203	−40.712918
ζ_s (au)	1.143846	3.057965	1.909168
ζ_p (au)	—	2.515332	2.008466
β_s (eV)	−5.911108	−29.472306	−11.194791
β_p (eV)	—	−28.515785	−11.985621
α (Å^{-1})	2.884915	4.404417	1.883237
G_{ss} (eV)	13.737453	14.234714	14.645747
G_{pp} (eV)	—	14.454530	11.694918
G_{sp} (eV)	—	14.539451	5.689654
G_{p^2} (eV)	—	12.942259	10.328696
H_{sp} (eV)	—	4.339705	1.175115
U_{dd} (eV)	—	—	−24.504161
ζ_d (au)	—	—	0.840667
β_d (eV)	—	—	−2.360095
$\tilde{\zeta}_s$ (au)	—	—	2.085120
$\tilde{\zeta}_p$ (au)	—	—	1.535336
$\tilde{\zeta}_d$ (au)	—	—	1.236266
ρ_{core} (au)	—	—	1.185130
G_{scale}[a]	1.000000	1.000000	0.353722
a_1 (unitless)	0.106238	0.288526	−0.344529
b_1 (Å^{-2})	5.735290	4.883265	3.034933
c_1 (Å)	1.261430	0.850910	1.134275
a_2 (unitless)	0.004043	0.061586	−0.021847
b_2 (Å^{-2})	7.080122	4.435791	1.684515
c_2 (Å)	2.084095	1.353681	2.716684
a_3 (unitless)	−0.002800	—	−0.036003
b_3 (Å^{-2})	0.739913	—	5.243357
c_3 (Å)	3.649474	—	1.924175

Source: Parameters from ref. 15.

[a] Scale factor of Gaussian core–core repulsion interactions.

barrier heights. (For further details, see ref. 15.) The final optimized parameters are presented in Table 1. The AM1/d-PhoT model has been demonstrated to offer significant improvement over the MNDO/d and AM1 methods relative to high-level density-functional results for geometries, charge distributions and relative energies of minima, transition states, and reactive intermediates of model phosphoryl transfer reactions, in addition to dipole moments, proton affinities, and other properties of compounds related to biological phosphoryl transfer.[15]

Hybrid QM/MM Calculations

Combined QM/MM calculations were performed with a locally modified version of the CHARMM molecular simulation software package[28] interfaced

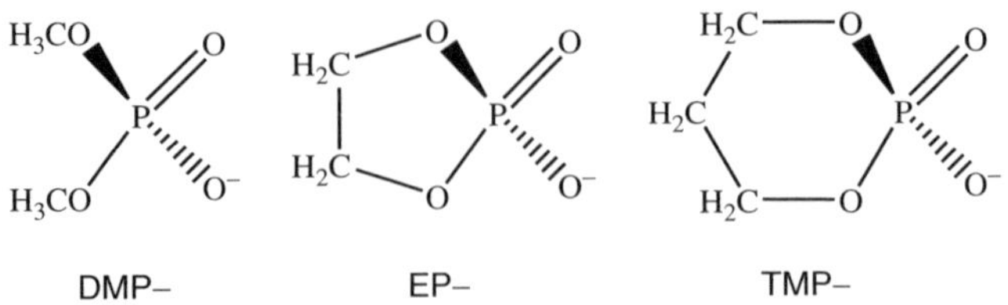

Figure 1 Dimethylphosphate (DMP^-), ethylene phosphate (EP^-), and trimethylene phosphate (TMP^-) molecules.

with the MNDO97[29] and MOPAC[30] programs and implemented using the QM/MM-Ewald method.[18] The dimethyl phosphate (DMP), ethylene phosphate (EP), and trimethylene phosphate (TMP) solutes (Figrue 1) were treated fully quantum mechanically at the semiempirical level and were solvated in a 40.0-Å cubic box of TIP3P water molecules,[31] resulting in a total of 2036, 2036, and 2035 waters, respectively. Internal water geometries were constrained using the SHAKE algorithm in all simulations.[32] A 10.0-Å group-based cutoff was used for van der Waals interactions and for evaluating the direct space term in the Ewald sums. The nonbonded list and crystal images were updated every 25 steps during molecular dynamics simulations. The Ewald κ value was chosen to be 0.340 $Å^{-1}$, the smooth particle mesh Ewald (PME) method was employed for reciprocal space summations between MM sites with an approximate grid size of 0.8 Å ($50 \times 50 \times 50$ FFT grid),[33,34] and two Na^+ ions were added to keep the system neutral. All simulations were propagated using the leapfrog Verlet algorithm with a 1-fs integration time step.[35] Periodic boundary conditions were used along with the isothermal–isobaric ensemble (NPT) at 1 atm and 298 K using the extended system pressure algorithm of Andersen[36] with effective mass of 500.0 amu and Hoover thermostat[37] with effective mass 1000.0 kcal/mol·ps^2, respectively.

The potential of mean force (PMF) profiles have been determined using umbrella sampling,[38] in which PMF is represented as a function of the reaction coordinate, defined as the difference in the leaving group–phosphorus and phosphorus–nucleophile distances ($R_{\text{LG-P}} - R_{\text{Nu-P}}$). After initial 200 ps of equilibration, 58 (DMP), 46 (EP), and 44 (TMP) separate umbrella sampling windows were executed to span the reaction coordinate from − 6.0 to 6.0 (DMP), 4.1 (EP), and 3.3 (TMP) Å, respectively, by applying a harmonic restraining potential centered at the center of the particular umbrella window. The spacing between neighboring windows was a function of the reaction coordinate values: 0.2- and 0.25-Å spacings were used for the reaction coordinate values ($R_c = R_{\text{LG-P}} - R_{\text{Nu-P}}$) in the range $|R_c| \leq 3.5$ Å and $|R_c| \geq 3.5$ Å respectively. The force constants used were chosen and adjusted based on the shape of the PMF profile for each system tested to guarantee sufficient overlap of the probability distribution with neighboring windows. (Force constant values ranged between 50.0 and 80.0 kcal/mol·Å near the transition state, 30.0 and 50.0 kcal/mol·Å in the region of intermediate reaction coordinate, and 20.0 and 30.0 kcal/mol·Å in the region of large separation between two reactants.) Each umbrella sampling window was equilibrated for 35 ps followed by 50 ps of production with data collected every step. The weighted histogram analysis method (WHAM)[39] was employed to compute the potential of mean force as a function of reaction coordinate R_c.

2. RESULTS AND DISCUSSION

The biological importance of phosphoryl transfer reactions has stimulated extensive theoretical and experimental investigations aimed at the identification and characterization of the underlying catalytic mechanisms.[6,7,40,41] It remains a challenge for experiment to determine unambiguously whether a particular phosphoryl transfer reaction proceeds through an associative or dissociative pathway (Figure 2). However, a vast array of experiments and calculated results supports the fact that, in solution, hydrolysis of phosphate diesters follow a bimolecular associative mechanism that proceeds via a pentacovalent intermediates and/or transition state with inversion around the phosphorus center. Although this general conclusion has been drawn for the reactions in solution, there is no consensus about the mechanism in enzymes and ribozymes. Unfortunately, kinetic models are not always able to discern between multiple mechanistic pathways that are able to fit experimental data equally well,[42–44] which underscores the need for the development and application of theoretical methods that are able to aid in the interpretation of mechanism.

Quantum mechanical electronic structure methods, in particular density functional theory (DFT), have been used widely to study phosphoryl transfer mechanism and to help interpret kinetic data.[42–65] Although these approaches provide insight into the nature of the reactions, there is a growing concern about the need for explicit models for the environment for reactions in aqueous solution and in the active sites of enzymes or ribozymes, where specific electrostatic interactions,

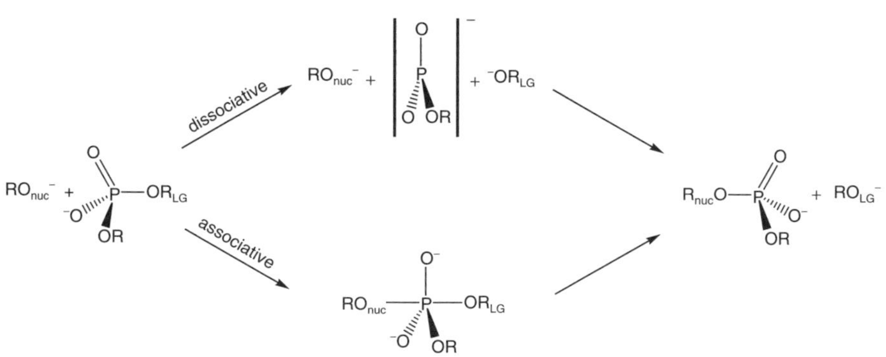

Figure 2 Generic pathways for phosphoryl transfer: dissociative and associative mechanisms.

hydrogen bonds, and solvent dynamics probably play a significant role.[66–73] Toward this end, semiempirical quantum models have proven invaluable tools when combined with molecular simulations in realistic environments using combined QM/MM potentials.[71,74–77]

In this section, the AM1, MNDO/d, and AM1/d-PhoT models are applied in QM/MM-Ewald simulations to the hydrolysis of acyclic and cyclic phosphates in aqueous solution. The hydrolysis of dimethyl phosphate (DMP^-), ethylene phosphate (EP^-), and trimethylene phosphate (TMP^-)(Figure 1) is examined. First we compare the AM1, MNDO/d, and AM1/d-PhoT models in the hydrolysis of DMP^- and EP^-. Then we analyze the effect of methoxide versus hydroxide ion nucleophile in attack on EP^- using the AM1/d-PhoT model. Finally, we compare the relative hydrolysis rates of DMP^-, EP^-, and TMP^- calculated with the AM1/d-PhoT model with experimental values.

Comparison of AM1, MNDO/d, and AM1/d-PhoT Models for Hydrolysis of DMP^- and EP^-

Tables 2 and 3 compare the gas-phase activation barriers and the transition-state geometries for hydrolysis (hydroxide attack) of DMP^-, EP^-, and TMP^- calculated with the DFT and with the AM1, MNDO/d, and AM1/d-PhoT models. The semiempirical models all underpredict the activation barriers relative to the DFT values in the gas phase. Overall, the AM1/d-PhoT model has the best agreement with the DFT values (MSE of − 3.9 kcal/mol), whereas the AM1 and MNDO/d models had MSE values of − 5.6 and 4.5 kcal/mol, respectively. This might seem at first to be encouraging. However, considering that the high reaction barriers for these reactions are to overcome the unfavorable Coulomb repulsion between two reactants, it is not surprising to predict similar barriers from the three semiempirical methods tested. On the other hand, closer inspection of the rate-controlling transition-state geometries reveals more significant problems. Table 3 compares the gas-phase transition-state geometries for hydrolysis (hydroxide attack) of DMP^-, EP^-, and TMP^- calculated with the DFT and with the AM1, MNDO/d,

TABLE 2. Comparison of the Gas-Phase Activation Barriers for Hydrolysis (Hydroxide Attack) of DMP^-, EP^-, and TMP^- Calculated with the DFT and with the AM1, MNDO/d, and AM1/d-PhoT Models[a]

Reaction	DFT	AM1	Error	MNDO/d	Error	AM1/d	Error
DMP^-	88.3	76.4	−11.9	80.9	−7.4	82.2	−6.1
EP^-	86.7	85.8	−0.9	82.3	−4.4	84.2	−2.5
TMP^-	89.0	84.8	−4.1	83.1	−5.9	86.0	−3.0

[a]Shown are the gas-phase activation energy barriers, $\Delta E^{\ddagger}$, in kcal/mol calculated from density functional theory (DFT) at the BLYP/6-311++G(3df,2p)//B3LYP/6-31++G(d,p) level (see ref. 16 for further details), calculated using the AM1, MNDO/d, and AM1/d-PhoT (referred to simply as AM1/d) models. Errors with respect to the DFT values are shown immediately to the right of the semiempirical $\Delta E^{\ddagger}$ values.

TABLE 3. Comparison of the Gas-Phase Transition-State Geometries for Hydrolysis (Hydroxide Attack) of DMP^-, EP^-, and TMP^- Calculated with the DFT and with the AM1, MNDO/d, and AM1/d-PhoT Models[a]

			Error		
Reaction		DFT	AM1	MNDO/d	AM1/d
DMP^-	r_1	2.186	0.505	0.641	0.061
	r_2	1.847	−0.055	−0.010	−0.030
	θ	169.0	1.9	−1.6	−1.9
EP^-	r_1	2.627	0.360	0.226	−0.074
	r_2	1.834	−0.101	0.025	−0.003
	θ	163.1	2.8	−1.6	−2.3
TMP^-	r_1	2.521	0.455	0.392	−0.045
	r_2	1.851	−0.124	0.005	−0.007
	θ	167.2	−1.6	−3.8	−4.4

[a]Shown are the gas-phase transition-state geometries along with geometric parameters $r_1 = \text{Nu} - \text{P}$ and $r_2 = \text{P} - \text{LG}$ distances (Å), and $\theta = \text{Nu-P-LG}$ angle (degrees), calculated from density functional theory (DFT) at the B3LYP/6-31++G(d,p) level (see ref. 16 for further details), calculated using the AM1, MNDO/d, and AM1/d-PhoT (referred to simply as AM1/d) models.

and AM1/d-PhoT models. It is clear that the $r_1 = \text{Nu} - \text{P}$ value is considerably too large for the AM1 and MNDO/d methods, whereas the AM1/d-PhoT model is quite close to the DFT value. This implies that for the AM1 and MNDO/d models, a very early transition state is predicted, particularly with AM1, which does not agree closely with the DFT values.

Figure 3 compares the PMF profiles for dimethyl phosphate hydrolysis with the AM1, MNDO/d, and AM1/d-PhoT models. There are several aspects of the profiles that become evident immediately. First, the AM1 and MNDO/d activation barriers for the hydrolysis of dimethyl phosphate, 13.3 and 18.5 kcal/mol, respectively, are considerably lower than that of the AM1/d-PhoT model (32.7 kcal/mol). As discussed in further detail below, the AM1/d-PhoT value is in close agreement with the activation barrier of 32 kcal/mol estimated from experiment.[55,78,79] This significant underestimation of the solvation barrier is due to overstabilization of the dianionic transition state relative to the separated monoanionic reactants, and leads to considerable artifacts in the simulations. It is of interest to note that the artificial stabilization is due mainly to the presence of an *earlier* transition state, predicted by the AM1 and MNDO/d models, which results in almost separated charge distribution close to hydroxide and DMP^- ions. Recall that the gas-phase barriers listed in Table 2 do not show sufficient differences between the semiempirical methods to account for the large differences in the barriers observed in aqueous solution. The primary origin of the differences occurs from the unrealistically high instability of the gas-phase hydroxide ion with the AM1 and MNDO/d methods, which leads to large solvent overstabilization in solution. This leads to an artificially *early* rate-controlling transition state with almost a full negative charge on the hydroxide. This negative charge is sensitive to solvent stabilization

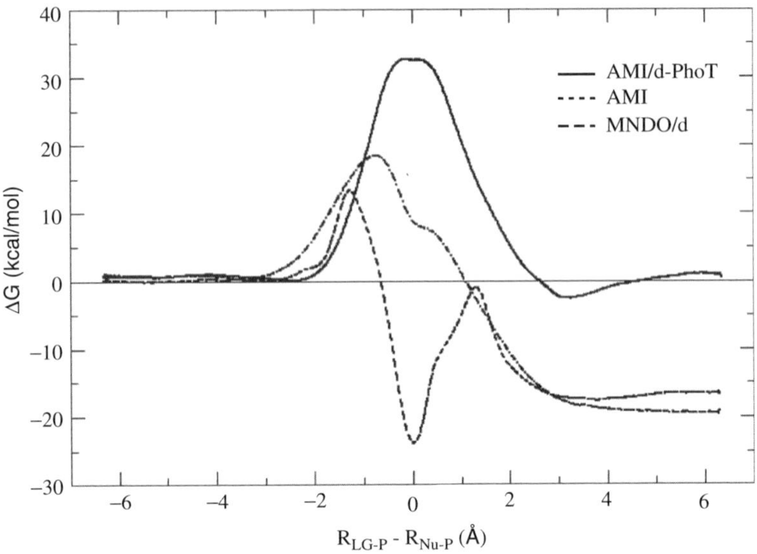

Figure 3 PMF profile for hydrolysis of dimethyl phosphate with the AM1, MNDO/d, and AM1/d-PhoT models.

and results in the AM1 and MNDO/d methods having solution free-energy barriers that are too low.

Another striking feature of the profiles is that for the AM1 and MNDO/d methods, the product free energy is considerably lower than that of the reactant state by approximately 17 to 20 kcal/mol. This lowering of the reaction free energy is in considerable error relative to what is expected based on the relative pK_a values of water (15.7) and methanol (15.5), which suggest that the reaction should be close to isoergic. This is due to the artificially high energy of the gas-phase hydroxide ion with AM1 and MNDO/d (i.e., the adiabatic proton affinity values for hydroxide are in error by 20.5 kcal/mol for AM1 and 30.6 kcal/mol for MNDO/d, whereas the AM1/d-PhoT model is in error by only 5.4 kcal/mol[15]). Being a small anion, the hydroxide ion is problematic for semiempirical methods that use a minimal valence basis. The AM1/d-PhoT model took special care in the parameterization to improve the stability of the hydroxide ion, arguably the most important nucleophile for phosphoryl transfer reactions.

The profile for the MNDO/d method predicts a single early transition state, corresponding to the approach of the hydroxide ion, and a shoulder (nonstationary point) in the region of the reaction coordinate corresponding to cleavage of the leaving group ($R_{\text{LG-P}} - R_{\text{Nu-P}}$ value of 0.5 Å). The early transition state is intrinsic, as can be seen from the gas-phase transition-state geometry (Table 3). The AM1 model, on the other hand, is qualitatively disastrous: There is a very early transition state that is followed by an intermediate phosphorane that is more stable than the reactant and product by approximately 24 and 7 kcal/mol, respectively. This profile is useless in predicting the phosphoryl transfer mechanism, where the lifetimes of intermediate states often have important consequences, such as phosphorane protonation and pseudorotation to form alternative products. The AM1/d-PhoT model predicts a profile with a single broad barrier centered around a reaction coordinate value of $R_{\text{LG-P}} - R_{\text{Nu-P}} = 0$. This is reasonable for the hydrolysis on DMP^-, where the nucleophile and leaving-group bonds are similar and the reaction is nearly isodesmic.

Figure 4 compares the PMF profiles for ethylene phosphate hydrolysis with the AM1, MNDO/d, and AM1/d-PhoT models. The main features of the dimethyl phosphate hydrolysis reaction are again observed. The AM1 and MNDO/d activation barriers for the hydrolysis of ethylene phosphate 12.1 and 13.5 kcal/mol, respectively, are considerably lower than that of the AM1/d-PhoT model (24.3 kcal/mol). The AM1 and MNDO/d activation barrier values are considerably too low with respect to the estimated experimental value of 24 kcal/mol,[55,78] whereas the AM1/d-PhoT model value is in very close agreement. Similar trends with respect to unreasonably large reaction free-energy values with the AM1 and MNDO/d models are observed, in addition to the prediction of a greatly overstabilized phosphorane intermediate in the case of AM1. The AM1/d-PhoT barriers and profile are in reasonable agreement with what is known experimentally about the reaction in solution (discussed further below).

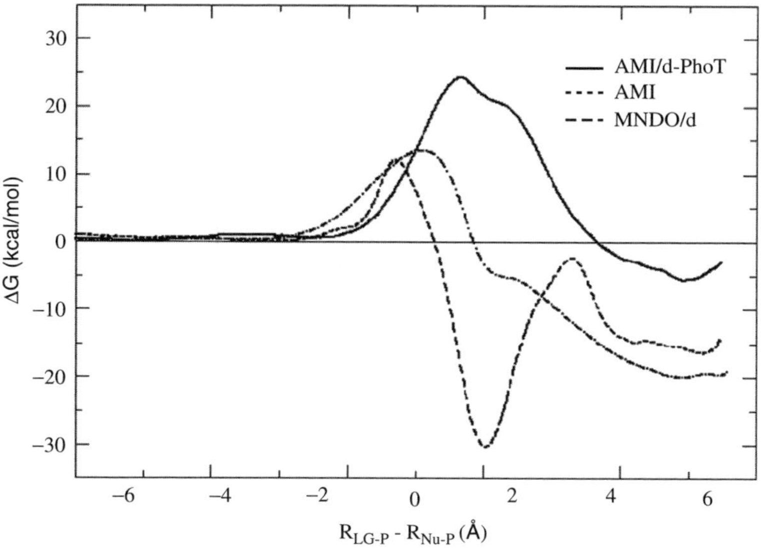

Figure 4 PMF profile for hydrolysis of ethylene phosphate with the AM1, MNDO/d, and AM1/d-PhoT models.

Comparison of OH^- and CH_3O^- Nucleophiles in Attack to EP

Cleavage of the phosphate backbone of RNA occurs via a transesterification reaction whereby the 2′ hydroxyl group becomes activated and engages in an in-line attack on the phosphorus of the scissile bond, resulting in departure of the 5′ alkoxide to form a 2′,3′-cyclic phosphate, which undergoes further hydrolysis to the final 3′-phosphate. This transesterification reaction is catalyzed by prototype RNA enzymes, or ribozymes, such as the hammerhead,[80,81] hairpin,[82,83] and hepatitis delta virus[84,85] ribozymes. Ethylene phosphate has served as a useful small-molecule model to mimic nonenzymatic RNA transesterification and hydrolysis.[7] Methanolysis of ethylene phosphate serves as a reverse reaction model for nonenzymatic RNA transesterification (i.e., the methoxide attack to the cyclic phosphate resulting in cleavage of the endocyclic bond is the reverse process to that of RNA transesterification).[47,68,69]

In Figure 5 we compare the PMF profiles for hydrolysis and methanolysis of ethylene phosphate with the AM1/d-PhoT model. As discussed briefly above, the hydrolysis reaction proceeds via an early transition state ($R_{\text{LG-P}} - R_{\text{Nu-P}}$ value of -0.35 Å) for the approach of the hydroxide followed by a shoulder (between $R_{\text{LG-P}} - R_{\text{Nu-P}}$ values of 0 and 0.5 Å) reminiscent of a $TS_{5'}$-type transition state,[47] corresponding to the departure of the cyclic alkoxide coincident with endocyclic P–O bond cleavage. The activation free-energy barrier is approximately 24.3 kcal/mol for hydrolysis. The methanolysis reaction, on the other hand, proceeds by a distinct, normally distributed single step with slightly early

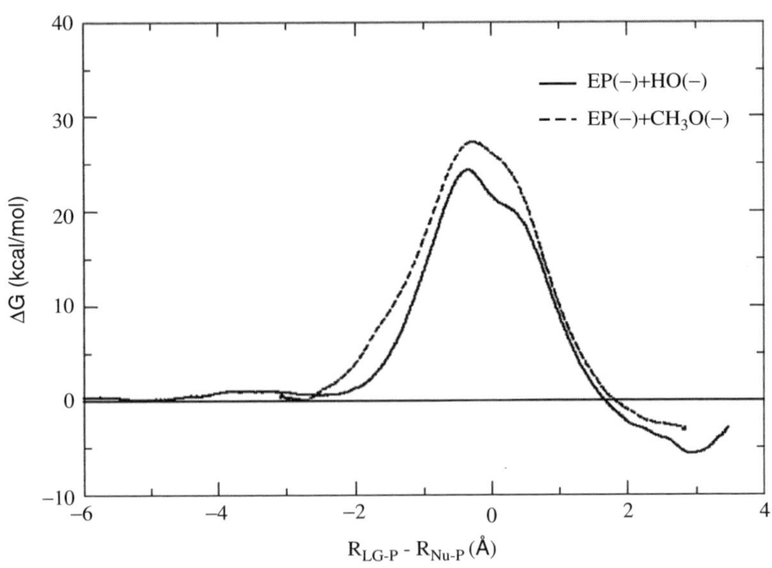

Figure 5 Comparison of the PMF profiles for hydrolysis and methanolysis of ethylene phosphate with the AM1/d-PhoT model.

transition state ($R_{LG\text{-}P} - R_{Nu\text{-}P}$ value of -0.27 Å) with a slightly elevated activation barrier (27.3 kcal/mol) relative to the hydrolysis reaction. The results are qualitatively reasonable but not quantitatively what would be expected based on the very similar pK_a values of water and methanol, the latter being higher by only 0.2 pK_a unit, which corresponds to about 0.3 kcal/mol. This small perceived inconsistency in the calculation underscores the need for continued development and testing to obtain increasingly accurate and robust methods.

Comparison of Hydrolysis Rates DMP⁻, EP⁻, and TMP⁻

A classic problem in phosphate reactivity is exemplified by the large differential hydrolysis rates of acyclic and certain cyclic phosphates first pointed out by Westheimer.[86] The prototype example involves the relative hydrolysis rates of DMP^- and EP^-. EP^- exhibits a remarkably enhanced hydrolysis rate relative to its acyclic analog, DMP^- (10^6- to 10^8-fold rate enhancement), which was first attributed to ground-state destabilization arising from ring strain of the cyclic structure. In contrast to EP^-, the hydrolysis rate for TMP^-, for example, is quite similar to that of DMP^-.[78] It was later argued that differential solvation effects could reasonably account for the majority of the difference in observed rate.[54,55] Subsequent studies, both theoretical and experimental, have generally arrived at the consensus that the origin of the rate enhancement arises mainly as a solvation effect.[47,56,64,87] Consequently, examination of the relative hydrolysis rates of DMP^-, EP^-, and TMP^- (Figure 1) is an important validation benchmark for the AM1/d-PhoT and QM/MM-Ewald methods.[18]

Figure 6 compares the PMF profiles for hydrolysis of dimethyl phosphate, ethylene phosphate, and trimethylene phosphate with the AM1/d-PhoT model. Each reaction proceeds via an associative mechanism, consistent with previous experimental and theoretical work for the alkaline hydrolysis of phosphate diesters,[7] and exhibits a single kinetically significant rate-controlling transition state. Table 4 summarizes the free-energy barriers calculated from the PMF profiles and estimated from the experimental rate constants (assuming unit transmission coefficient and no tunneling effects). The AM1/d-PhoT barriers are in very good agreement with those obtained from experiment. The largest deviation occurs for the TMP^- barrier, calculated to be approximately 3.8 kcal/mol below that for DMP^-, whereas the corresponding experimental barriers are more similar. The AM1/d-PhoT model very reliably predicts the relative hydrolysis rate for DMP^- and EP^-. The rates calculated correlate with the $r_1 = \mathrm{Nu} - \mathrm{P}$ value, which becomes smaller with increasing barrier size (decreasing rate). Stated alternatively, in the series DMP^-, TMP^-, and EP^-, a steady decrease in rate-controlling activation barrier is accompanied by the occurrence of a successively earlier transition state (having a longer $r_1 = \mathrm{Nu} - \mathrm{P}$ distance). The origin for the possible small error in the activation barrier for TMP^- appears subtle, and will be a topic of further study and characterization. Overall, however, the results for the AM1/d-PhoT model are highly encouraging, especially with respect to other semiempirical alternatives, and hence the AM1/d-PhoT model may have

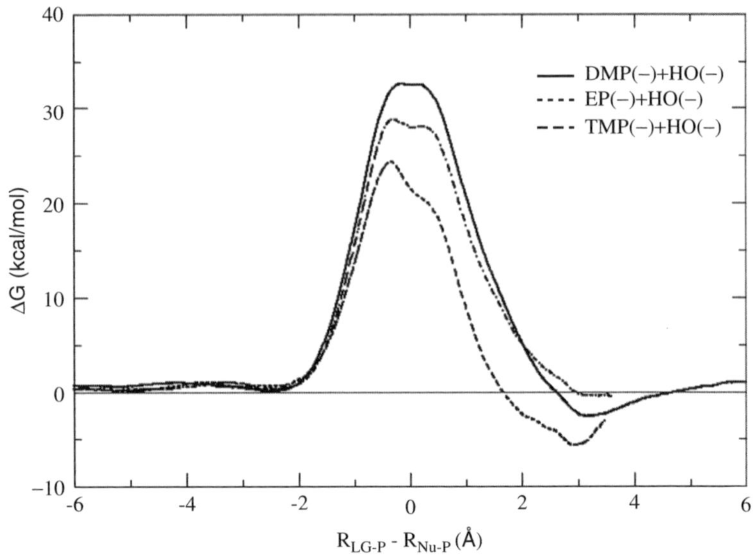

Figure 6 Comparison of the PMF profiles for hydrolysis of dimethyl phosphate, ethylene phosphate, and trimethylene phosphate with the AM1/d-PhoT model.

TABLE 4. Comparison of the Aqueous Activation Barriers and the Average Transition-State Geometries for Hydrolysis (Hydroxide Attack) of DMP^-, EP^-, and TMP^- from QM/MM-Ewald Simulation with the AM1/d-PhoT Model[a]

Reaction	r_1	r_2	θ	$\Delta G^{\ddagger}$	Expt.
DMP^-	1.978	1.812	167.8	32.7	~32
EP^-	2.146	1.768	164.5	24.3	~21–24
TMP^-	2.084	1.779	170.5	28.9	~32

[a] Shown are the calculated free-energy barrier ($\Delta G^{\ddagger}$) in kcal/mol from the PMF profile, along with average geometric parameters $r_1 = \text{Nu} - \text{P}$ and $r_2 = \text{P} - \text{LG}$ distances (Å), and $\theta = \text{Nu} - \text{P} - \text{LG}$ angle (degrees). Also shown is the free-energy barrier in kcal/mol estimated from the experimental rate (Expt.) assuming unit transmission coefficient and no tunneling effects.

promise as a convenient tool to study biological phosphoryl transfer reactions in solution, enzymes, and ribozymes.

3. CONCLUSIONS

A modified AM1/d Hamiltonian, AM1/d-PhoT, has been used with a new QM/MM-Ewald method to study prototype phosphate hydrolysis reactions in solution. The hydrolysis of dimethyl phosphate and ethylene phosphate was considered first. The AM1 and MNDO/d methods were both observed to perform very poorly for these systems, whereas the AM1/d-PhoT model provides PMF profiles that were in good agreement with experiment. The effect of nucleophile in the attack in ethylene phosphate was studied by comparing the PMF profiles for hydrolysis and methanolysis. The profiles suggest that the methanolysis reaction proceeds via a single step with a slightly higher barrier than for the hydrolysis reaction, which proceeds via two transition states separated by a kinetically insignificant intermediate. The AM1/d-PhoT model was validated further by comparison of the relative hydrolysis rates of DMP^-, EP^-, and TMP^-. Results are in excellent agreement with experiment and accurately reproduce the known enhanced reactivity of cyclic versus acyclic phosphates discussed previously in the literature. Overall, the AM1/d-PhoT model and QM/MM-Ewald method have been demonstrated to have considerable promise as reliable methods for the study of phosphoryl transfer reactions in aqueous solution. It is the hope that this method can lend more quantitative insight into problems of biological phosphoryl transfer in enzymes and ribozymes.

Acknowledgments

This work was supported partially by grants from the National Institutes of Health (GM62248 and GM46736), and the Office of Naval Research (ONR) under grant N00012-05-01-0538. Computational resources were provided by the Minnesota Supercomputing Institute.

REFERENCES

1. Ahn, N.*Chem. Rev.* 2001, 101: 2207–2208.
2. Boyer, P. D.; Cross, R. L.; Momsen, W. *Proc. Natl. Acad. Sci. USA*, 1973, 70: 2837–2839.
3. Admiraal, S.; Herschlag, D. *Chem. Biol.* 1995, 2: 729–739.
4. Blackburn, G. M.; Gait, M. J., Eds. *Nucleic Acids in Chemistry and Biology*, Oxford University Press, New York, 1996.
5. Bloomfield, V. A.; Crothers, D. M.; Tinoco, I., Jr. *Nucleic Acids: Structures, Properties, and Functions*, University Science Books, Sausalito, CA, 2000.
6. Oivanen, M.; Kuusela, S.; Lönnberg, H. *Chem. Rev.*, 1998, 98: 961–990.
7. Perreault, D. M.; Anslyn, E. V. *Angew. Chem. Int. Ed.*, 1997, 36: 432–450.
8. Scott, W. G. *Curr. Opin. Struct. Biol.*, 1998, 8: 720–726.
9. Butcher, S. E. *Curr. Opin. Struct. Biol.*, 2001, 11: 315–320.
10. Takagi, Y.; Ikeda, Y.; Taira, K. *Top. Curr. Chem.*, 2004, 232: 213–251.
11. Thiel, W.; Voityuk, A. A. *Theor. Chim. Acta*, 1992, 81: 391–404.
12. Thiel, W.; Voityuk, A. A. *J. Phys. Chem.*, 1996, 100: 616–626.
13. York, D. M.; Darden, T.; Pedersen, L. G. *J. Chem. Phys.*, 1993, 99: 8345–8348.
14. Sagui, C.; Darden, T. A. *Annu. Rev. Biophys. Biomol. Struct.*, 1999, 28: 155–179.
15. Nam, K.; Cui, Q; Gao, J.; York, D. M. *J. Chem. Theory Comput.*, 2007, 3: 486–504.

16. Giese, T. J.; Gregersen, B. A.; Liu, Y.; Nam, K.; Mayaan, E.; Moser, A.; Range, K.; et al. *J. Mol. Graph. Model.*, 2006, 25: 423–433.
17. QCRNA. http://theory.chem.umn.edu/Database/QCRNA.
18. Nam, K.; Gao, J.; York, D. M. *J. Chem. Theory Comput.*, 2005, 1: 2–13.
19. Stewart, J. J. P. *Rev. Comput. Chem.*, 1990, 1: 45–81.
20. Thiel, W. In *Advances in Chemistry and Physics*,Vol. 93, Prigogine, I.; Rice, S. A., Eds., Wiley, New York, 1996, pp. 703–757.
21. Winget, P.; Selçuki, C.; Horn, A.; Martin, B.; Clark, T. *Theor. Chem. Acc.*, 2003, 110: 254–266.
22. Cramer, C. J. *Essentials of Computational Chemistry: Theories and Models*, 2nd ed., Wiley, Chichester, England, 2004.
23. Dewar, M. J. S.; Thiel, W. *J. Am. Chem. Soc.*, 1977, 99: 4899–4907.
24. Giese, T. J.; Sherer, E. C.; Cramer, C. J.; York, D. M. *J. Chem. Theory Comput.*, 2005, 1: 1275–1285.
25. Stewart, J. J. P. *J. Comput. Chem.*, 1989, 10: 209–220.
26. Press, W. H.; Teukolsky, S. A.; Vetterling, W. T.; Flannery, W. P. *Numerical Recipes in Fortran*, 2nd ed., Cambridge University Press, Cambridge, 1992.
27. Giese, T. J.; York, D. M. In preparation.
28. Brooks, B. R.; Bruccoleri, R. E.; Olafson, B. D.; States, D. J.; Swaminathan, S.; Karplus, M. *J. Comput. Chem.*, 1983, 4: 187–217.
29. Thiel, W. *MNDO97, Version 5.0*, University of Zurich, Zurich, Switzerland, 1998.
30. Stewart, J. J. P. *J. Comput.-Aided Mol. Des.*, 1990, 4: 1–105.
31. Jorgensen, W. L.; Chandrasekhar, J.; Madura, J. D.; Impey, R. W.; Klein, M. L. *J. Chem. Phys.*, 1983, 79: 926–935.
32. Ryckaert, J. P.; Ciccotti, G.; Berendsen, H. J. C. *J. Comput. Phys.*, 1977, 23: 327–341.
33. Darden, T.; York, D.; Pedersen, L. *J. Chem. Phys.*, 1993, 98: 10089–10092.
34. Essmann, U.; Perera, L.; Berkowitz, M. L.; Darden, T.; Hsing, L.; Pedersen, L. G. *J. Chem. Phys.*, 1995, 103: 8577–8593.
35. Allen, M. P.; Tildesley, D. J. *Computer Simulation of Liquids*, Oxford University Press, Oxford, 1987.
36. Andersen, H. C. *J. Chem. Phys.*, 1980, 72: 2384–2393.
37. Hoover, W. G. *Phys. Rev. A*, 1985, 31: 1695–1697.
38. Torrie, G. M.; Valleau, J. P. *J. Comput. Phys.*, 1977, 23: 187–199.
39. Kumar, S.; Bouzida, D.; Swendsen, R. H.; Kollman, P. A.; Rosenberg, J. M. *J. Comput. Chem.*, 1992, 13: 1011–1021.
40. Zhou, D.-M.; Taira, K. *Chem. Rev.*, 1998, 98: 991–1026.
41. Hengge, A. C. *Acc. Chem. Res.*, 2002, 35: 105–112.
42. Florián, J.; Åqvist, J. A.; Warshel, A. *J. Am. Chem. Soc.*, 1998, 120: 11524–11525.
43. Florián, J.; Warshel, A. *J. Phys. Chem. B*, 1998, 102: 719–734.
44. Åqvist, J.; Kolmodin, K.; Florián, J.; Warshel, A. *Chem. Biol.*, 1999, 6: R71–R80.
45. Liu, Y.; Gregersen, B. A.; Hengge, A.; York, D. M. *Biochemistry*, 2006, 45: 10043–10053.
46. Liu, Y.; Lopez, X.; York, D. M. *Chem. Commun.*, 2005, 31: 3909–3911.

47. Liu, Y.; Gregersen, B. A.; Lopez, X.; York, D. M. *J. Phys. Chem. B*, 2005, 109: 19987–20003.
48. López, C. S.; Faza, O. N.; R. de Lera, A.; York, D. M. *Chem. Eur. J.*, 2005, 11: 2081–2093.
49. Lim, C.; Karplus, M. *J. Am. Chem. Soc.*, 1990, 112: 5872–5873.
50. Uchimaru, T.; Tanabe, K.; Nishikawa, S.; Taira, K. *J. Am. Chem. Soc.*, 1991, 113: 4351–4353.
51. Lim, C.; Tole, P. *J. Phys. Chem.*, 1992, 96: 5217–5219.
52. Mercero, J. M.; Barrett, P.; Lam, C. W.; Fowler, J. E.; Ugalde, J. M.; Pedersen, L. G. *J. Comput. Chem.*, 2000, 21: 43–51.
53. Arantes, G. M.; Chaimovich, B. *J. Phys. Chem. A*, 2005, 109: 5625–5635.
54. Dejaegere, A.; Lim, C.; Karplus, M. *J. Am. Chem. Soc.*, 1991, 113: 4353–4355.
55. Dejaegere, A.; Karplus, M. *J. Am. Chem. Soc.*, 1993, 115: 5316–5317.
56. Tole, P.; Lim, C. *J. Phys. Chem.*, 1993, 97: 6212–6219.
57. Tole, P.; Lim, C. *J. Am. Chem. Soc.*, 1994, 116: 3922–3931.
58. Chang, N.-Y.; Lim, C. *J. Phys. Chem. A*, 1997, 101: 8706–8713.
59. Florián, J.; Warshel, A. *J. Am. Chem. Soc.*, 1997, 119: 5473–5474.
60. Hu, C.-H.; Brinck, T. *J. Phys. Chem. A*, 1999, 103: 5379–5386.
61. Lopez, X.; Dejaegere, A.; Karplus, M. *J. Am. Chem. Soc.*, 2001, 123: 11755–11763.
62. Lopez, X.; Schaefer, M.; Dejaegere, A.; Karplus, M. *J. Am. Chem. Soc.*, 2002, 124: 5010–5018.
63. Lopez, X.; Dejaegere, A.; Leclerc, F.; York, D. M.; Karplus, M. *J. Phys. Chem. B*, 2006, 110: 11525–11539.
64. Chen, X.; Zhan, C.-G. *J. Phys. Chem. A*, 2004, 108: 6407–6413.
65. Xu, D.; Guo, H.; Liu, Y.; York, D. M. *J. Phys. Chem. B*, 2005, 109: 13827–13834.
66. Karplus, M. *J. Phys. Chem. B*, 2000, 104: 11–27.
67. Warshel, A. *Annu. Rev. Biophys. Biomol. Struct.*, 2003, 32: 425–443.
68. Gregersen, B. A.; Lopez, X.; York, D. M. *J. Am. Chem. Soc.*, 2003, 125: 7178–7179.
69. Gregersen, B. A.; Lopez, X.; York, D. M. *J. Am. Chem. Soc.*, 2004, 126: 7504–7513.
70. López, C. S.; Faza, O. N.; Gregersen, B. A.; Lopez, X.; R. de Lera, A.; York, D. M. *Chem. Phys. Chem.*, 2004, 5: 1045–1049.
71. Garcia-Viloca, M.; Gao, J.; Karplus, M.; Truhlar, D. G. *Science*, 2004, 303: 186–195.
72. Garcia-Viloca, M.; Truhlar, D. G.; Gao, J. *J. Mol. Biol.*, 2003, 327: 549–560.
73. Garcia-Viloca, M.; Poulsen, T. D.; Truhlar, D. G.; Gao, J. *Protein Sci.*, 2004, 13: 2341–2354.
74. Warshel, A.; Levitt, M. *J. Mol. Biol.*, 1976, 103: 227–249.
75. Field, M. J.; Bash, P. A.; Karplus, M. *J. Comput. Chem.*, 1990, 11: 700–733.
76. Gao, J. *Rev. Comput. Chem.*, 1995, 7: 119–185.
77. Gao, J. *Curr. Opin. Struc. Biol.*, 2003, 13: 184–192.
78. Cox, J. R., Jr.; Ramsay, O. B.*Chem. Rev.*, 1964, 64: 317–351.
79. Chen, X.; Zhan, C.-G. *J. Phys. Chem. A*, 2004, 108: 3789–3797.
80. Scott, W. G.; Murray, J. B.; Arnold, J. R. P.; Stoddard, B. L.; Klug, A. *Science*, 1996, 274: 2065–2069.

81. Scott, W. G. *Q. Rev. Biophys.*, 1999, 32: 241–294.
82. Walter, N. G.; Burke, J. M. *Curr. Opin. Chem. Biol.*, 1998, 2: 24–30.
83. Rupert, P. B.; Massey, A. P.; Sigurdsson, S. T.; Ferré-D'Amaré, A. R. *Science*, 2002, 298: 1421–1424.
84. Shih, I.-H.; Been, M. D. *EMBO J.*, 2001, 20: 4884–4891.
85. Shih, I.-H.; Been, M. D. *Annu. Rev. Biochem.*, 2002, 71: 887–917.
86. Westheimer, F. H. *Acc. Chem. Res.*, 1968, 1: 70–78.
87. Lopez, X.; York, D. M.; Dejaegere, A.; Karplus, M. *Int. J. Quantum Chem.*, 2002, 86: 10–26.

13

MODELING THE THERMAL DECOMPOSITION OF LARGE MOLECULES AND NANOSTRUCTURES

MARC R. NYDEN
National Institute of Standards and Technology, Gaithersburg, Maryland

STANISLAV I. STOLIAROV
SRA International, Egg Harbor Township, New Jersey

VADIM D. KNYAZEV
Department of Chemistry, The Catholic University of America, Washington, DC

Considerable effort has been directed toward understanding the mechanisms of chemical transformations in complex multi-component systems. For example, scientists have been able to resolve the combustion of gaseous fuels and chemical processes in the earth's atmosphere in terms of elementary chemical reactions with defined kinetic parameters.[1,2] Although the quantitative kinetic models are still far from being perfect, our knowledge of the underlying gas-phase chemistry has reached the point where it can be used to make realistic predictions of the behavior of such systems. Our understanding of the complex processes that govern the chemistry of the condensed-phase systems, however, is not so well developed.

The thermal degradation of polymers and polymer nanocomposites is one area, in particular, where our knowledge of the underlying chemical mechanisms is

Multiscale Simulation Methods for Nanomaterials, Edited by Richard B. Ross and Sanat Mohanty

insufficient to make accurate quantitative predictions. Experimental information is usually limited to data obtained from thermogravimetric analysis (TGA) or TGA in conjunction with mass spectrometry (TGA/MS), which can only provide rate constants for the kinetics of the overall mass loss and yields of the final gaseous products. Statistical reaction-rate theories (together with quantum chemical calculations of potential energy surfaces) have been quite successful when used to predict the rate constants of elementary reactions in the gas phase.However, they are not directly applicable to condensed-phase environments because of the absence of a consistent and physically realistic method for handling the coupling between all of the degrees of freedom. At the same time, the need to develop a molecular-level understanding of the thermal decomposition of polymers is becoming increasingly important in areas of science and technology associated with the combustion and high-temperature processing of these materials.

In this chapter we review progress that has been made in the development of an extension of classical force field–based molecular dynamics (MD) to the simulation of chemical reactions involving large molecules and nanostructures. With this technique, which we refer to as *reactive molecular dynamics* (RMD), it is possible to reach nanometer-length scales while accounting for the effects of intermolecular interactions in condensed phases. It represents one possibility of how to span the power and limitations of static 0 K high-accuracy few-atom solutions of the time-independent electronic Schrödinger equation (ab initio quantum chemistry) and dynamic real-temperature low-chemical-accuracy many-atom molecular simulations (molecular dynamics and ensemble Monte Carlo methods).

We have made substantial progress in development of one variant of RMD implemented in the computer program called MD_REACT.[3–7] Two related approaches are MD based on the reactive empirical bond-order (REBO)[8] and ReaxFF force fields.[9] Although both the REBO and ReaxFF force fields were developed from scratch to accommodate chemical reactivity, MD_REACT employs a conventional (nonreactive) force field for the description of atomic structures near their equilibrium geometries and performs dynamic modifications of the force field to model potential energy surfaces associated with chemical transformations. The development of all three computational approaches is driven by a common motivation: to enable a detailed analysis of chemical processes that have length/time scales that are too small for experimental methods and too large for quantum chemical methods.

1. DESCRIPTION OF THE MD_REACT APPROACH TO RMD

In its conventional formulation, molecular dynamics consists of solving the classical equations of motion for the time-dependent positions (and velocities) of the atoms that comprise the system under investigation. The objective is usually to determine the evolution of a molecular structure after it is subjected to a perturbation due to thermal excitation and/or external forces. The perturbation is introduced in the form of a velocity distribution for the constituent atoms. This results in a distortion of the structure, which, in turn, affects the potential energy.

Typically, the potential energy is represented by simple analytical functions, which in conjunction with parameters that are differentiated on the basis of atom types, comprise the force-field. One of the most important advantages of employing a force-field description is that it provides a mechanism for describing the potential energy of a large system in terms of components that can be parameterized from experimental information and/or from the results of quantum calculations on smaller model compounds and fragments. This makes it possible, at least in principle, to obtain accurate representations for nanoscale structures, which might otherwise be outside the realm of computational chemistry.

In our approach, the interactions between atoms are defined by analytical potentials that are derived from the consistent valence force field (CVFF)[10]. The general structure of the CVFF formulation is given by

$$V = \sum^{N\text{bonds}} V_{\text{bond}} + \sum^{N\text{angles}} V_{\text{angle}} + \sum^{N\text{torsions}} V_{\text{torsion}} + \sum^{N\text{pairs}} V_{\text{nonbond}} \tag{1}$$

What follows is a description of the basic components of this potential, including how the force field is modified to account for chemical reactivity.

The potential energy for stretching a covalent bond is represented by a Morse function,

$$V_{\text{bond}} = D\left[1 - \exp(-\alpha(r - r_e))\right]^2$$
$$\alpha = \sqrt{\frac{k_b}{2D}} \tag{2}$$

where D is the bond dissociation energy, r the distance between the bonded atoms, r_e the equilibrium bond length, and k_b the force constant associated with stretching the bond. The potential energy resulting from a change in the bond angle involving atoms a, b, and c is

$$V_{\text{angle}} = S_{ab} S_{bc} k_\theta (\theta - \theta_e)^2 \tag{3}$$

where θ_e is the equilibrium angle, k_θ the angular force constant, and S_{ab} and S_{bc} are switching functions, which are defined below. Rotations about covalent bonds are restricted by torsional potentials of the form

$$V_{\text{torsion}} = S_{ab} S_{bc} S_{cd} k_\varphi [1 + \cos(n\varphi - \varphi_e)] \tag{4}$$

The dihedral angle, φ, is defined by the three bonds between four adjacent atoms: a, b, c, and d (planes abc and bcd). Parameters k_ϕ, n, and φ_e determine the height, multiplicity, and position of the barrier to internal rotation.

The V_{nonbond} terms in Eq. (1) are comprised of electrostatic and van der Waals interactions between each pair of atoms that are not directly bonded to each other, provided that they are separated by at least three covalent bonds (i.e., one to four interactions) when they are part of a continuously bonded segment. These interactions are defined by the sum of the Coulomb and Lennard-Jones 6-12 potentials as described in refs. 10 and 11. The switching functions, S, are fractional bond orders defined as

$$S = \begin{cases} 1 & r \leq r_e \\ 1 - \frac{V_{\text{bond}}}{D} & r > r_e \end{cases} \tag{5}$$

where V_{bond} is the bond energy [described by Eq. (2)]. The purpose of the switching functions is to simulate the decay of the bending and twisting forces as the covalent bonds are stretched.

Chemical transformations are modeled in MD_REACT by the following algorithm. After every time step of molecular dynamics, fractional bond orders [defined by Eq. (5)] are computed for every covalent bond in the system. In the case of double bonds, the bond orders of π bonds are computed. The bond orders are compared with a predefined bond-dissociation criterion (BDC). If a fractional bond order is less than or equal to BDC, the bond is eliminated and the atoms that had been connected by the eliminated bond are labeled as chemically active. Next, a set of new bonds is generated, consisting of all possible covalent interactions between the chemically active atoms. The most energetically favorable subset of the new bonds [which corresponds to the highest value of $\sum^{N}{}_{\text{bonds}}(D - V_{\text{bond}})$] that complies with the rules if atomic valence is selected. If the total number of bonds to an atom is equal to its valence and the fractional bond order of each of its bonds is higher than BDC, the chemically active label is removed from the atom. Otherwise, if an atom retains its chemically active status, it is allowed to form one additional bond, provided that the sum of the bond orders associated with the (valence + 1) bonds to that atom does not exceed its valence. For example, a chemically active carbon atom can participate in five bonds as long as the sum of the bond orders associated with these bonds does not exceed four. This feature is used to describe chemical reactions that occur via so-called "hypervalent" transition states, such as disproportionation or abstraction reactions. Once the bond analysis is complete and the structural information is updated, the next time step of molecular dynamics is executed.

It should be noted that as a result of the bond-breaking/bond-making procedure described above, the bonds between atoms are not formally removed unless they are replaced by new, more energetically favorable bonds. For the purpose of quantifying chemical events, a covalent bond between atoms is considered to be broken when its energy is within $0.25RT$ (R is the ideal gas constant and T is the simulation temperature) of the dissociation energy. The bond-dissociation criterion determines whether covalently bonded atoms are eligible to participate in chemical reactions.

As is the case with most of the traditional force fields, the parameters depend on the types of atoms that are involved in the interactions. Several atom types are used for each chemical element to account for the influence of the chemical environment on the interatomic potentials. During the bond-breaking/bond-making procedure, the atom types of chemically active atoms are updated in accordance with the new arrangement of covalent bonds. When the "chemically active" label is removed from an atom, the types of atoms bonded to the former chemically active atom are also updated. The "chemically active" label does not itself alter properties of the atoms.

The current version of MD_REACT can be used to model any chemical reactions involving σ and/or π bonds, with the exception of triple bonds and aromatics, in a carbon–hydrogen–oxygen molecular system. However, only homolytic bond dissociation is permitted (i.e., no ions can be formed). The partial charges on the atoms participating in breaking and making of covalent bonds are adjusted to accommodate changes in their bonding environments.

MD_REACT is implemented as a Fortran/C computer code that is interfaced to Discover 95. The latter is a commercially available molecular dynamics computer program offered by Accelrys Inc. (formerly Molecular Simulations Inc.).[†] The function of the MD_REACT program is to compute the reactive force field, while Discover 95 updates the molecular geometry on the basis of the solution of the equations of motion. A separate computer program, called Molecview, has been developed to visualize the results of the MD_REACT simulations in three dimensions. Typically, the simulations are carried out in two stages. An equilibration stage is performed at constant (atmospheric) pressure and temperature (NPT ensemble). The structure and velocity distribution from the equilibration stage are then used to initiate the RMD simulations, which are performed at constant volume and temperature (NVT ensemble).

2. CALIBRATION OF THE FORCE FIELD

The atom types (see Table 1) and most of the potential energy parameters used in MD_REACT were taken from version 2.3 of the CVFF, which is the default force field for Discover. However, to improve the parametric description of reactive potential energy surfaces, dissociation energies (D) and equilibrium bond lengths (r_e) used in the CVFF force field were modified on the basis of results obtained from high-level quantum chemical calculations performed on small model compounds. What follows is a brief description of the results of the quantum chemical calculations and how these results were used to obtain the new force-field parameters.

Energetics for the reactions of model compounds were calculated using the CBS-QB3 method[12–15] as implemented in the Gaussian 98 computational chemistry package.[16] CBS-QB3 is a hybrid approach consisting of a series of density-functional and ab initio calculations that determine the geometry of the molecule and extrapolate the energy of the molecule at this geometry to the complete basis

[†]Certain commercial equipment, instruments, materials or companies are identified in this chapter to adequately specify the experimental procedure. This in no way implies endorsement or recommendation by NIST.

TABLE 1. Atom Types and Morse Parameters Used in the CVFF

Atom Type	Description
H	hydrogen atom
C_g	generic sp^3 carbon
C	sp^3 carbon bonded to 4 heavy atoms
C_1	sp^3 carbon bonded to 1 hydrogen and 3 heavy atoms
C_2	sp^3 carbon bonded to 2 hydrogens and 2 heavy atoms
C_3	sp^3 carbon bonded to 3 hydrogens and 1 heavy atom
C=	nonaromatic doublybonded carbon
CC=	sp^3 carbon atom adjacent to C=
C′	carbon in C=O group
CC′	sp^3 carbon atom adjacent to C
O	sp^3 oxygen
O′	oxygen in C=O group

set limit.[12–14] The mean absolute deviation of the $\Delta H^o(T = 0\,\text{K})$ values computed using this method from the experimental values is less than 4.2 kJ/mol.[15]

The energy differences between products and reactant were obtained for the set of bond dissociation and β-scission reactions.[5] The model reactions were selected to represent chemical transformations of importance in the thermal decomposition of poly(methyl methacrylate) (PMMA). Every reactant molecule was represented by its lowest energy conformation. The geometries of the products were obtained by dividing the reactant molecule into the fragments of interest and optimizing the structures of each fragment. Although experimental reaction enthalpies are available for a few of the model reactions, many of the required values have not been measured. In most cases, even when there are experimental values, they have not been evaluated critically, so their accuracies are uncertain. By using the CBS-QB3 method to compute all of the reaction enthalpies used in the force field, we have reasonable assurance that the errors will be both small and uniform across the reaction set.

The energy differences between the products and reactants obtained for the dissociation reactions were assigned to the corresponding bond dissociation energies (D) in the CVFF force field (Table 2). In an effort to account for the effects of quantum mechanics to first order, these energy differences were corrected for zero-point energy contributions [i.e., the D parameters were set equal to the $\Delta H^o(T = 0\,\text{K})$ of the corresponding reactions]. Computed enthalpies of β-scission reactions were used to determine the energies of π bonds. The π-bond energies were calculated by subtracting the $\Delta H^o(T = 0\,\text{K})$ of the β-scission reaction from the energy of the σ bond broken as a result of this process. Bond dissociation energies of the σ bonds in the radical reactants were assumed to be equal to those in the corresponding nonradical species. The D parameters for double bonds were set equal to the sums of energies of the corresponding σ and

TABLE 2.

Bond	r_e[a](Å)	D(kcal/mol)	α[b]($Å^{-1}$)	Comment[c]
1. C_3–C_3	1.53	88.5	1.915	$D = \Delta H(R14)$
2. C_2–C_2	1.53	86.9	1.915	$D = \Delta H(R18)$
3. C_2–C_3	1.53	87.5	1.915	$D = \Delta H(R16)$
4. C–C_2	1.55	84.5	1.915	$D = \Delta H(R19)$
5. C–C_3	1.55	85.1	1.915	$r_e = r_e(4)$ $D = D(4) + (D(3) - D(2))$
6. C–C	1.57	82.1	1.915	$r_e = r_e(4) + [r_e(4) - r_e(2)]$ $D = D(4) - [D(2) - D(4)]$
7. C_1–C_3	1.54	86.3	1.915	Parameters of C_1–X bonds are taken to be equal to the average between C_2–X and C–X bond parameters
8. C_1–C_2	1.54	85.7	1.915	
9. C_1–C_1	1.55	84.5	1.915	
10. C_1–C	1.56	83.3	1.915	
11. C_g–C_1	1.54	85.7	1.915	Parameters of C_g–X bonds are taken to be equal to C_2–X bond parameters
12. C_g–C_2	1.53	86.9	1.915	
13. C_g–C_3	1.53	87.5	1.915	
14. C_g–C	1.55	84.5	1.915	
15. C_g–C_g	1.53	86.9	1.915	
16. H–C_3	1.09	99.9	1.771	$D = \Delta H(R15)$
17. H–C_2	1.10	97.1	1.771	$D = \Delta H(R17)$
18. H–C_1	1.10	94.3	1.771	$r_e = r_e(17)$ $D = D(17) - [D(16) - D(17)]$
19. H–C_g	1.10	97.1	1.771	$r_e = r_e(17)$ $D = D(17)$
20. CC′–C_2	1.55	78.7	1.915	$D = \Delta H(R1)$
21. CC′–C_3	1.54	79.4	1.915	$D = \Delta H(R5)$
22. CC′–C_1	1.56	77.5	1.915	$r_e = r_e(20) + [r_e(8) - r_e(2)]$ $D = D(20) - [D(2) - D(8)]$
23. CC′–C	1.57	76.3	1.915	$r_e = r_e(20) + [r_e(4) - r_e(2)]$ $D = D(20) - [D(2) - D(4)]$
24. CC′–C_g	1.55	78.7	1.915	$r_e = r_e(20)$ $D = D(20)$
25. CC′–H	1.10	88.9	1.771	$D = D(17) - [D(2) - D(20)]$
26. C′–CC′	1.53	91.5	1.915	$D = \Delta H(R2)$
27. C′–C_g	1.53	91.5	1.915	$r_e = r_e(26)$ $D = D(26)$
28. C′–H	1.11	97.1	1.771	$D = D(19)$
29. C=–C=	1.33	146.2	2.000	$D = D(15) + [\Delta H(R13) - \Delta H(R9)) + \Delta H(R1) - \Delta H(R12)) + (\Delta H(R15) - \Delta H(R8)) + (\Delta H(R5) - \Delta H(R11)) + (\Delta H(R2) - \Delta H(R10))]/5$
30. C=–C_g	1.50	98.4	2.000	$D = \Delta H(R20)$
31. C=–C′	1.49	103.0	2.000	$D = D(30) + (D(27) - D(15))$
32. C=–CC′	1.52	90.2	2.000	$r_e = r_e(30) + [r_e(24) - r_e(15)]$ $D = D(30) - [D(15) - D(24)]$
33. C=–H	1.09	106.1	1.771	$D = \Delta H(R22)$
34. CC=–C=	1.50	98.4	2.000	$r_e = r_e(30)$ $D = D(30)$

(*continued overleaf*)

TABLE 2. (*Continued*)

Bond	r_e[a](Å)	D(kcal/mol)	α[b]($Å^{-1}$)	Comment[c]
35. CC=–C_3	1.54	73.4	1.915	$D = \Delta H$(R21)
36. CC=–C_2	1.54	72.8	1.915	$r_e = r_e(35)$ $D = D(35) - [D(13) - D(12)]$
37. CC=–C_1	1.55	71.6	1.915	$r_e = r_e(35) + [r_e(11) - r_e(13)]$ $D = D(35) - [D(13) - D(11)]$
38. CC=–C	1.56	70.4	1.915	$r_e = r_e(35) + [r_e(14) - r_e(13)]$ $D = D(35) - [(D(13) - D(14)]$
39. CC=–C_g	1.54	72.8	1.915	$r_e = r_e(36)$ $D = D(36)$
40. CC=–C′	1.54	77.4	1.915	$r_e = r_e(35)$ $D = D(35) + [D(27) - D(13)]$
41. CC=–CC′	1.56	64.6	1.915	$r_e = r_e(35) + [r_e(24) - r_e(13)]$ $D = D(35) - [D(13) - D(24)]$
42. CC=–H	1.10	83.0	1.771	$D = D(19) - [D(15) - D(39)]$
43. CC=–CC=	1.55	58.7	1.915	$r_e = r_e(39) + [r_e(39) - r_e(15)]$ $D = D(39) - [D(15) - D(39)]$
44. C′–C′	1.53	96.1	1.915	$r_e = r_e(27)$ $D = D(27) + [D(27) - D(15)]$
45. CC′–CC′	1.57	70.5	1.915	$r_e = r_e(24) + [r_e(24) - r_e(15)]$ $D = D(24) - [D(15) - D(24)]$
46. H–H	0.74	104.4	1.956	$D = \Delta H$(R23)
47. O–C_3	1.42	86.9	2.000	$D = \Delta H$(R3)
48. O–C_2	1.42	86.0	2.000	$D = \Delta H$(R26)
49. O–C_1	1.43	84.8	2.000	$r_e = r_e(48) + [r_e(11) - r_e(12)]$ $D = D(48) - [D(12) - D(11)]$
50. O–C	1.44	83.6	2.000	$r_e = r_e(48) + [r_e(14) - r_e(12)]$ $D = D(48) - [D(12) - D(14)]$
51. O–C_g	1.42	86.0	2.000	$r_e = r_e(48)$ $D = D(48)$
52. O–C′	1.35	100.3	2.000	$D = \Delta H$(R4)
53. O–CC′	1.44	77.8	2.000	$r_e = r_e(51) + [r_e(24) - r_e(15)]$ $D = D(51) - [D(15) - D(24)]$
54. O–C=	1.39	97.5	2.000	$r_e = r_e(51) - [r_e(15) - r_e(30)]$ $D = D(51) + [D(30) - D(15)]$
55. O–CC=	1.43	71.9	2.000	$r_e = r_e(51) + [r_e(39) - r_e(15)]$ $D = D(51) - [D(15) - D(39)]$
56. O–H	0.96	104.1	2.280	$D = \Delta H$(R24)
57. O–O	1.47	39.6	1.718	$D = \Delta H$(R25) α is derived from the corresponding calculated vibrational frequency
58. O′–C′	1.21	194.8	2.060	$D = D(51) + [(\Delta H(R2) - \Delta H(R6)) + (\Delta H(R3) - \Delta H(R7))]/2$

[a] r_e parameters are obtained from the B3LYP/CBSB7 optimized geometries of the model molecules unless stated otherwise in the Comment section.

[b] α parameters are taken from version 2.3 of the CVFF provided by Accelrys Inc. unless stated otherwise in the Comment section.

[c] (R#) is the reference to the reaction number in Table I in reference 3, (#) is the reference to the bond number in this table.

π bonds. Equilibrium bond-length (r_e) parameters were obtained directly from the optimized structures of the reactants and products.

The set of the Morse parameters determined from the quantum chemical calculations was expanded by performing a series of simple linear extrapolations to account for different atom types. For example, the energy of the bond between C, an sp^3 carbon attached to four nonhydrogens, and C_3, an sp^3 carbon attached to three hydrogens and one nonhydrogen, was computed as

$$D(\mathrm{C} - \mathrm{C}_3) = D(\mathrm{C} - \mathrm{C}_2) + D(\mathrm{C}_2 - \mathrm{C}_3) - D(\mathrm{C}_2 - \mathrm{C}_2) \qquad (6)$$

where C_2 is the sp^3 carbon attached to two hydrogens and two nonhydrogens.

3. THERMAL DECOMPOSITION OF POLY(METHYL METHACRYLATE)

The thermal degradation of PMMA is a well-studied process. Although it is known that PMMA depolymerizes almost exclusively to the monomer when heated, there are still unanswered questions about the nature of the initiation reaction and the relative reactivities of the primary and tertiary radicals formed in the degradation process.

Simulations were performed on a model of PMMA consisting of a single 15-unit polymer chain using periodic boundary conditions (Figure 1) to account for the nonbonded interactions between the reacting chain and the bulk polymer. The chains were terminated by an H atom at one end and a CH_3 group at the other end. The simulations were performed at 1000, 1200, and 1500 K. The total time for each reactive dynamics simulation was chosen in such a way that the degree of conversion would be 20 to 30%, determined as the ratio of the mass of

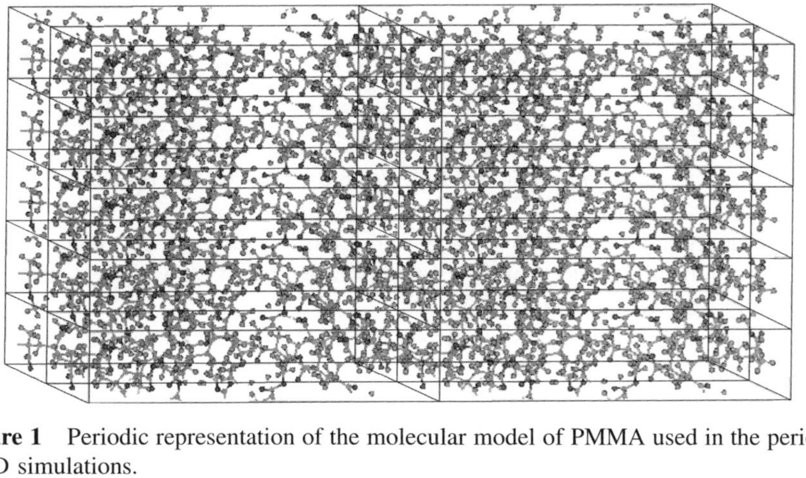

Figure 1 Periodic representation of the molecular model of PMMA used in the periodic RMD simulations.

volatile products versus initial mass of the polymer. For this purpose, products with masses less than the mass of methyl methacrylate dimer were considered volatile. The value of 1000 K as the lowest temperature was dictated by the necessity to observe reactive events within computationally feasible simulation times. The average density of the equilibrated PMMA systems showed a weak temperature dependence decreasing from 1.04 g/cm^3 at the lowest simulation temperature to 1.02 g cm^{-3} at the highest. To accumulate statistics on chemical reactions, 10 independent simulations were done at every temperature.

The simulations indicate that the decomposition of the polymer is initiated by random scissions of the main chain. The resulting tertiary and primary macroradicals then "unzip," undergoing β-scission reactions to give the monomer as indicated in reactions 1 and 2:

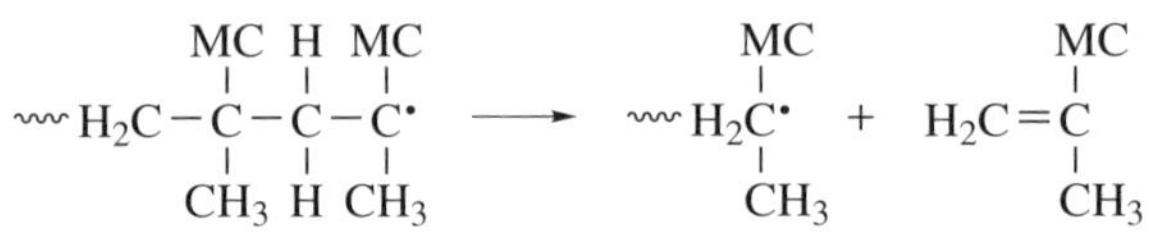

(reaction 1)

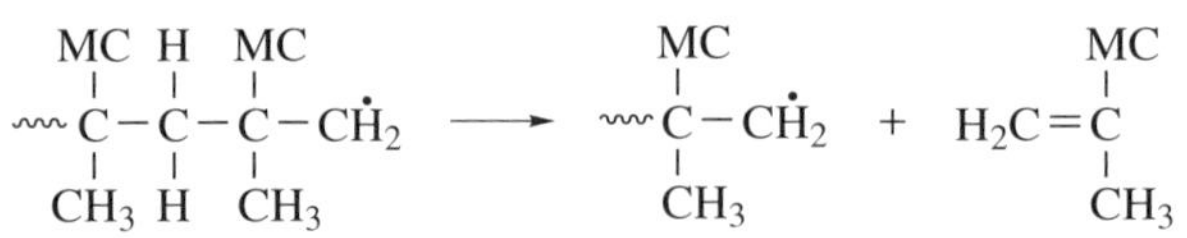

(reaction 2)

For the tertiary macroradical, reaction 1 is essentially the only reaction channel observed in the simulations. However, the primary macroradicals also undergo the alternative β-scission reactions:

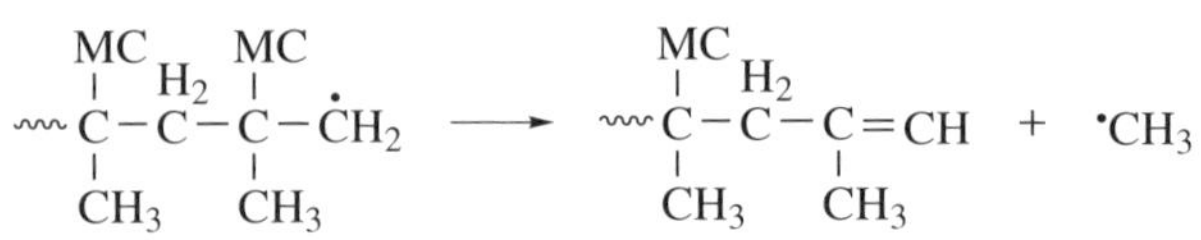

(reaction 3)

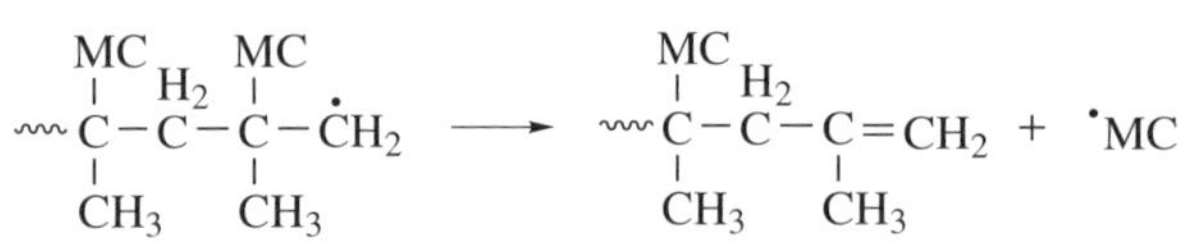

(reaction 4)

These reactions result in first-order termination of the depolymerization process. This difference in behavior of the tertiary and primary macroradicals was first postulated by Kashiwagi et al.[17] These authors proposed the β-scission of a pendant methyl carboxylate (MC) group in order to provide a mechanistic explanation of the results of their measurements of the molecular-weight distributions of partially degraded PMMA samples.

Our simulations show that reactions 1 and 2 produce monomers at similar rates. At the lowest simulation temperature (1000 K), several monomers are generated by the primary radical site before termination occurs by reaction 3 or 4. However, as the temperature increases, the number of monomers produced by primary macroradicals decreases. At 1500 K, less than 0.5 monomer (on average) are generated by the primary macroradical before termination. Thus, while the rates of reactions 1 and 2 increase with increasing temperature, the primary radical channel is effectively shut down by the competing termination reactions, resulting in the formation of oligomers with an unsaturated end.

The simulations indicate that methyl methacrylate (monomer) is the dominant volatile product of the thermal degradation of PMMA. The average mass fraction of monomer (with respect to the total mass of volatile products) decreased from 90% to 80% as the temperature was increased from 1000 to 1500 K. The generation of monomer was accompanied by formation of a number of low-molecular-weight stable species (H_2, CO, CO_2, CH_4, C_2H_4, C_2H_6, $HCOOCH_3$) in trace amounts. Experimental analysis of the volatile products of the decomposition of PMMA[18] indicates that 81 to 83% of the product is monomer at 1070 K. This value, along with the experimental results of analysis of trace volatile products, is in good agreement with the results of the MD_REACT simulations. However, at 1470 K the experimentally determined fraction of methyl methacrylate decreases to about 13%. This decrease is attributed to cracking of the monomer in secondary reactions. Although some decomposition of the monomer was observed in the simulations, full-scale modeling of such secondary processes would require much longer simulation times and was not the primary goal of this study.

4. THERMAL DECOMPOSITION OF POLY(BISPHENOL A CARBONATE)

Like PMMA, poly(bisphenol A carbonate) (PC) is transparent to visible light and is therefore used for many of the same applications. However, because it is more thermally stable, PC is used in place of PMMA when thermal stability and fire resistance are important. Simulations were performed on single chains consisting of 25 bisphenol A carbonate monomers using periodic boundary conditions. The simulation temperatures ranged from 2000 to 2500 K. These high temperatures were needed because of the exceptional thermal stability of this polymer, which failed to react in reasonable simulation times at the lower temperatures used in the PMMA simulations described above. The predominate reactions observed in the simulations were side-chain scissions of the methyl groups and main-chain scissions of the C–O bonds followed by the transfer of a hydrogen (from a nearby methyl group), resulting in the formation of oligomeric fragments terminated by OH. A still frame from a representative simulation is shown in Figure 2.

Experimental measurements of the thermal decomposition products were made by matrix-assisted laser desorption/ionization mass spectrometry (MALDI-MS), which was performed on PC samples before and after partially decomposing the sample by heating it for several minutes in a TGA at 400°C.[19] The samples were irradiated with a standard nitrogen laser in a matrix of 2,5-dihydroxybenzoic acid (DHB) and sodium trifluoroacetate to provide sodium cations.

Representative spectra from samples before and after heating are compared in Figure 3. Notice the reduction in the intensities of the peaks at masses of

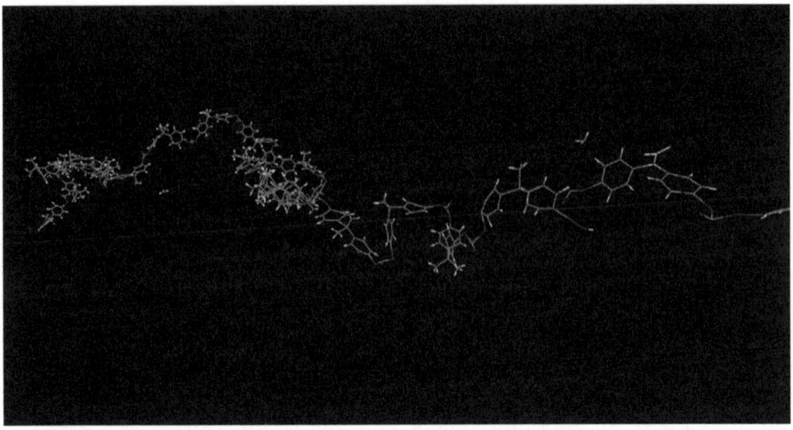

Figure 2 Still frame from RMD simulations showing the formation of an OH-terminated fragment (oxygen atoms are colored red and hydrogens silver) and free methyl groups resulting from side-chain scission reactions. (*See insert for color representation figure*.)

about 3040 and 3080 amu and the enhancement of intensities at 3170 and 3310 amu. These changes correspond to a decrease in the concentrations of linear and cyclic polymers and a corresponding increase in linear polymer terminated by OH on one side (and phenyl on the other) and by OH on both sides, which are consistent with the mechanism observed in the RMD simulations. In addition,

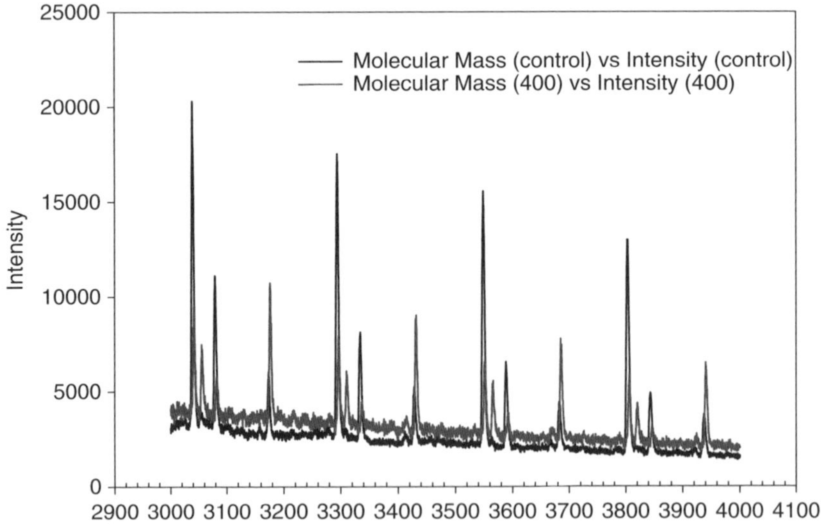

Figure 3 Comparison of the MALDI-MS spectra of the partially decomposed (upper) and original (lower) samples.

analysis of the gases evolved during the thermal decomposition of the samples by Fourier transform infrared spectroscopy confirmed the presence of methane, which would be the expected product of the side-chain scissions (of the methyl groups) observed in the simulations.

5. THERMAL DECOMPOSITION OF POLY(ISOBUTYLENE)

Polymer debinding is involved in many powder metallurgy processes, where clean removal of the binder is essential to ensure a sound product. Poly(isobutylene) is under consideration for applications as a fugitive binder because of its favorable rheological properties.[20] RMD simulations were performed to gain insights into the mechanisms of the thermal decomposition of PIB and to determine whether the presence of titanium affects the thermal decomposition process.[21] The simulations were performed on molecular models of the polymers consisting of a single chain with 50 monomers terminated by H atoms. In an effort to determine the effects of the Ti–polymer interactions on the thermal decomposition of the polymer, simulations were performed with and without the nonbond interactions between the polymer and a <100> titanium surface consisting of 190 Ti atoms. The surface was kept immobile and nonreactive. Periodic boundary conditions were used, but the cell was elongated in the direction perpendicular to the titanium surface to provide a quasi two-dimensional model of a polymer chain interacting with a surface. As a consequence of this construction, the densities of the model polymers in these simulations were unrealistically low.

The simulations indicate that the mechanism of the thermal decomposition of PIB is similar to that of PMMA. The major reactions are illustrated in a series of still frames presented in Figure 4. Although these particular frames were taken from a simulation without the Ti surface, we did not observe any systematic differences when the nonbonded interactions between the polymer and the surface were included. Thus, the following description of the reactions applies equally well to both sets of simulations. A random scission of the polymer chain is evident in the first frame of Figure 4 at 0.20 ps ($1\,ps = 10^{-12}$ s). A monomer forms by scission of the C–C bond β to the site of the primary radical at 0.23 ps. Another random scission and several more β-scission reactions can be seen to occur at another site in the sequence of frames from 0.33 to 0.39 ps. At 0.60 ps, the propagation of monomers from the second site is terminated by the β-scission of a methyl group from a primary radical (indicated by arrow).

Taken as a whole, the 112 simulations performed during this investigation indicate that monomer is the primary volatile compound produced in the thermal decomposition of PIB. In fact, they account for an average (over all seven temperatures) of about 65% of the mass of all the fragments containing 12 carbons or less produced in the simulations. This observation is consistent with experimental measurements reported by Madorsky.[18] At the high temperatures used in the simulations, however, it would be expected that once formed, a large fraction of the monomers would decompose into smaller hydrocarbons. Although we did

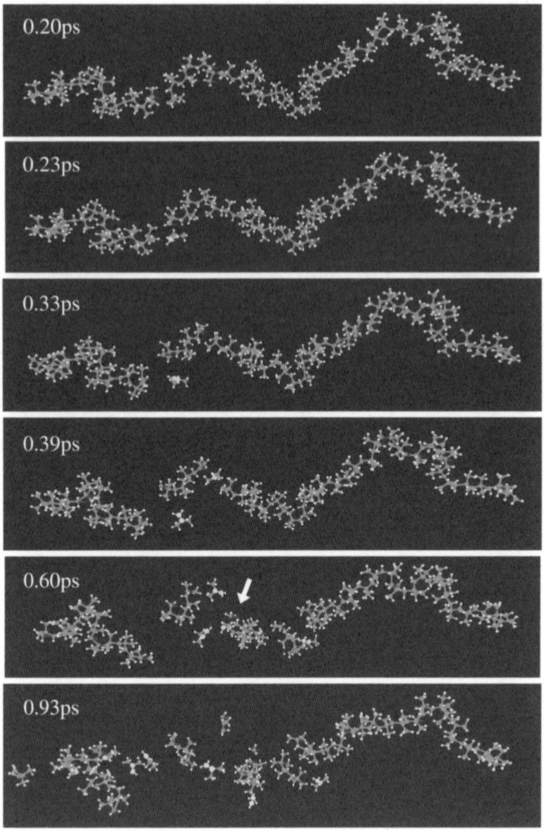

Figure 4 Still frames from a PIB simulation at 2000 K showing chain scission followed by propagation and termination (gray atoms are hydrogen, green indicates single-bonded carbon bonds, yellow indicate's double-bonded carbon bonds, and blue indicates radical sites). (*See insert for color representation figure*.)

observe secondary reactions resulting in the formation of CH_3 and C_3H_6 from the thermal decomposition of monomers, the monomer yields remained substantial even in the highest-temperature simulations, presumably because the simulation times ($\sim 10^{-11}$ s) are not sufficient for the effects of these secondary reactions on the product distributions to become evident.

The formation of trace amounts of C_8H_{16} (cyclic dimer) was also observed in some simulations. An example of this reaction is shown in Figure 5. The circled area at 2.28 ps shows the formation of a monomer from β-scission of the primary radical. Propagation continued at the same site to produce another isobutylene monomer at 2.36 ps. With the formation of the second monomer, the primary radical site reacted with the tertiary radical site at the other end of the chain, forming the cyclic dimer as shown in the frame at 2.47 ps. This structure was stable for the next 7 ps of the simulation.

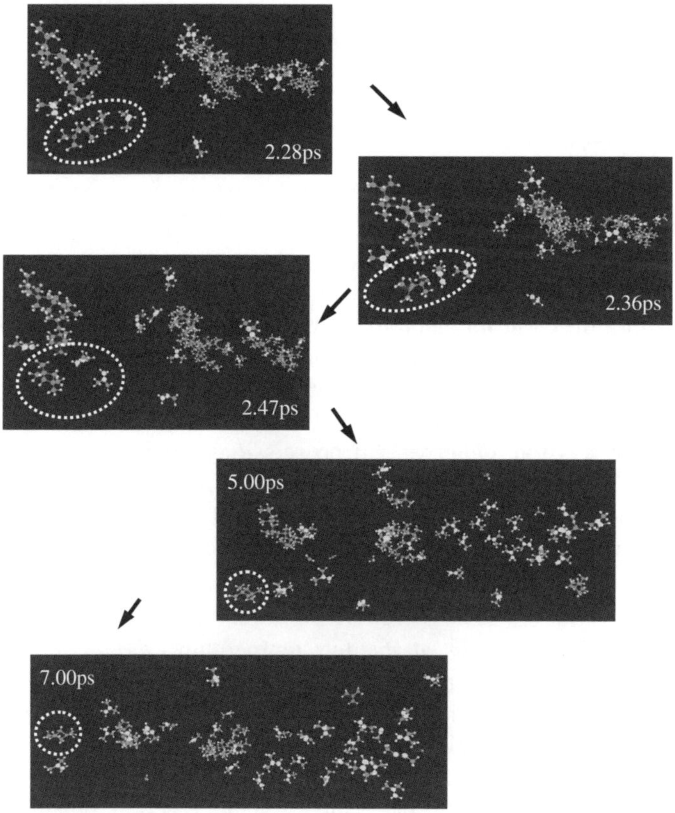

Figure 5 Still frames obtained from a simulation of PIB at 2000 K showing the formation of cyclic dimer (in the white circle). (*See insert for color representation figure.*)

Hydrogen transfer reactions, which are thought to be important at relatively low temperature, did not occur with high frequency in any of the simulations. These reactions usually have lower activation energies than β-scission reactions, but they also tend to have much smaller preexponential factors. Thus, although they may compete effectively with β-scission reactions at low temperatures, at some point (presumably, at much lower temperatures than we can simulate), the rate of β-scission overtakes the rate of hydrogen transfer, and the formation of monomer predominates.

Rate constants for random scission of the main chain were computed for each model (i.e., with and without the surface) at seven values of the temperature between 1300 to 1600 K from the formula

$$K_s(T) = \frac{\sum_i n_i}{\sum_i t_i} \tag{7}$$

TABLE 3. Arrhenius Parameters for Backbone Scission Reactions in PIB and Ti/PIB

Model	ln(Å/s)	E_a(kJ/mol)
PIB	35 ± 5 (39.6 ± 1)	241 ± 7 (247 ± 8)
Ti/PIB	36 ± 5 (41.6 ± 4)	246 ± 8 (258 ± 20)

In this equation, i is an index used to distinguish the independent simulations (initiated with a different momentum distribution consistent with the Boltzmann distribution at the specified temperature) performed at each temperature. The values of n_i and t_i were obtained as follows. If at least one scission was observed during the ith simulation (at the specified temperature), then $n_i = 1$ and t_i is the time to the first scission. Otherwise, if no scissions were observed, $n_i = 0$ and t_i is the total simulation time. Activation energies, E_a, and preexponential factors, A, were obtained from the slopes and intercepts of Arrhenius plots in a manner consistent with

$$k(T) = Ae^{-E_a/RT} \tag{8}$$

where $k(T)$ is the rate constant at absolute temperature, T.

At the relatively low thermal decomposition temperatures that prevail in TG experiments (620 to 730 K), the rate of mass loss should be controlled by the rate of the random scission initiation reactions because the activation energy for this reaction is much higher than the activation energies of the hydrogen transfer, and β-scission reactions. The kinetic parameters for this reaction, obtained from both sets of simulations (i.e., with and without the surface), are reported in Table 3. The activation energies obtained from the simulations are in good agreement with the experimental TG results for PIB (listed in parentheses in Table 3). The preexponential factors from the simulations are also consistent with the values from the TG measurements if they are scaled to correct for the difference between the number of monomers in the molecular models (50) and the average degree of polymerization of the PIB in the experimental samples ($\sim$7000). Moreover, since the values obtained with and without the surface are nearly identical, the current simulation results support the experimental observation that the mass-loss kinetics of PIB is unaffected by the presence of the Ti.

6. RATES OF SCISSION REACTIONS FOR HOMOLOGOUS POLYMERS

Additional RMD simulations were performed on the homologous series of vinyl polymers: polyethylene (PE), polypropylene (PP), and PIB to determine the rates of backbone scission reactions. The kinetic parameters obtained from the simulations are listed in Table 4. These data suggest that the kinetics of reactions

TABLE 4. Arrhenius Parameters for Backbone Scission Reactions for PE, PP, and PIB

Polymer	ln(Å/s)	E_a(kJ/mol)
PE	39.0	275
PP	34.5	185
PIB	34.5	170

involving bond dissociations in polymers can differ substantially from the analogous reactions in smaller molecules. At the high temperatures employed in the simulations, the activation energies are substantially smaller than the corresponding bond energies. Moreover, a comparison of the Arrhenius parameters obtained from the simulations on PE, PP, and PIB reveals that the activation energies tend to decrease with increasing substitution. The magnitude of this effect is much larger than the relatively small (~20 kJ/mol) decrease in bond dissociation energies attributable to the resonance stabilization of the tertiary radical fragment that forms as a result of breaking the $RCH_2{-}C(CH_3)_2R$ bond in PIB.

The effect of the length of the polymer chains on the Arrhenius parameters was also explored. The results from simulations performed on a series of PIB oligomers ranging in length from 4 to 150 monomers is summarized in Table 5. Based on these data, it seems clear that whatever is responsible for the anomalously low activation energies in PIB is related to the length of the polymer chain. Indeed, our simulations on the isobutylene dimer indicate that the activation energy for chain scission in this molecule is on the order of the bond dissociation energy (~345 kJ/mol) obtained from the quantum mechanical calculations on model compounds.

One aspect of the problem that requires further examination is the variation of bond energies along the polymer backbone. In fact, we think that it is likely that the scission reactions may be occurring at weak links resulting from steric interactions between the bulky pendant groups. This may explain why the activation energies obtained from RMD simulations using low-density

TABLE 5. Arrhenius Parameters for Backbone Scission Reactions for PIB Oligomers

Polymer	ln(Å/s)	E_a(kJ/mol)
PIB4	39.8 ± 1.5[a]	239 ± 19[a]
PIB14	37.9 ± 1.7[a]	224 ± 21[a]
PIB50	35.7 ± 0.7[a]	189 ± 9[a]
PIB150	34.6 ± 0.8[a]	170 ± 9[a]

[a]The uncertainties are one standard deviation (1σ). The uncertainties were calculated from deviations between the rate constant data and the least-squares fit.

models of PIB (Table 3) are significantly higher than those obtained at more realistic densities (Table 5). We investigated this hypothesis by performing quantum chemical calculations using the multilayer ONIOM method[22–25] available in the Gaussian software package.[26] More specifically, high-level calculations [density functional, B3LYP/6-31G(d,p)] were used for the carbon atoms of the decomposing C–C bond and all atoms bonded to these carbon atoms; low-level calculations (molecular mechanics with Amber[27] force field) were used for the rest of the macromolecule. Also, a series of pure molecular mechanics calculations were performed using a modified Amber force field with incorporated Morse stretches used to describe bond breaking. In these calculations, extensive potential energy surface (PES) scans were performed for polymer globules (500-monomer polyethylene and 300-monomer polyisobutylene) following initial structure preparation via molecular dynamics simulations performed at 600 K. In these PES scans, positions of dissociating bonds were randomly selected along the polymer chains; C–C distances of the dissociating bonds were then incremented successively by small amounts while all the remaining internal coordinates of the macromolecule were optimized (i.e., potential energy was minimized).

The results demonstrated significant effects of internal polymer strain and attractive dispersion forces on the PES of reactions. In particular, a distribution of effective activation energies for C–C backbone scission was observed. Most C–C bonds in the backbone chain have dissociation energies clustered around the C–C bond strength of an isolated prototype gas-phase molecule (e.g., *n*-octane for PE). However, a fraction of these bonds have significantly smaller dissociation energies. The corresponding C–C bond scans yielded PES profiles with sharp potential energy drops caused by sudden rearrangements of the polymer chain on a multimonomer scale (i.e., the scale of 5 to 50 monomer units). Presumably, in these cases steric strain is released due to the increased mobility of the polymer chains resulting from bond scissions. Such PES profiles are illustrated in Figure 6.

7. COMPARISON OF RATE CONSTANTS OBTAINED FROM RMD AND TST CALCULATIONS

In an attempt to evaluate the applicability of existing statistical rate theories to reactions of macromolecules in polymer melt, a systematic comparison of polyethylene chain dissociation rates obtained in transition-state theory (TST) and RMD modeling was performed.[28] These calculations were carried out for isolated single molecules (i.e., molecules in the gas phase). Polyethylene chains ranging in length from PE-1 to PE-1000 were studied. Here, the number corresponds to the number of monomer units in the polymer chain: PE-1 is ethane, PE-5 is decane, and so on. Calculations were performed using the same representation of the potential energy surface in both the TST and the RMD modeling. Since the main purpose of the study was investigation of the hypothetical effects of chain length, rather simplistic potential energy surfaces (PES's) given by the all-atom OPLS-AA force field[29] (which is very similar to that of Amber) with

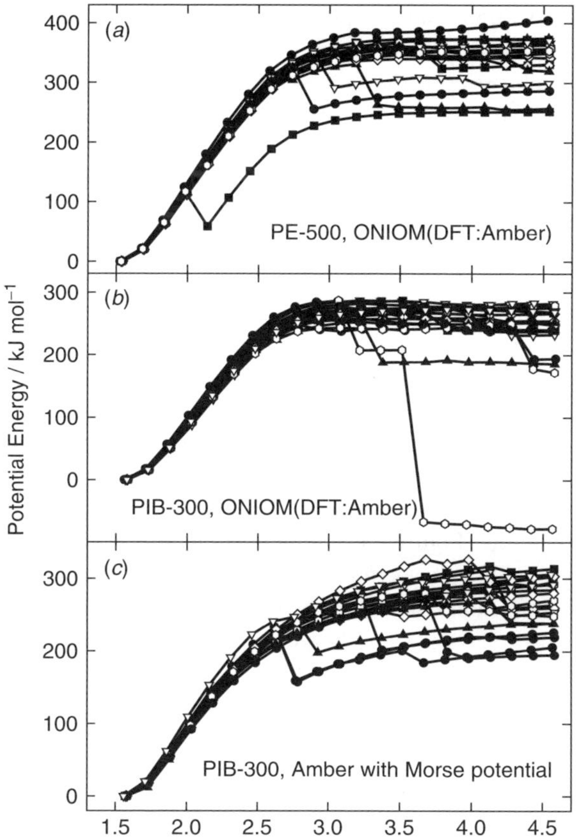

Figure 6 Results of a series of relaxed PES scans of C–C bond scission in (a) a 500-monomer-unit PE globule and (b,c) a 300-monomer-unit PIB globule. Different symbols and lines represent different individual C–C bonds within the same globule. The results demonstrate the existence of a distribution of effective activation energies for C–C backbone scission. Calculations were performed using the ONIOM(B3LYP/6-31G(d,p):Amber) (a,b) and Amber molecule mechanics (c) methods.

C–C stretch terms replaced by the Morse function were used. One feature of such a force field that immediately draws attention as unrealistic is that the force constants of bending degrees of freedom do not change with elongation of adjacent bonds. For this reason, the resulting modeling cannot be expected to reproduce the effects of "loose" transition states of C-C bond dissociation (e.g., refs. 30 and 31. Nevertheless, since the same PES is used in both types of modeling, this deficiency should not affect comparisons of TST- and RMD-based rate constants.

TST calculations were performed for dissociation of two molecules: PE-1 ($C_2H_6 \rightarrow CH_3 + CH_3$) and PE-5 ($nC_{10}H_{22} \rightarrow nC_5H_{11} + nC_5H_{11}$). Potential energy profiles and projected vibrational frequencies[32] were calculated along the reaction path using a Gaussian03 program package.[26] For these calculations, parameters of the Amber force field included in Gaussian03 were modified to most closely match those of the OPLS-AA force field. The canonical variational transition-state theory approach[31] was used to obtain the rate constant values. Classical (nonquantum) partition functions were used in all TST calculations to maintain compatibility with RMD modeling, which is inherently classical.

RMD calculations were performed using the GROMACS program package.[33] For each polyethylene molecule, a series of 10 to 30 calculations were performed at each temperature. Each MD calculation was continued until one carbon–carbon bond dissociated and the time required to achieve dissociation was recorded. Per-bond rate constants were calculated using formula (7) (with all $n_i = 1$) and dividing by the number of bonds in the chain. The temperatures used in modeling ranged from 2100 to 3000 K; such high temperatures were necessary to make the RMD rates observable within the practicable range of times achievable in molecular dynamics simulations.

The results demonstrate the effects of chain length on the rate of polymer backbone dissociation (Figure 7). Although transition-state theory does not predict any increase in the dissociation rate constant with increasing chain length, RMD calculations demonstrated significantly faster backbone scission (in terms of rate constants expressed per C–C bond, k_{CC}) for longer molecules. RMD results for PE-1 (ethane) agree with those obtained in TST calculations. However, this agreement breaks down for larger molecules, for which RMD predicts larger rate constants. The dependence of k_{CC} on the chain length N in PE-N displays a fast growth at low N ($N = 1$–5) and saturation at larger N values ($N = 25$ to 1000). On average, RMD rates of C–C bond dissociation of long-chain molecules are ~10 times larger than those of ethane. The difference between the results of TST and RMD calculations is also close to a factor of 10 for these long-chain molecules.

Even larger effects of reaction acceleration were found for macromolecules that were forced to remain in extended configuration by fixing the positions of the ends. Three series of RMD calculations were performed for PE-250 molecules. In one of the series, the molecule was not restricted in any way and thus was free to undergo changes in configuration during molecular dynamics simulations. In the second series, the distance between the terminal C atoms was fixed at 90% of its equilibrium value in extended PE-250 configuration. In the third series, this distance was kept at 100% of the equilibrium value. As can be seen from the plot in Figure 8, such restrictions imposed on the molecule conformation resulted in significant further increases in the per-bond rates of dissociation: by approximately a factor of 5 in the 90% extended chain case and approximately a factor of 50 in the 100% extended chain case. The restricted distance models bear similarity to condensed-phase (polymer melt) situations where motion of polymer chains is limited by interactions with neighboring chains. In this respect, the observed effects of conformational restrictions resulting in increases in the chain dissociation rates may contribute to the fact that dissociation of polymers

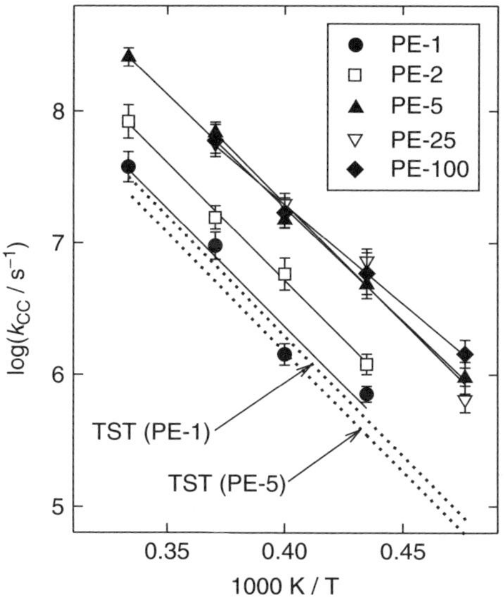

Figure 7 Per-bond rate constants k_{CC} for C–C backbone scission obtained in TST (dotted lines) and RMD (symbols and fitted solid lines) simulations. The plot shows the effect of chain length: rate constants increase with increasing chain length; this dependence is saturated for polyethylene chains longer than PE-5.

in their melts occurs with rates that are significantly larger than the reaction rates of similar (but smaller) hydrocarbons in the gas phase.

The increases observed in the per-bond rate constant with increasing chain length and with conformational restrictions imposed on the macromolecule were found to correlate with corresponding increases in average C–C bond lengths. Average C–C bond-length values observed in molecular dynamics simulations were larger than the equilibrium value even if harmonic (nonreactive) potential was used; the differences (up to 0.047 Å) increased with temperature and chain length. This correlation brings forth a hypothesis that the increases in reaction rates relative to the case of small molecules may be, at least partially, caused by the distortion of the effective potential for C–C dissociation due to chain elongation as described by Crist et al.[34] and later by Hageman et al.[35] and Krisyuk.[36]

8. DEVELOPMENT OF A NEW FORCE FIELD FOR RMD

As indicated above, the force field used in performing the calculations reported in this chapter is dynamic in that the atom types are updated at every time step to accommodate changes in the chemical environment caused by bond breaking or bond making. This feature can result in artificial jumps in the potential energy, which although partially offset by the temperature control in the RMD/NVT

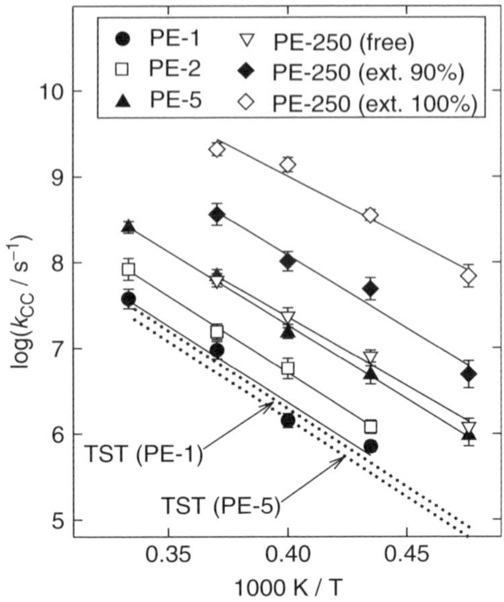

Figure 8 Per-bond rate constants k_{CC} for C–C backbone scission obtained in TST (dotted lines) and RMD (symbols and fitted solid lines) simulations. The plot shows the effect of restrictions imposed on chain motions: rate constants increase in the case of chains forced to remain in 90% or 100% extended configurations (see the text).

simulations, can nevertheless lead to undesirable effects. This problem is circumvented in a new approach, RMDff (RMD force field), which is currently being developed in collaboration with the University of Massachusetts.[37] The basic idea behind RMDff is the same as that of MD_REACT: Potential energy surfaces of chemical reactions are constructed from functions and parameters of a traditional (nonreactive) force field. However, rather than predicting reactions on the basis of a bond breaking/bond making algorithm, RMDff is being designed as a tool for incorporation of detailed potential energy surfaces (obtained from quantum calculations and/or experiments on small molecules) of predefined reaction types into a large-scale molecular simulation.

In RMDff, each type of chemical reaction (C–C single-bond scission, H abstraction from a hydrocarbon by Cl, etc.) is described by a reaction coordinate (RC), which is a function of coordinates of atoms involved in the reaction. RC is chosen in such a way that it defines the progress of reaction and corresponds to the minimum energy path from reactants to products. Continuous switching functions (SW) that depend on the reaction coordinate are used to interpolate between nonreactive energy terms describing the state of reactants (V_{rcts}), transition state (V_{ts}), and the state of products (V_{prods}). The presence of a transition state in this description is optional. A smooth transition is accomplished by using a linear combination of the energy terms:

$$V_{\text{react}} = \sum_i \text{SW}^i_{\text{rcts}}(\text{RC})V^i_{\text{rcts}} + \sum_j \text{SW}^j_{\text{ts}}(\text{RC})V^j_{\text{ts}} + \sum_k \text{SW}^k_{\text{prods}}(\text{RC})V^k_{\text{prods}} \quad (9)$$

and/or by switching the values of the term parameters (α):

$$V_{\text{react}} = \sum_i V_i(\text{SW}^i_{\text{rcts}}(\text{RC})\alpha^i_{\text{ts}} + \text{SW}^i_{\text{prods}}(\text{RC})\alpha^i_{\text{prods}} \cdots \quad (10)$$

The switching functions turn on (SW = 1) for values of RC that are in the vicinity of the corresponding state and turn off (SW = 0) for values of RC that correspond to geometries that are far from that state. The minimum energy differences between reactants, transition state, and products are accounted for by constant terms (which are also multiplied by the corresponding switching functions) that are added to the sums of the transition state and product energies. At first glance, Eq. (9) suggests that during a reactive event, the number of force-field calculations triples. However, in a large molecular system, only a small fraction of potential energy terms (those that correspond to the atoms participating in reactions) would undergo this transition at any given time.

An important feature of RMDff is that one reaction potential can be built on the basis of another. This is realized by making the reaction coordinate be not only a function of atomic coordinates but also of another reaction coordinate. Extension of the RC definition makes it possible to use simple reaction potentials in the construction of more complex ones. For example, a C–H bond scission potential can be used as a basis for an H abstraction potential. An additional benefit of such hierarchical approach is that more complex reactive configurations need to be examined only when the corresponding basis reactions are detected (i.e., when the RC values of the basis reactions deviate considerably from that of the reactants).

9. CONCLUSIONS

The results of reactive molecular dynamics simulations of the thermal degradation of poly(isobutylene), poly(methyl methacrylate), and poly(bisphenol A carbonate) were compared to experimental measurements of product distributions and mass-loss kinetics. The mechanistic information and product distributions obtained from these simulations are in agreement with experimental observations for all three polymers. The kinetic parameters for the random scission initiation reactions obtained from the RMD simulations of the thermal decomposition of PIB and Ti/PIB are consistent with experimental values obtained from mass-loss measurements and confirm that the presence of Ti does not have a significant effect on the thermal decomposition kinetics of PIB. These favorable comparisons are evidence of the predictive power of RMD and instill confidence in its application for future investigations of the thermal decomposition mechanisms of large molecules and nanostructures.

The simulations also reveal interesting and meaningful information that cannot be obtained directly from experimental measurements. For example, our simulations of the thermal decomposition of a series of structurally related polymers indicate that the kinetics of polymer scission reactions differs substantially from that of bond dissociations in small gas-phase molecules, apparently because of conformational rearrangements in the fragmenting polymer chains. This is significant because conventional kinetic modeling of macromolecules and nanostructures relies, at least in part, on the presumption that rate constants for the elementary reactions can be described in terms of activation energies and preexponential factors obtained from measurements (and/or calculations) on smaller (typically, gas-phase) molecules. The results of our RMD simulations suggest that this practice will result in unrealistic predictions. Furthermore, since the mobility of the constituent atoms appears to be responsible for this disparity, we suspect that the differences will become more significant with increasing size and complexity of the system under investigation.

Acknowledgments

Significant contributions to the content of this chapter were made by our collaborators, Philip Westmoreland and Ken Smith from the Department of Chemical Engineering, University of Massachusetts–Amherst, and Xiao Guo and Caroline Jee from the Department of Materials, Queen Mary, University of London. The authors would also like to acknowledge the long-term support of the Federal Aviation Administration's Fire Safe Aircraft Cabin Materials Program, and its manager, Richard Lyon.

REFERENCES

1. Gardiner, W. C., Jr., Ed. *Combustion Chemistry*, Springer-Verlag; New York, 1984.
2. Seinfeld, J. H.; Pandis, S. N. *Atmospheric Chemistry and Physics*, Wiley, New York, 1998.
3. Nyden, M. R.; Forney, G. P.; Brown, J. E. Molecular modeling of polymer flammability: application to the design of flame-resistant polyethylene, *Macromolecules* 1992, 25: 1658.
4. Nyden, M. R.; Coley T. R.; Mumby, S. Applications of molecular dynamics to the study of thermal degradation in aromatic polymers: I. Polystyrene, *Polym. Eng. Sci*. 1997, 37: 1496.
5. Stoliarov, S. I.; Westmoreland, P. R.; Nyden, M. R.; Forney, G. P. A reactive molecular dynamics model of thermal decomposition in polymers: I. Poly(methyl methacrylate), *Polymer*, 2003, 44: 883–894.
6. Nyden, M. R.; Stoliarov, S. I.; Westmoreland, P. R.; Guo, Z. X.; Jee, C. S. Y. Applications of reactive molecular dynamics to the study of the thermal decomposition of polymers and nanoscale structures, *Mater. Sci. Eng. A*, 2004, 365: 114–121.

7. Stanislav I.; Stoliarov, S. I.; Lyon, R. E.; Nyden, M. R. A reactive molecular dynamics model of thermal decomposition in polymers: II. Polyisobutylene, *Polymer*, 2004, 45: 8613–8621.
8. Brenner, D W. *Mater. Res. Soc. Bull*., 1996, 21: 36.
9. Strachan, A.; van Duin, A. C. T; Chakraborty, D.; Dasgupta, S.; Goddard, W. A., III. *Phys. Rev. Lett.*, 2003, 91: 098301.
10. Dauber-Osguthorpe, P.; Roberts, V. A.; Osguthorpe, D. J.; Wolff, J.; Genest, M.; Hagler, A. T. *Struct. Funct. Genet*., 1988, 4: 31.
11. Discover 95: User Guide, Biosym/MSI, San Diego, CA, 1995.
12. Petersson, G. A.; Tensfeldt, T. G.; Montgomery, J. A., Jr. A complete basis set model chemistry: III. The complete basis set–quadratic configuration interaction family of methods, *J. Chem. Phys*., 1991, 94: 6091.
13. Montgomery, J. A., Jr.; Ochterski, J. W.; Petersson, G. A. A complete basis set model chemistry: IV. An improved atomic pair natural orbital method, *J. Chem. Phys*., 1994, 101: 5900.
14. Ochterski, J. W.; Petersson, G. A.; Montgomery, J. A., Jr. A complete basis set model chemistry: V. Extensions to six or more heavy atoms, *J. Chem. Phys*., 1996, 104: 2598.
15. Petersson, G. A. In *Computational Thermochemistry: Prediction and Estimation of Molecular Thermodynamics*, Irikura, K. K.; Frurip, D. J.; Eds., ACS Symposium Series, American Chemical Society, Washington, DC, 1998.
16. Frisch, M. J.; Trucks, G. W.; Schlegel, H. B.; Scuseria, G. E.; Robb, M. A.; Cheeseman, J. R.; Zakrzewski, V. G.; *Gaussian 98, Revision A.9*, Gaussian, Inc., Pittsburgh, PA, 1998.
17. Kashiwagi, T.; Inaba, A.; Hamins, A. Behavior of primary radicals during thermal degradation of poly(methyl methacrylate), *Polym. Degrad. Stabil*., 1989, 26: 161.
18. Madorsky, S. L. *Thermal Degradation of Organic Polymers*, Interscience, New York, 1964.
19. Nyden, M. R., Wallace, W. E., Awad, W. H. An application of MALDI-MS and reactive molecular dynamics in determining the mechanisms of char-formation during the thermal and oxidative degradation of polymers, ACS PMSE Preprints, 2003.
20. Lobley, C. M.; Guo, Z. X. Analysis of binders and binder burnout in tape cast Ti/Sic Mmcs, presented at the 3rd Pacific Rim International Conference on Advance Materials and Processing (PRICM 3), Minerals, Metals and Materials Society, Warrendale, PA, 1998.
21. Jee, C. S. Y.; Guo, Z. X.; Stoliarov, S. I.; Nyden, M. R. Experimental and molecular dynamics studies of thermal decomposition of a polyisobutylene binder, Acta Mater. In press.
22. Maseras, F.; Morokuma, K. IMOMM: a new integrated ab initio + molecular mechanics geometry optimization scheme of equilibrium structures and transition states, *J. Comput. Chem*., 1995, 16: 1170–1179.
23. Humbel, S.; Sieber, S.; Morokuma, K. The IMOMO method: integration of different levels of molecular orbital approximations for geometry optimization of large systems—test for *N*-butane conformation and SN2 reaction: $RCl + Cl^-$, *J. Chem. Phys*., 1996, 105: 1959–1967.

24. Svensson, M.; Humbel, S.; Froese, R. D. J.; Matsubara, T.; Sieber, S.; Morokuma, K. ONIOM: a multilayered integrated MO + MM method for geometry optimizations and single point energy predictions.—a test for Diels–Alder reactions and $Pt(P(t\text{-}Bu)_3)_2 + H_2$ oxidative addition, *J. Phys. Chem.*, 1996, 100: 19357–19363.

25. Vreven, T.; Morokuma, K. On the application of the IMOMO (integrated molecular orbital + molecular orbital) method, *J. Comput. Chem.*, 2000, 21: 1419–1432.

26. Frisch, M. J.; Trucks, G. W.; Schlegel, H. B.; Scuseria, G. E.; Robb, M. A.; Cheeseman, J. R.; Montgomery, J. A., Jr.; et al. *Gaussian03, Revision C.02*, Gaussian, Inc., Wallingford, CT, 2004.

27. Cornell, W. D.; Cieplak, P.; Bayly, C. I.; Gould, I. R.; Merz, K. M.; Ferguson, D. M.; Spellmeyer, D. C.; Fox, T.; Caldwell, J. W.; Kollman, P. A. A second generation force-field for the simulation of proteins, nucleic-acids, and organic-molecules, *J. Am. Chem. Soc.*, 1995, 117: 5179–5197.

28. Knyazev, V. D. Manuscript in preparation, 2006.

29. Jorgensen, W. L.; Maxwell, D. S.; Tirado-Rives, J. Development and testing of the OPLS all-atom force field on conformational energetics and properties of organic liquids, *J. Am. Chem. Soc.*, 1996, 118: 11225–11236.

30. Benson, S. W. *Thermochemical Kinetics*, 2nd ed., Wiley, New York, 1976.

31. Gilbert, R. G.; Smith, S. C. *Theory of Unimolecular and Recombination Reactions*, Blackwell, Oxford, 1990.

32. Miller, H. W.; Handy, N. C.; Adams, J. E. Reaction path Hamiltonian for polyatomic molecules, *J. Chem. Phys.*, 1980, 72: 99–112.

33. Lindahl, E.; Hess, B.; van der Spoel, D. GROMACS 3.0: a package for molecular simulation and trajectory analysis, *J. Mol. Mod.*, 2001, 7: 306–317.

34. Crist, B.; Oddershede, J.; Sabin, J. R.; Perram, J. W.; Ratner, M. A. Polymer fracture: a simple-model for chain scission, *J. Polym. Sci. B Polym. Phys.* 1984, 22: 881–897.

35. Hageman, J. C. L.; de Wijs, G. A.; de Groot, R. A.; Meier, R. J. Bond scission in a perfect polyethylene chain and the consequences for the ultimate strength, *Macromolecules*, 2000, 33: 9098–9108.

36. Krisyuk, B. E. Reactivity of stressed molecules: calculation of the effect of chain deformation on scission for macromolecules and middle macroradicals, *J. Mol. Struct. (Theochem)*, 2004, 677: 77–83.

37. Smith, K. D.; Stoliarov, S. I.; Nyden, M. R.; Westmoreland, P. R. RMDff: a smoothly transitioning, forcefield-based representation of kinetics for reactive molecular dynamics simulations, *Mol. Simal.*, 2007, 33: 361–368.

14

PREDICTING DYNAMIC MESOSCALE STRUCTURE OF COMMERCIALLY RELEVANT SURFACTANT SOLUTIONS

FIONA CASE
Case Scientific, Essex Junction, Vermont

Low-molecular-weight surfactants are used in a huge variety of consumer and industrial products[1,2]: shampoos and cosmetics, detergents, paints and inks, foods; and in industrial processes such as emulsion polymerization, fiber spinning, and enhanced oil recovery.[3] Surfactant molecules can self-assemble. They spontaneously form spheres or rods or lamella (or more complicated structures); they can also assemble themselves at interfaces—for example, at a surface or at the boundary between two fluids (see Figure 1). It is these soft, flexible mesoscale (or nanoscale) structures that determine properties and behavior of the surfactant-containing material.[2]

Surfactants are classified according to the chemical nature of their *head group* (the hydrophilic end of a surfactant used in water). Anionic surfactants (with negatively charged head groups) are the most widely used, usually in combination with nonionic surfactants.[1,4] Optimum performance almost always requires a mixture of several different surfactants in systems with carefully controlled levels of salt.[5–7]

Experimental characterization of soft dynamic submicrometer structures in fluid systems is challenging, especially in relatively concentrated systems containing many different materials. Computer modeling has the potential to predict the mesoscale structure and dynamics of surfactant-containing materials and to revolutionize the design of these commercially important materials. In Section 1

Multiscale Simulation Methods for Nanomaterials, Edited by Richard B. Ross and Sanat Mohanty

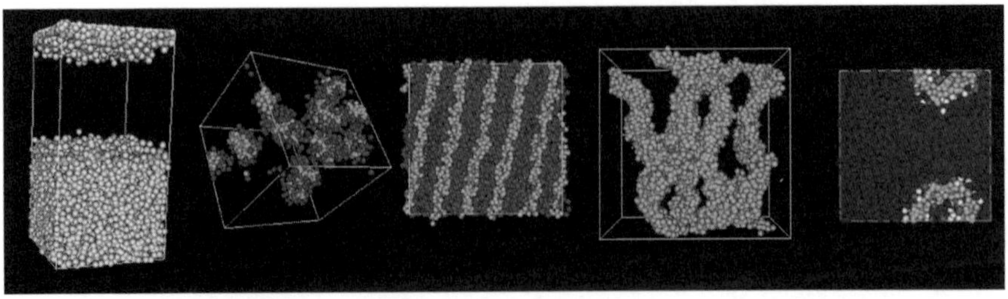

Figure 1 Mesoscale structures for surfactant solutions predicted using dissipative particle dynamics: surfactants self-assembled at an oil–water interface, spherical micelles, a lamella structure, entangled rods, and a slice through a single-layer vesicle.

we discuss mathematical modeling techniques that are already an established part of the toolkit of the industrial formulation chemist, used to predict the most thermodynamically stable mesoscale structures for individual surfactants or simple mixtures in dilute solutions, and as an indicator of trends in more complicated or concentrated systems. We then discuss simulation-based approaches, in particular dissipative particle dynamics, which provide complementary capabilities, including prediction of dynamic behavior, nonequilibrium structures, and modeling of more concentrated systems containing mixtures of surfactants.

1. CALCULATING EQUILIBRIUM MESOSCALE STRUCTURES OF SURFACTANT SOLUTIONS

The simplest approaches to predicting mesoscale structures for surfactants are based on geometrical arguments (Figure 2).[8] A surfactant packing parameter can

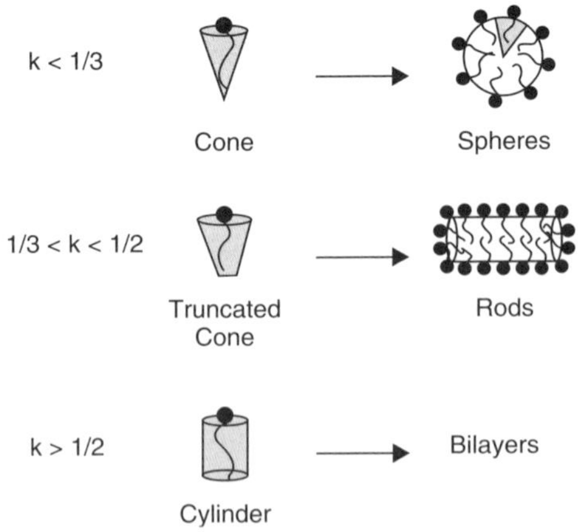

Figure 2 Surfactant packing parameter.

be defined: $k = V/al$, where V is the volume of the surfactant tail, l the length of the surfactant tail, and a the cross-sectional area of the head group at the surface of the aggregate.

Surfactant molecules with large head groups (relative to their tails) have packing parameters of less than $\frac{1}{3}$ and are predicted to form spherical micelles. Decreasing the average head-group size (e.g., by adding a cosurfactant with a smaller head size, or adding salt to shield repulsive electrostatics interactions between like-charged head groups) increases the packing parameter, reduces the curvature micelle surface, and results in a rodlike or wormlike structure. Further increase in the packing parameter leads to a lamella structure and to inverse micelles when $V/al > 1$. A similar concept, the hydrophobic/hydrophilic balance (HLB) uses the relative solubility of the head and tail groups in two solvents (e.g., an oil and water) to predict the mesoscale structure in emulsions.[9,10] These empirical approaches explain trends in behavior for single surfactant systems or simple mixtures.

The various molecular thermodynamic models provide more quantitative prediction of equilibrium mesoscale structure in dilute systems.[11–16] The free-energy changes associated with each process contributing to self-assembly are estimated by fitting to experimental data or to other models. For example, in the approach proposed by Tanford[17] and extended by Nagarajan,[12,18] the free energy of micellization to form a aggregate containing N_{agg} surfactants is estimated as follows:

$$\frac{\Delta\mu^0_{N_{agg}}}{kT} = \left(\frac{\Delta\mu^0_{N_{agg}}}{kT}\right)_{tr} + \left(\frac{\Delta\mu^0_{N_{agg}}}{kT}\right)_{def} + \left(\frac{\Delta\mu^0_{N_{agg}}}{kT}\right)_{in} + \left(\frac{\Delta\mu^0_{N_{agg}}}{kT}\right)_{head} \quad (1)$$

where $(\Delta\mu^0_{N_{agg}}/kT)_{tr}$ is the free-energy change due to the escape of the hydrophobic surfactant tails from water into the micelle core (the driving force for micelle formation); $(\Delta\mu^0_{N_{agg}}/kT)_{def}$ is the deformation energy—the result of packing the tail into the hydrophobic core (this term is different for different aggregate shapes); $(\Delta\mu^0_{N_{agg}}/kT)_{int}$ accounts for energetically unfavorable interactions between water and those tail groups that are not completely shielded by the surfactant heads (a desire to reduce the area of this unfavorable tail–water interface leads to an increase in N_{agg}); and $(\Delta\mu^0_{N_{agg}}/kT)_{head}$ is the contribution due to interactions between head groups at the surface of the micelle (repulsive steric and electrostatic interactions between head groups act to limit the growth in micelle size). The total free-energy change involved in forming a sphere, a lamella structure, or a rodlike micelle of a given length, is calculated. The most stable structure is predicted to form. Molecular thermodynamics has been able to predict mesoscale structure for ionic, nonionic, and zwitterionic surfactants[12,14,19] and to predict some properties for mixtures of surfactants,[15,20] emulsions,[21] and systems containing both surfactants and polymers.[22] For example, in their 1991 paper, Nagarajan and Ruckenstein[21] used an analytical solution to the Poisson–Boltzmann equation for spherical and cylindrical micelles[23] to calculate the electrostatic contribution to $(\Delta\mu^0_{N_{agg}}/kT)_{head}$ for ionic surfactants as a function of the aggregate shape, the surfactant concentration, the counterion concentration, and the concentration of added salt. Figures 3 and 4 show their results for sodium dodecyl sulfate (SDS) and the results of early experimental studies of SDS solutions close to the critical micelle concentration (8.2×10^{-3} M).[24] There is reasonable agreement between predicted and experimental aggregate size (Figure 3) and of the Nagarajan and Ruckenstein ln K parameter, which quantifies the probability of transition from spherical to rodlike micelle (Figure 4).[12]

Figure 5 shows another molecular thermodynamics example.[21] The radius of emulsion droplets stabilized by SDS is predicted to change as pentanol is

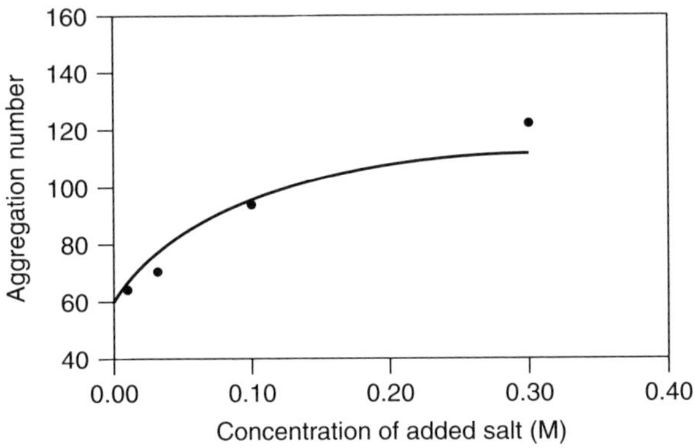

Figure 3 Influence of added NaC1 concentration on the average aggregation number and sphere to rod transition probability for SDS micelles. The line shows predicted values and the points experimental data at concentrations close to the CMC (8.2×10^{-3} M). (Replotted using data from ref. 12.)

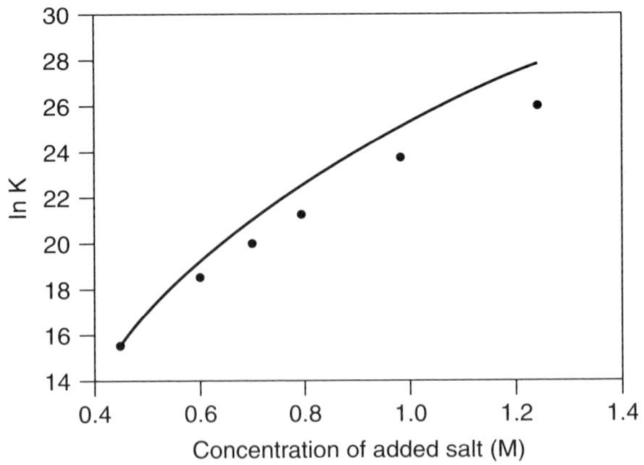

Figure 4 Influence of added NaC1 concentration on the sphere to rod transition probability for SDS micelles. The line includes the values predicted and the points show experimental data. (Replotted using data from ref. 12).

added. As indicated in Figure 6, the alcohol intersperses the SDS molecules at the oil–water interface, changing the balance between the average volume of the heads and tails. This leads to a change in the packing parameter and in the curvature of the interface. At higher levels of alcohol the interface starts to curve in the opposite direction (the packing parameter is greater than 1).

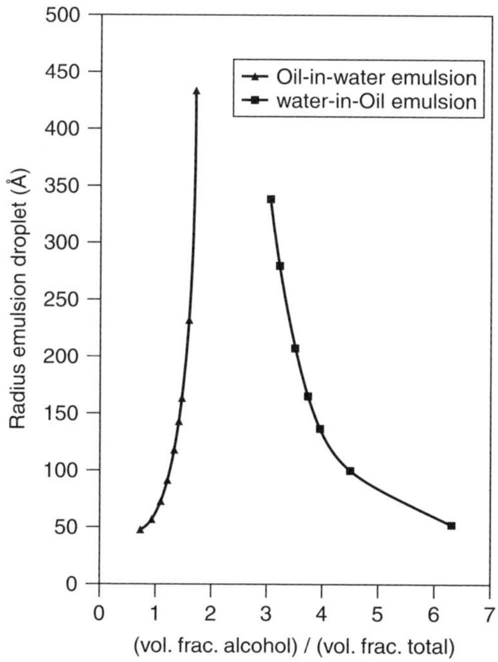

Figure 5 Radii of microemulsion droplets predicted as a function of the volume ratios of alcohol to surfactant in microemulsions. The predictions are for a system consisting of sodium dodecyl sulfate (SDS), 1-pentanol, 0.3 M NaC1, water, and cyclohexane. The total volume fraction of the surfactant in the microemulsions is 0.01. (Replotted using data from ref. 21.)

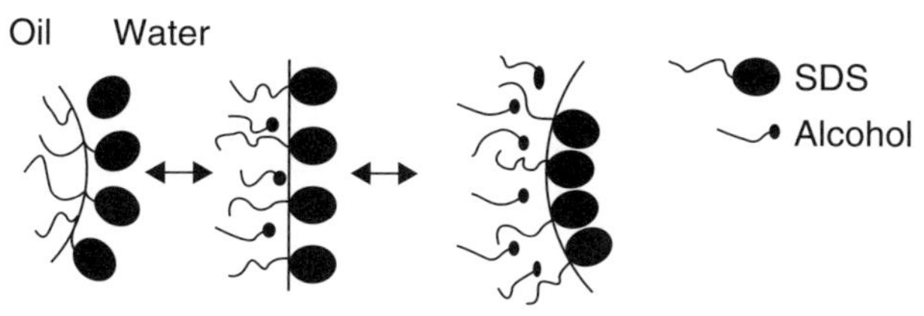

Figure 6 Effect of 1-pentanol on the structure of the surfactant layer at the boundary between oil and water.

2. SIMULATIONS OF SURFACTANT SOLUTIONS

Molecular thermodynamics approaches can be useful within an industrial setting for predicting numerical data about the equilibrium structure of dilute solutions.[25,26] They cannot, however, provide information about dynamic behavior, or predict nonequilibrium structures, or model more concentrated systems in which interactions between mesoscale structures become significant. Simulation-based approaches provide complementary capabilities.[27] Techniques such as molecular dynamics[28–30] and Monte Carlo,[31–33] particle or bead-based approaches such as dissipative particle dynamics (DPD) and lattice boltzmann,[34,35] or field-based simulations for larger-scale systems[36,37] can provide predictions of dynamic and nonequilibrium behavior and are most efficient at higher concentrations. The ability to generate three-dimensional models and images of the mesoscale structure, in addition to numerical data, is also important and can provide considerable insight into properties and behavior of meso- or nanostructured fluids. In the next sections we focus on the application of DPD.

Brief Description of Dissipative Particle Dynamics

Using dissipative particle dynamics[38] (DPD) the behavior of a soft meso- (or nano)-structured material can be modeled by simulating the behavior of soft interpenetrating beads, each of which represents a portion of a molecule (perhaps the tail of a surfactant) or a "fluid element" (perhaps five or six water molecules) (see Figure 1). Springs are defined between beads to model larger molecules.[39,40]

The force experienced by an individual bead (not bonded to other beads) in a dissipative particle dynamics simulation without electrostatics interactions, f_i, is

$$f_i = \sum_{i \neq j} (F_{ij}^C + F_{ij}^D + F_{ij}^R)$$

where F_{ij}^D and F_{ij}^R are the dissipative (or drag) force and the random force. It is important that these interdependent forces are set correctly since they provide the thermostat for the system.[41] In principle, different values for F_{ij}^D and F_{ij}^R could be defined for every pair of bead types in the system, and this will be necessary for prediction of rheological or flow behavior.[42,43] However, here we use the default values for F_{ij}^D and F_{ij}^R in the Culgi DPD package[41,44,45] and focus on the third contribution, F_{ij}^C, the conservative contribution to the force. It is the F_{ij}^C parameter that captures the "chemistry" of a particular system and determines the mesoscale (or nanoscale) structure. The effect of short-range van der Waals interactions, dipole–dipole interactions, and hydrogen bonding between the atoms in the molecules represented by each bead is bundled into this one parameter. Longer-range electrostatic interactions are handled separately (see later).

DPD beads are soft; they can interpenetrate. The repulsive F_{ij}^C force keeps them apart. Since there is no unique length scale in DPD (the DPD bead can represent an arbitary amount of material—this choice determines how coarse-grained the model is), the cutoff radius for the bead–bead force calculation r_c is referred to as one DPD length unit. For two beads representing two different types of material, A and B, less than r_c apart:

$$F^C_{\mathrm{AB}} = a_{\mathrm{AB}}\left(1 - \frac{r_{\mathrm{AB}}}{r_c}\right) \tag{2}$$

To parameterize DPD we need to obtain DPD a_{AB} parameters to quantify the energy of interaction for every pair of bead types in the system.

The Flory–Huggins equation is commonly used to estimate energies of mixing for polymers represented as beads on a lattice. Groot and Warren were the first to point out the resemblance between Flory–Huggins theory and DPD (which, they suggested, could be considered a continuous version of the lattice model[44]) and the relationship between the DPD a_{AB} parameter and the Flory–Huggins χ_{AB} parameter. For a mixture of poly(A) and poly(B) the Flory–Huggins free energy of mixing per lattice site is

$$\frac{\Delta G_{\mathrm{mix}}}{k_B T} = \frac{\phi_{\mathrm{A}}}{N_{\mathrm{A}}}\ln\phi_{\mathrm{A}} + \frac{\phi_{\mathrm{B}}}{N_{\mathrm{B}}}\ln\phi_{\mathrm{B}} + \chi_{\mathrm{AB}}\phi_{\mathrm{A}}\phi_{\mathrm{B}}$$

where ϕ_{A} is the volume fraction of poly(A) in the mixture, and N_{A} is the number of beads per poly(A) polymer. The first two terms capture the combinatorial entropy of mixing (which always favors mixing but is small in polymer systems). The χ_{AB} parameter captures the chemistry of the interaction between A and B beads. When χ_{AB} is small or negative, a mixture of A and B will be miscible. When it is positive, and larger that the critical value, the mixture will phase separate into two phases, an A-rich phase and a B-rich phase. The volume fraction of A in the B-rich phase, ϕ_{A}, will then be related to the χ_{AB} parameter as follows[44]:

$$\chi_{\mathrm{AB}} N_{\mathrm{A}} = \frac{\ln[(1 - \phi_{\mathrm{A}})/\phi_{\mathrm{A}}]}{1 - 2\phi_{\mathrm{A}}} \tag{3}$$

To derive the relationship between a_{AB} and χ_{AB}, DPD simulations of mixtures of A and B beads can be carried out using different values of a_{AB}. The simulation can either be started from homogeneous mixture of A and B (which will then phase-separate if a_{AB} is larger than the critical value), or it can be started from a structure that is completely phase-separated (which will then mix if a_{AB} is smaller than the critical value). Both approaches were used in this study and gave similar results. Figure 7 shows the predicted distribution of A beads for 50:50 mixtures of A and B with $a_{\mathrm{AB}} = 34$, 40, 45, and 50 in a $12 \times 24 \times 12$ box, with a bead density of 3 (a total of 10,368 beads). In all cases the system has phase-separated; the bottom layer contains the A-rich phase and the top layer is B-rich. Use of a noncubic box encourages the phase boundary to occur perpendicular to the y-axis. Note that although the boxes are all shown with the A-rich phase at the bottom for clarity, there is no gravity in this simulation, and the beads have the same mass, so "top" and "bottom" have no real meaning. An

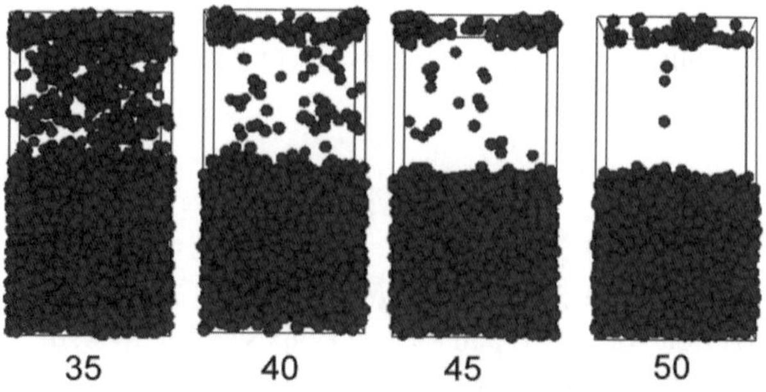

Figure 7 Predicted distribution of A beads in mixtures with $a_{AB} = 34$, 40, 45, and 50. In all cases the system has phase-separated. The bottom layer contains the A-rich phase, the top layer is B-rich (use of a noncubic box encourages the phase boundary to occur perpendicular to the x-axis).

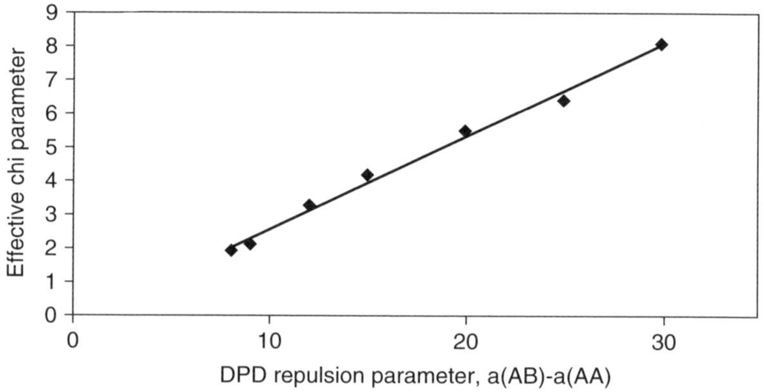

Figure 8 Correlation predicted between DPD ($a_{AB} - a_{AA}$) and the Flory–Huggins χ_{AB} parameter for a 50:50 mixture of A beads and B beads.

effective χ_{AB} for each system is calculated from the predicted concentrations in each phase using Eq. (3). The graph in Figure 8 shows the results.

The relationship obtained from this study,

$$\chi_{AB} = (0.267 \pm 0.028)(a_{AB} - a_{AA})$$

is in reasonable agreement with the results of the 1997 study by Warren and Groot[44] using a system size of $8 \times 8 \times 10$ and a bead density of 3:

$$\chi_{AB} = (0.286 \pm 0.002)(a_{AB} - a_{AA})$$

In the following studies, system sizes of $25 \times 25 \times 25$ DPD length units were used unless noted. A bead density of 3.0 was used throughout, and all systems were equilibrated for at least 100,000 simulation steps (with a time-step size of 0.05) before data collection.

Modeling Nonionic Surfactant Systems

Emulsions Almost all of the published DPD simulations are for nonionic systems, and most focus on predicting trends in behavior rather than on obtaining quantitative results. One of the first validation studies I carried out, using DPD code supplied by Accelrys, was a simple test to see if DPD could predict the relationship between the shape of a surfactant and the mesoscale structure of the oil–water–surfactant emulsion it stabilizes.[46] Emulsions are commercially important[47] and are an obvious candidate for mesoscale modeling techniques such as DPD.[48] The stability of an emulsion is affected by processing conditions (how it was made) and by the solubility of one phase in the other (e.g., the solubility of the oil in water determines the rate of Ostwald ripening in an oil-in-water emulsion).[49,50] But surfactant molecules can direct the shape and size of the emulsion droplets, controlling the nanostructure as described by molecular thermodynamics (Figure 5). Figure 9 shows more recent DPD simulation results, carried out using the Culgi DPD code, and with larger system sizes than were practical in 2003. The parameterization of the model nonionic surfactant is a simple one (Table 1). The oil is an alkane, so the tail groups and oil groups have the same parameters. The head group is assumed to have the same parameters as water. The a_{OW} parameter for oil and water is set at 80, a highly repulsive value used in previous surfactant simulations[48,51–53] (the predicted mesoscale structure is not sensitive to small changes in this parameter); the water–water interaction, a_{WW}, is set to 25.[44]

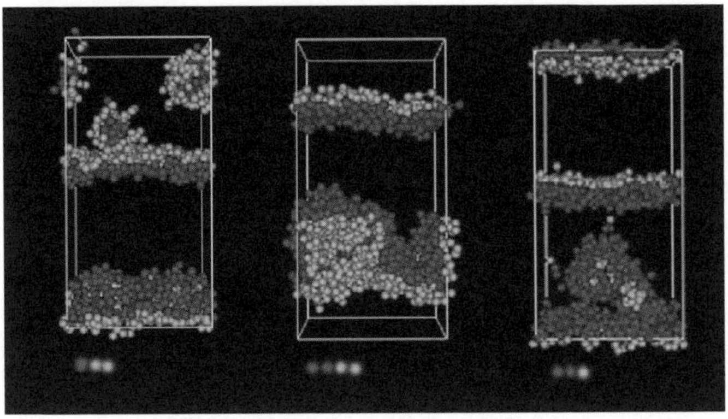

Figure 9 Structures predicted for mixtures of oil, water, and surfactants with different relative head and tail size. An example of one surfactant molecule is shown below each simulation box. Hydrophilic head beads are dark gray, hydrophobic tail beads are light gray. The oil and water beads are not displayed.

TABLE 1. Repulsion Parameters for the Model Nonionic Surfactant Used in Emulsion Simulations

DPD a_{AB}	Head	Tail	Oil	Water
Head	25	80	80	25
Tail	80	25	25	80
Oil	80	25	25	80
Water	25	80	80	25

When the head of the surfactant is larger than the tail (surfactant packing parameter less than 1), the system self-assembles into two phases: an oil phase and a phase containing an oil-in-water emulsion (classified as a Winsor I system). When the tail is larger than the head (surfactant parameter greater than 1), the system self-assembles into the Winsor II structure: a water phase and a phase containing a water-in-oil emulsion. When the surfactant is "balanced," with equal head and tail and a packing parameter of 1, the surfactant does not wish to stabilize either an oil-in-water or a water-in-oil emulsion and forms a flexible layered structure. Experimental evidence for this bicontinuous structure in systems containing balanced surfactants (the Winsor II structure) is provided by nuclear magnetic resonance diffusion studies.[10]

Nonionic Surfactant Solutions To predict mesoscale structure and behavior for specific types of nonionic surfactants, we need to derive values for their DPD bead–bead repulsion parameters, particularly for interactions involving the head–bead interaction parameters, a_{HW} and a_{HH}, since these will have the greatest effect on the aggregate shape. In our 2003 work[46] we used an approximate model for an ethoxylate surfactant with an average of six ethylene oxide repeat units as head: $C_{12}H_{23}(OCH_2CH_2)_6OH$, or $C_{12}E_6$. The DPD repulsion parameters are given in Table 2.

The model reproduces the experimentally observed trends in $C_{12}E_6$ ethoxylate surfactant phase behavior.[54] At very low concentrations, small spherical or ellipsoidal micelles are formed, but as concentration increases, the structure grows quickly, forming rodlike or wormlike micelles, which branch and break

TABLE 2. Repulsion Parameters for the Simple $C_{12}E_6$ Ethoxylate Surfactant Model

DPD a_{AB}	E_3	Tail	Water
E_3	27	60	27
Tail	60	25	80
Water	27	80	25

and re-form over time. At high surfactant concentration, a lamella structure is formed. With this model we see entangled worms at moderate concentrations rather than the ordered-looking hexagonal structure reported by Jury et al. in 1999.[55] This could be due in part to system size effects; using a $10 \times 10 \times 10$ box size as in the 1999 paper results in more "perfect"-looking structures.

Simulation of Oil Solubilization One of the key properties of surfactants is their ability to solubilize hydrophobic materials. A detergent cleans by solubilizing oily dirt and holding it in the cleaning solution; a surfactant in a drug formulation can prevent a hydrophobic drug molecule from precipitating out of solution during and injection. Figure 10 shows part of a recent study using Culgi DPD to test the ability of the DPD surfactant model to solubilize oil. Seven layers of hydrophobic oil beads are placed along the $Y = 0$ face of a $20 \times 40 \times 20$ simulation box. The center layer is fixed in place, creating a permeable wall. A solution containing 10% randomly distributed nonionic surfactant molecules in water is placed in contact with the oil beads to create the starting structure. Surfactant parameters are as in Table 2, parameters for oil and water as in Table 1. As the DPD simulation progresses, the surfactants self-assemble into micelles and start to diffuse through the solution. Occasionally, they interact with the oil layer. When this happens, some of the oil molecules are absorbed into the micelles and taken away from the surface when the micelles leave. The nonionic surfactant model is able to solubilize some of the oil, although not all of it (mixtures of nonionic and ionic surfactants are used in cleaning because they show better oil-solubilization capabilities than do nonionics alone[5]).

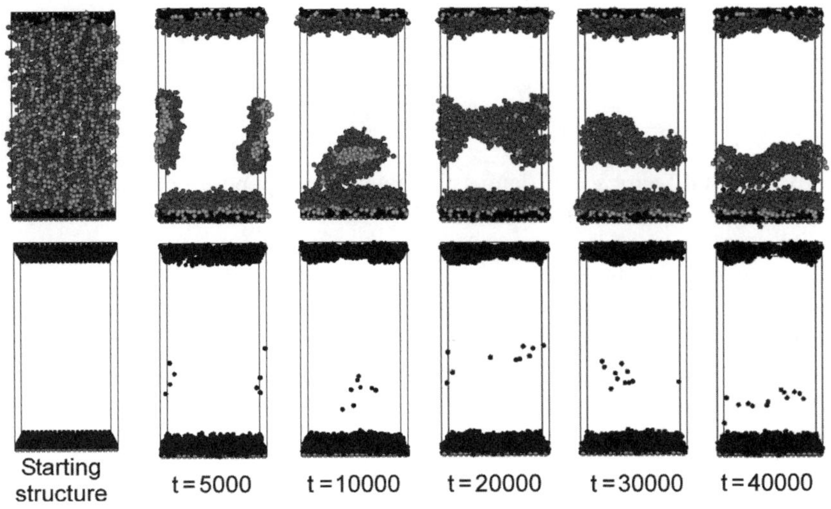

Figure 10 Evolution of the solubilization simulation over time. The top row shows both the nonionic surfactant and the oil; the bottom row of images show just the oil beads. The water beads are not displayed.

Modeling Ionic Surfactants: Sodium Dodecyl Sulfate

A method for including electrostatic interactions in DPD was published by Groot in 2003.[56] The Culgi software uses a broadly similar approach.[57] From a given concentration distribution c_B of charged beads, the electrostatic potential ψ is obtained by iteration using linearized Poisson–Boltzmann theory or the Debye–Hückel model:

$$\frac{\nabla^2\psi - \psi}{D^2} = \frac{-ez_Bc_B}{\varepsilon}$$

where e is the charge on an electron, z_B the valence of the beads, ε the permittivity of the fluid, and D, the Debye length, is defined as

$$D \equiv \sqrt{\frac{\varepsilon kT}{2e^2c}}$$

In this model, the electrostatic potential around a point charge of valence z is known analytically and is given by

$$\psi(r) = ze\frac{e^{-r/D}}{4\pi\varepsilon r}$$

The potential is alternatively described by

$$\frac{\psi(r)}{kT/e} = zl_B\frac{e^{-r/D}}{r}$$

where the Bjerrum length is defined as

$$l_B \equiv \frac{e^2}{4\pi\varepsilon_0 kT}$$

The exponential factor $e^{-r/D}$ damps the potential, depending on the value of the Debye length. The background electrolyte is said to "screen" the electrostatic interactions. In a larger-scale model (where each bead represents a larger amount of material) it would be reasonable to vary D to model the effect of background electrolyte. A simple rule of thumb is that the Debye length in water is given by $D = 1/\sqrt{10c}$, where c is the concentration of 1–1 electrolyte, but in this work all the ions are modeled explicitly as charged beads and D is set to a large value (100). The charge from a given bead is smeared over neighboring grid cells with Gaussian weights; then all charges on the vertex points are collected, the potential is calculated, and the electrostatic force on a bead at any given point is calculated by interpolation and used in the next step of the simulation. Figure 11 shows the pair correlation functions predicted for a 10% solution of anions (A-bead, with a charge of -1) and cations (B-beads, with a charge of $+1$) in water (beads with no charge). All the DPD repulsion parameters are set to 25.

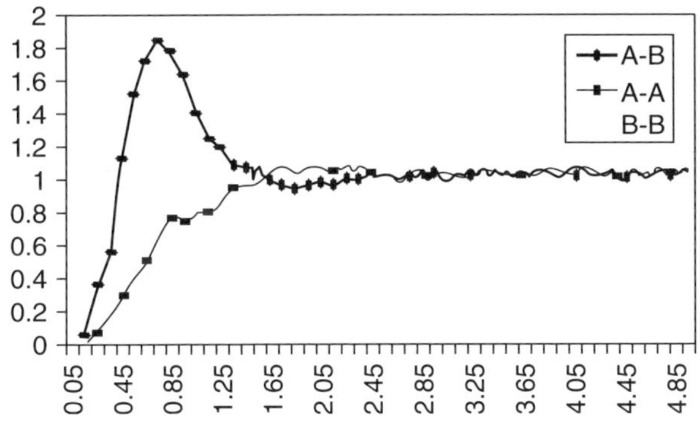

Figure 11 Pair correlation functions predicted for a 10% solution of anions (A-bead, with a charge of −1) and cations (B-beads, with a charge of +1) in water (beads with no charge).

Experimental and modeling work has shown that ions in water are surrounded by a layer of bound water molecules.[58,59] In this work the charged beads represent both the ion and its bound water layer. However, atomistic-scale simulations of the potentials of mean force between, for example, Na^+ and Cl^-, suggest that the association between the ions takes place in two or more steps, leading to a final contact pair in which the ions are in close contact without bound water between them.[59,60] To model this behavior and to allow oppositely charged ion beads to approach more closely once they have come within the DPD cutoff distance, a DPD $a_{\text{anion-cation}}$ repulsion parameter of less than 25 is used.

Given the commercial importance of ionic surfactants it is perhaps surprising that so few researchers have attempted to include electrostatics in their DPD simulations and that there has been scant application of this type of approach beyond the original validation work.[56] Figure 12 and Table 3 show the DPD model for sodium dodecyl sulfate (SDS) used in the current study. Each tail bead represents six methyl (CH_2 or CH_3) groups. This is the same length scale used in the previous nonionic surfactant simulations, and the same parameters are

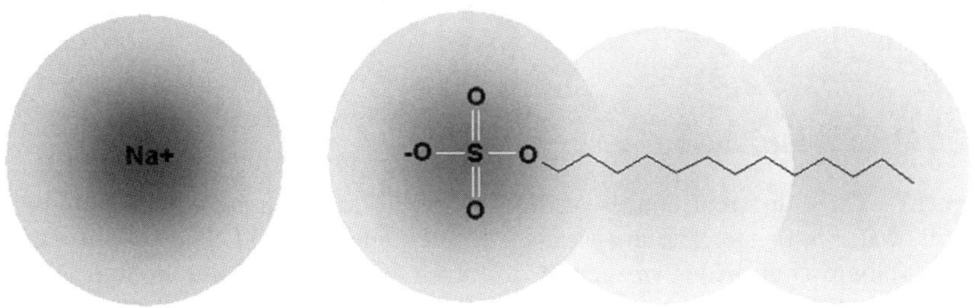

Figure 12 DPD model for sodium dodecyl sulfate used in the current study. Each tail bead represents six methyl (CH_2 or CH_3) groups.

TABLE 3. SDS Repulsion Parameters

DPD a_{AB}	Head	Tail	Water	Sodium Ion	Chloride Ion
Head	25	60	25	18	25
Tail	60	25	80	60	60
Water	25	80	25	25	25
Sodium ion	18	60	25	25	18
Chloride ion	25	60	25	18	25

used for the tail beads (maintaining the transferability of parameters is important for future simulations of mixtures). The head bead represents the charged SO_4^- head group and its bound water. Sodium and chloride ions are modeled as charged beads, which include bound water.

The distance between SDS head groups at their maximum packing density at an oil–water interface has been calculated using experimental surface tension data[61] and provides a length scale for this model: The average head–head distance in a simulation of this system (Figure 13) is 0.8 DPD length units and the experimental distance is 0.632 nm (twice the reported radius of 0.316 nm), so 1 DPD length unit is 0.79 nm. In simulations of concentration ranges where the SDS forms spherical micelles, the average length of the SDS molecule (including the sodium counterion) is 3.24 DPD units, or 2.56 nm. This is in reasonable agreement with published experimental and modeling results. The average number of surfactants per micelle for a 0.19 volume fraction solution of SDS in the absence of added salt is $N_{agg} = 116$. This is in reasonable agreement with the experimental value of 111 for this concentration obtained recently using time-resolved fluorescence quenching (TRFQ)[62] and the value of $N_{agg} = 112$ obtained by analysis

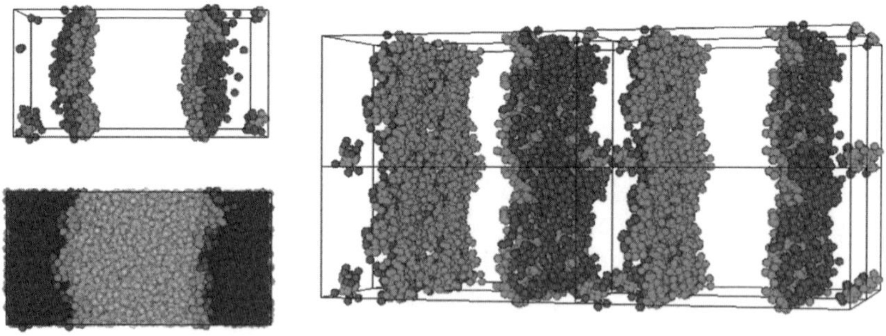

Figure 13 Three different images of SDS molecules at their maximum packing density at an oil–water interface. The image at the top left shows the SDS surfactant molecules with the sodium counterions (magenta) displayed. The image at the bottom left shows the oil (yellow), the water (blue), and the SDS molecules. The image on the right shows $2 \times 2 \times 2$ simulation cells. Only the SDS molecules are displayed (heads are red, tails are green). (*See insert for color representation figure.*)

of a number of older studies employing a range of techniques[63] but smaller than the value of $N_{agg} = 136$ obtained from small-angle neutron scattering (SANS).[62]

Sensitivity of the Model to Changes in Input Parameters In the current DPD implementation the same value of l_B is used throughout the system (a constant dielectric constant). Fortunately, experimental data are available giving dielectric constant at the surface of a SDS micelle, the location where correct modeling of electrostatic interactions is most critical. The experimental dielectric constant[64] of 31 gives $l_B = 1.9$ nm (2.4 DPD units), used in this study. For pure water ($\varepsilon_0 = 78$), $l_B = 0.7$ nm.

The aggregation number predicted is also sensitive to changes in the hydrophobicity of the counterion (the a_{SW} interaction parameter) and the repulsive interaction between the charged surfactant head group and the counterion (as determined by the a_{SH} interaction parameter). This is reasonable since these parameters define the chemical nature and size of the counterion. The experimentally observed variation in dodecyl sulfate aggregation number with different counterions (Li, Na, K, Rb, Cs) has been shown to be correlated directly with the size and hydrophobicity of the counterion.[65]

Predicting the Effect of Salt on the SDS Aggregation Number Addition of salt causes a dramatic change in the mesoscale structure of SDS solutions: The spherical micelles found at low salt concentrations become elliptical and then turn into extended wormlike structures as salt is added.[66] The phenomenon has been studied extensively: for example, using light scattering,[24] small-angle neutron scattering,[62] electron paramagnetic resonance,[62,67] positron annihilation lifetime spectroscopy,[68] and fluorescence quenching.[69] Figure 3 showed the results of a molecular thermodynamic calculation for this system. Unfortunately, many of the experimental studies have been carried out at very low SDS concentrations. For example, scattering studies have typically been for SDS concentrations of around 0.07 M. To simulate a system containing 2000 SDS molecules at this concentration using our current model would require a $46 \times 46 \times 46$ DPD length unit simulation. Figure 14 shows the change in aggregation number for 0.25 and 0.51 M SDS with increasing salt predicted using the DPD parameters from Table 4. The mesoscale structures are shown in Figure 15. The model has predicted the change in size and shape of the micelles as salt is added. Analysis of the head–head pair correlation function (Figure 16) reveals that the increasing salt permits more of the head groups to pack closely together. The reduction in the average head–head separation distance leads to an increase in the packing parameter (Figure 2).

At higher salt concentrations the DPD simulation predicts that the wormlike structures become branched (Figure 15). Branched structures have been observed in cryo-TEM studies of cationic surfactants under conditions of high salt.[70] Branch formation has been proposed as an explanation for the rheological behavior of ionic surfactant solutions as salt is added: initially, the viscosity increases (as the wormlike micelles get longer and entangle[71]), but at high salt concentrations the viscosity decreases[72] because a connection can slide along the micelle

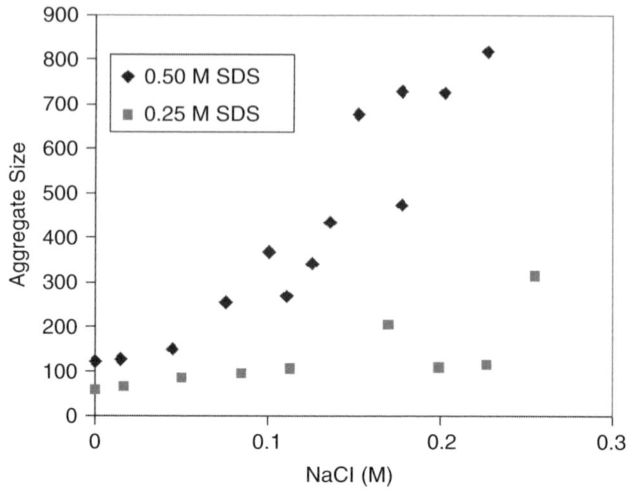

Figure 14 Predicted aggregation numbers for 0.5 and 0.25 M SDS as a function of added salt.

TABLE 4. Examples of Experimentally Derived β Parameters for Surfactant Mixtures

Surfactant Mixture	Mixture Type	Medium	β
$SDS/C_{12}E_8$	Anionic/nonionic	Water	−3.9
$SDS/C_{12}E_8$	Anionic/nonionic	0.5 M NaCl	−2.6
$C_{14}TABr/C_8E_4$	Cationic/nonionic	0.05 M NaBr	−1.8
$C_{10}E_3/C_{12}E_8$	Nonionic/nonionic	Water	−0.4
$SDS/C_{12}TABr$[a]	Anionic/cationic	Water	−25.5

Source: Data from refs. 2 and 78.

[a] $C_{12}TABr$ is the cationic surfactant dodecyltrimethylammonium bromide.

in response to a viscous flow to relax the stress. The mechanism of junction formation in entangled wormlike micelles has been studied using particle-based simulations[73] and has recently been described using molecular thermodynamics,[74] but this is the first time the full range of salt-dependent behavior has been seen in a simulation.

Mixed Surfactant Systems

The simplest description of the behavior of mixed surfactant systems is provided by regular solution theory (RST),[16,75–78] where deviation from ideal mixing is parameterized by a single interaction parameter, β. For a mixture of surfactant A

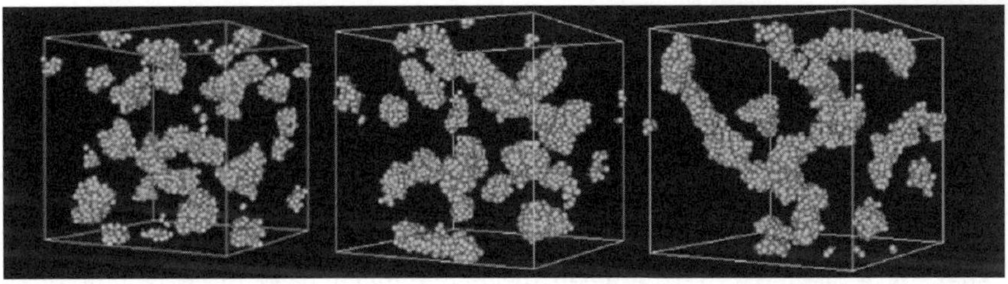

Figure 15 Mesoscale structure for 0.25 M SDS with 0 M, 0.085 M, and 0.227 M added salt. Only the hydrophobic core of the micelles (the surfactant tails) are displayed.

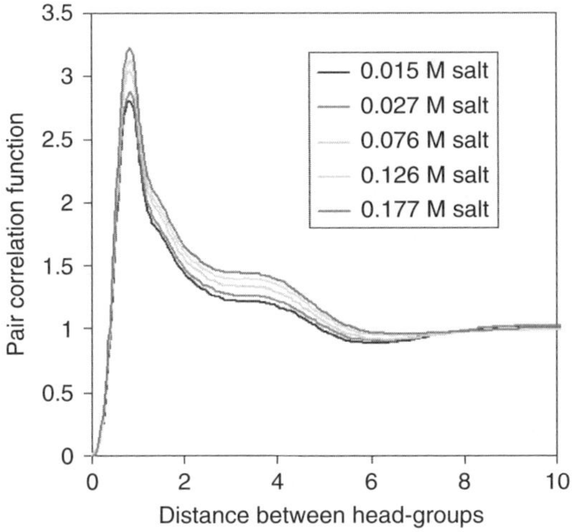

Figure 16 Head–head pair correlation function for 0.5 M SDS with added salt.

and surfactant B,

$$\beta = \frac{N(W_{AA} + W_{BB} - 2W_{AB})}{RT}$$

where W_{AA} and W_{BB} are the self-interactions and W_{AB} the energy of interaction between surfactant A and surfactant B in the mixed micelle.

The β parameter can be obtained by fitting to experimental data. For example, in a mixture of surfactants A and B the variation in the critical micelle concentration of the mixed system, CMC_{mix}, with composition (mole fraction) m is

$$\frac{1}{CMC_{mix}} = \frac{m}{f_A CMC_B} + \frac{1-m}{f_B CMC_B}$$

$$f_A = \exp[\beta(1-x)^2]$$

$$f_B = \exp[\beta(x)^2]$$

where f_A is the activity of surfactant A and x is composition of the micelle (which is often assumed to equal m). Extensive tables of experimentally derived β values are available,[2,4,78] and these highlight the significant nonideality of surfactant mixtures (see Table 4). In particular, there are strongly synergistic interactions between anionic and nonionic surfactants (which depend on salt concentration) and between anionic and cationic surfactants. β can also be estimated using molecular thermodynamics.[12,16,79]

There is a clear parallel between the β parameter used in surfactant science and the χ parameter used to characterize mixing in polymers. One approach to modeling mixtures of surfactants with chemically similar tail groups using DPD might be to correlate the α_{AB} interaction energy parameter for the head beads of surfactants A and B [Eq. (2)] with the β value for that surfactant mixture (obtained from experiment or from molecular thermodynamics). This would be appropriate for mixtures of nonionic materials. However, the strongly synergistic interactions between ionic and nonionic surfactants or between ionic surfactants, and zwitterionic surfactants, are likely to be due to electrostatic interactions, and in this work electrostatic interactions are calculated rather than entered as semiempirical parameters. It would be useful if DPD were able to predict structures of mixtures including ionic surfactants without the need for input of a head–head interaction parameter (or β parameter) from either experiment or molecular thermodynamics since RST is known to break down for some of these systems. A single interaction parameter is not sufficient to describe their behavior.[80] To test the ability of the Culgi DPD program to predict synergistic interactions, mixtures of SDS and $C_{12}E_6$ ethoxylate were created using the parameters in Tables 2 and 3. (This is one of the surfactant mixtures that has been shown not to obey RST.[81]) The repulsion parameter describing the interaction between the SDS and ethoxylate surfactant head groups, a_{HE}, was left at the default value of 25. The predicted aggregation numbers are shown in Figure 17. Structures of the mixed micelles are shown in Figure 18. It is clear that the SDS surfactants are breaking up the rodlike structures. The experimental data in Figure 19 are at a different set of conditions,[82] at a surfactant concentration that is too low for convenient simulation but shows the same trends.

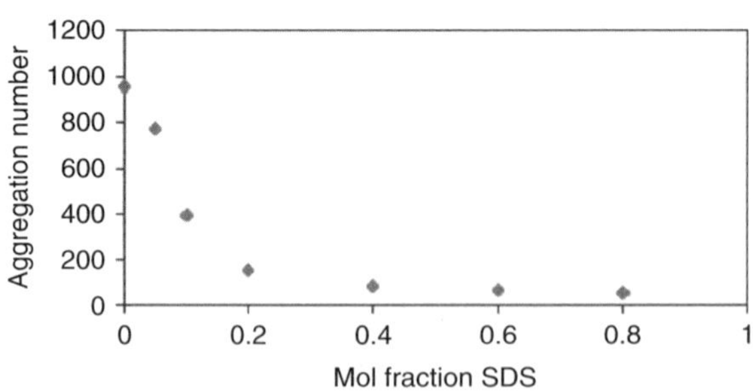

Figure 17 Aggregation number predicted for mixtures of SDS and $C_{12}E_3$ ethoxylate surfactant.

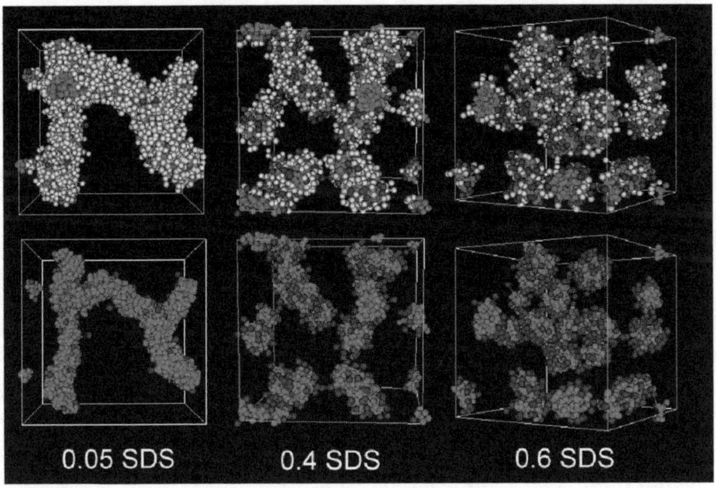

Figure 18 Mesoscale structure predicted for mixtures of $C_{12}E_6$ ethoxylate and SDS for different mole fractions of SDS. Only the surfactant tails and SDS head groups are shown in the second row of images. (*See insert for color representation figure*.)

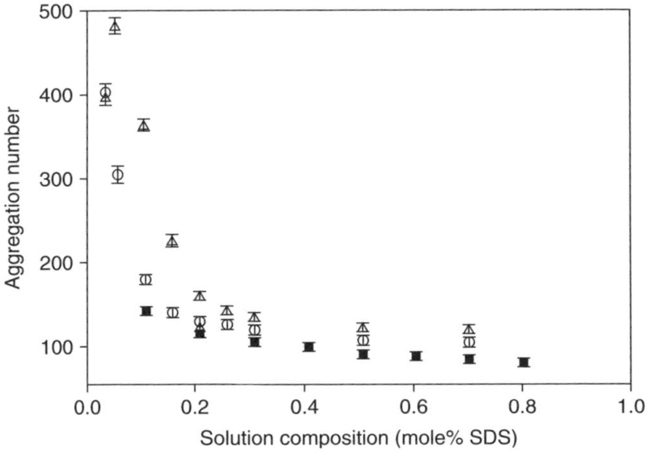

Figure 19 Variation in micelle aggregation number with solution composition for 25 mM SDS/$C_{12}E_6$: (●) in D_2O; (○) in 0.01 M NaCl; (△) in 0.05 M NaCl. Obtained using small-angle neutron scattering (SANS). (Reprinted with permission from ref. 82. Copyright © 2005, American Chemical Society.)

Correct modeling of synergistic interactions is important for property prediction. Jost, Leiter, and Schwuger showed that mixtures of SDS and $C_{16}E_3$ nonionic surfactants gave more that twice as much foam as either surfactant alone and showed an enhanced ability to disperse MnO_2 and zeolite.[83] Moore

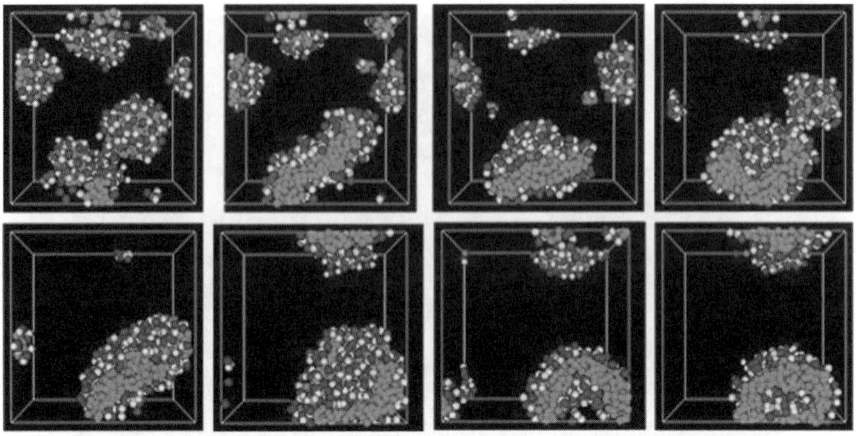

Figure 20 Self-assembly of a mixture of anionic and cationic surfactants to form a vesicle. (*See insert for color representation figure.*)

et al. showed that addition of $C_{12}E_6$ decreases the amount of SDS penetrating the epidermis (skin); they suggest that this is the result of the change in the mesoscale structure,[84] and Acharya et al. recently quantified the dramatic changes in rheological behavior as the two surfactants were mixed.[85]

Mixtures of Anionic and Cationic Surfactants As a final test of the new ionic surfactant model, mixtures of anionic and cationic surfactants were created. Experimental evidence from small-angle neutron scattering (SANS) and static light scattering (SLS) shows that equal molar mixtures of anionic SDS and a cationic surfactant (DTAB) with identical hydrocarbon C_{12} tail self-assemble to form vesicles in water.[86] Figure 20 shows the predicted mesoscale structure evolution for a 10% solution of 50:50 anionic and cationic surfactants with no added salt (using repulsion parameters from Table 3). The surfactants initially form mixed spherical micelles, but as they fuse together they transform into a lamella sheet, which then folds to create a vesicle. The ability of simulation approaches such as DPD to predict dynamic behavior and the mechanism of mesoscale structure formation, in addition to prediction of the final mesoscale structure, could provide useful insight into the behavior of surfactant-structured fluids during processing and manufacture and during their use as commercial materials.

3. CONCLUSIONS

The dissipative particle dynamics (DPD) method is capable of predicting the mesoscale structure of nonionic surfactant solutions, as has been demonstrated in numerous published studies. In this chapter it was shown that in a simulation of an emulsion, the switch from Winsor I to Winsor II and Winsor III is predicted as the relative size of surfactant head and tail was varied. A simple model of a $C_{12}E_6$ ethoxylate surfactant solution shows the correct trends in mesoscale structure, changing from spheres to long rods (or worms) as the surfactant concentration increases and forming a lamella structure at high concentrations. This surfactant model is able to solubilize oil, removing it from a surface—a simple model of one aspect of cleaning by detergents. However, most commercial surfactant systems consist of surfactant mixtures, including ionic surfactants and salt. A strategy for modeling ionic system in DPD has been proposed. The model is capable of predicting the effect of added salt on the mesoscale structure of SDS, which changes from spheres into wormlike micelles of increasing length. Initial results suggest that the DPD + electrostatics model is also capable of predicting synergistic interactions in mixed surfactant systems.

REFERENCES

1. Hamley, I. W. *Introduction to Soft Matter*, Wiley, Chicester, England, 2000.
2. Jonsson, B.; Lindman, B.; Holmberg, K.; Kronberg, B. *Surfactants and Polymers in Aqueous Solution*, Wiley, New York, 1999.
3. Maitland, G. C. Oil and gas production, *Curr. Opin. Colloid Interface Sci.*, 2000, 5(5–6): 301–311.
4. Rosen, M. J. *Surfactants and Interfacial Phenomena*, Wiley, New York, 1989.
5. Lim, W. H.; Salmiah, A. Dishwashing performance of mixed palm stearin sulfonated methyl esters–nonylphenol ethoxylate alcohol, *J. Surfactants Detergents*, 2004, 7(1): 53–58.
6. Scamehorn, J. F.; Agino, K.; Abe, M. *Mixed Surfactant Systems*, Marcel Dekker, New York, 1992.
7. Schwuger, M. J. Interfacial and performance properties of sulfated polyoxyethylenated alcohols, *ACS Symp. Ser.*, 1984, 253: 3–26.
8. Israelachvili, J.; Mitchell, D. J.; Ninham, B. W. *J. Chem. Soc. Faraday Trans. 2*, 1986, 72: 1525–1568.
9. Griffin, W. C. *J. Soc. Cosmet. Chem.*, 1949, 1: 311.
10. Olsson, U.; Shinoda, K.; Lindman, B. Change of the structure of microemulsions with the hydrophile lipophile balance of nonionic surfactant as revealed by NMR self-diffusion studies, *J. Phys. Chem.*, 1986, 90(17): 4083–4088.
11. Hines, J. D. Theoretical aspects of micellisation in surfactant mixtures, *Curr. Opin. Colloid Interface Sci.*, 2001, 6(4): 350–356.
12. Nagarajan, R.; Ruckenstein, E. Theory of surfactant self-assembly: a predictive molecular thermodynamic approach, *Langmuir*, 1991, 7(12): 2934–2969.
13. Srinivasan, V.; Blankschtein, D. Effect of counterion binding on micellar solution behavior: 2. Prediction of micellar solution properties of ionic surfactant–electrolyte systems, *Langmuir*, 2003, 19(23): 9946–9961.
14. Srinivasan, V.; Blankschtein, D. Effect of counterion binding on micellar solution behavior: 1. Molecular–thermodynamic theory of micellization of ionic surfactants, *Langmuir*, 2003, 19(23): 9932–9945.

15. Nagarajan, R. Micellization of binary surfactant mixtures: theory, *ACS Symp. Ser.*, 1992, 501: 54–95.
16. Hines, J. D. Theoretical aspects of micellisation in surfactant mixtures, *Curr. Opin. Colloid Interface Sci.*, 2001, 6(4): 350–356.
17. Tanford, C. *The Hydrophobic Effect*, Wiley-Interscience, New York, 1973.
18. Nagarajan, R. Molecular packing parameter and surfactant self-assembly: the neglected role of the surfactant tail, *Langmuir*, 2002, 18(1): 31–38.
19. Goldsipe, A.; Blankschtein, D. Molecular–thermodynamic theory of micellization of ph-sensitive surfactants, *Langmuir*, 2006, 22(8): 3547–3559.
20. Mulqueen, M.; Blankschtein, D. Theoretical and experimental investigation of the equilibrium oil–water interfacial tensions of solutions containing surfactant mixtures, *Langmuir*, 2002, 18(2): 365–376.
21. Nagarajan, R.; Ruckenstein, E. Molecular theory of microemulsions, *Langmuir*, 2000, 16(16): 6400–6415.
22. Nagarajan, R. Polymer-induced structural transitions in microemulsions, *Langmuir*, 1993, 9(2): 369–375.
23. Evans, D. F.; Ninham, B. W. Ion binding and the hydrophobic effect, *J. Phys. Chem.*, 1983, 87(24): 5025–5032.
24. Mysels, K. J.; Princen, L. H. Light scattering by some sodium laurel sulfate solutions, *J. Phys. Chem.*, 1959, 63: 1695–1700.
25. Blankschtein, D. Computer programs: Predict, mix, mix2, surf, and dynamic, available at http://web.mit.edu/dbgroup/computer/index.html, 2003.
26. Blankschtein, D.; Shiloach, A.; Zoeller, N. User-friendly computer programs to predict surfactant solution behavior. *J. Soc. Cosmet. Chem.*, 1997, 48(1): 71–72.
27. Shelley, J. C.; Shelley, M. Y. Computer simulation of surfactant solutions, *Curr. Opin. Colloid Interface Sci.*, 2000, 5(1–2): 101–110.
28. Saiz, L.; Bandyopadhyay, S.; Klein, M. L. Towards an understanding of complex biological membranes from atomistic molecular dynamics simulations, *Biosci. Rep.*, 2002, 22(2): 151–173.
29. Srinivas, G.; Nielsen, S. O.; Moore, P. B.; Klein, M. L. Molecular dynamics simulations of surfactant self-organization at a solid–liquid interface, *J. Am. Chem. Soc.*, 2006, 128(3): 848–853.
30. deMoura, A. F.; Freitas, L. C. G. Molecular dynamics simulation of the sodium octanoate micelle in aqueous solution, *Chem. Phys. Lett.*, 2005, 411(4–6), 474–478.
31. Cheong, D. W.; Panagiotopoulos, A. Z. Monte Carlo simulations of micellization in model ionic surfactants: application to sodium dodecyl sulfate, *Langmuir*, 2006, 22(9): 4076–4083.
32. Chen, B.; Siepmann, J. I. Microscopic structure and solvation in dry and wet octanol, *J. Phys. Chem. B*, 2006, 110(8), 3555–3563.
33. Firetto, V.; Floriano, M. A.; Panagiotopoulos, A. Z. Effect of stiffness on the micellization behavior of model h4t4 surfactant chains, *Langmuir*, 2006, 22(15): 6514–6522.
34. Nekovee, M.; Coveney, P. V. Lattice-Boltzmann simulations of self-assembly of a binary water-surfactant system into ordered bicontinuous cubic and lamellar phases, *J. Am. Chem. Soc.*, 2001, 123(49): 12380–12382.

35. van derSman, R. G. M.; van derGraaf, S. Diffuse interface model of surfactant adsorption onto flat and droplet interfaces, *Rheol. Acta*, 2006, 46(1): 3–11.
36. Fredrickson, G. H. *The Equilibrium Theory of Inhomogeneous Polymers*, Clarendon Press, Oxford, 2006.
37. Fraaije, J.; Zvelindovsky, A. V.; Sevink, G. J. A. Computational soft nanotechnology with mesodyn. *Mol. Simul.*, 2004, 30(4): 225–238.
38. Hoogerbrugge, P. J.; Koelman, J. Simulating microscopic hydrodynamic phenomena with dissipative particle dynamics, *Europhys. Lett.*, 1992, 19(3): 155–160.
39. Kong, Y.; Manke, C. W.; Madden, W. G.; Schlijper, A. G. Effect of solvent quality on the conformation and relaxation of polymers via dissipative particle dynamics. *J. Chem. Phys.*, 1997, 107(2): 592–602.
40. Schlijper, A. G.; Hoogerbrugge, P. J.; Manke, C. W. Computer-simulation of dilute polymer-solutions with the dissipative particle dynamics method, *J. Rheol.*, 1995, 39(3): 567–579.
41. Espanol, P. Hydrodynamics from dissipative particle dynamics, *Phys. Rev. E*, 1995, 52(2): 1734–1742.
42. Chen, S.; Phan-Thien, N.; Fan, X. J.; Khoo, B. C. Dissipative particle dynamics simulation of polymer drops in a periodic shear flow, *J. Non-Newtonian Fluid Mech.*, 2004, 118(1): 65–81.
43. Clark, A. T.; Lal, M.; Ruddock, J. N.; Warren, P. B. Mesoscopic simulation of drops in gravitational and shear fields, *Langmuir*, 2000, 16(15): 6342–6350.
44. Groot, R. D.; Warren, P. B. Dissipative particle dynamics: bridging the gap between atomistic and mesoscopic simulation, *J. Chem. Phys.* 1997, 107(11): 4423–4435.
45. Fraaije, J. G. E. M. *Chemistry Unified Language Interface Modelling Library*, Version 2.1, Culgi, 2006.
46. Broze, G.; Case, F. H. Impact of mesoscale structure and phase behavior on rheology and performance in superwetting cleaners. in *Mesoscale Phenomena in Fluid Systems*, ACS symposium series, Vol. 861, Case, F. H.; Alexandridis, P., Eds., American Chemical Society, Washington, DC, 2003, pp. 376–389.
47. Sjoblom, J. *Enclycopedic Handbook of Emulsion Technology*, Marcel Dekker, New York, 2001.
48. Rekvig, L.; Hafskjold, B.; Smit, B. Molecular simulations of surface forces and film rupture in oil/water/surfactant systems, *Langmuir*, 2004, 20(26): 11583–11593.
49. Pays, K.; Mabille, C.; Schmitt, V.; Leal-Calderon, F.; Bibette, J. Understanding the stability and lifetime of emulsions, *J. Dispers. Sci. Technol.*, 2002, 23(1–3), 175–186.
50. Jiao, J.; Burgess, D. J. Ostwald ripening of water-in-hydrocarbon emulsions, *J. Colloid Interface Sci.*, 2003, 264(2): 509–516.
51. Groot, R. D. Mesoscopic simulation of polymer–surfactant aggregation, *Langmuir*, 2000, 16(19): 7493–7502.
52. Kranenburg, M.; Smit, B. Simulating the effect of alcohol on the structure of a membrane, *FEBS Lett.*, 2004, 568(1–3): 15–18.
53. Horsch, M. A.; Zhang, Z. L.; Iacovella, C. R.; Glotzer, S. C. Hydrodynamics and microphase ordering in block copolymers: Are hydrodynamics required for ordered phases with periodicity in more than one dimension? *J. Chem. Phys.*, 2004, 121(22): 11455–11462.

54. Mitchell, D. J.; Tiddy, G. J. T.; Waring, L.; Bostock, T.; McDonald, M. P. Phase-behavior of polyoxyethylene surfactants with water: mesophase structures and partial miscibility (cloud points), *J. Chem. Soc. Faraday Trans. 1*, 1983, 79: 975–1000.
55. Jury, S.; Bladon, P.; Cates, M.; Krishna, S.; Hagen, M.; Ruddock, N.; Warren, P. Simulation of amphiphilic mesophases using dissipative particle dynamics, *Phys. Chem. Chem. Phys.*, 1999, 1(9), 2051–2056.
56. Groot, R. D. Electrostatic interactions in dissipative particle dynamics-simulation of polyelectrolytes and anionic surfactants, *J. Chem. Phys.*, 2003, 118(24): 11265–11277.
57. Fraaije, J. G. E. M.; vanMale, J. Electrostatics in culgi, Documentation for Culgi, Version 2.1, 2005,
58. Ohtaki, H.; Radna, T. Structure and dynamics of hydrated ions, *Chem. Rev.*, 1993, 93: 1157–1204.
59. Koneshan, S.; Rasaiah, J. C. Computer simulation studies of aqueous sodium chloride solutions at 298 K and 683 K, *J. Chem. Phys.*, 2000, 113(18): 8125–8137.
60. Chialvo, A. A.; Cummings, P. T.; Cochran, H. D.; Simonson, J. M.; Mesmer, R. E. *J. Chem. Phys.*, 1995, 103: 9379.
61. Kralchevsky, P. A.; Danov, K. D.; Kolev, V. L.; Broze, G.; Mehreteab, A. Effect of nonionic admixtures on the adsorption of ionic surfactants at fluid interfaces: 1. Sodium dodecyl sulfate and dodecanol, *Langmuir*, 2003, 19(12): 5004–5018.
62. Griffiths, P. C.; Paul, A.; Heenan, R. K.; Penfold, J.; Ranganathan, R.; Bales, B. L. Role of counterion concentration in determining micelle aggregation: evaluation of the combination of constraints from small-angle neutron scattering, electron paramagnetic resonance, and time-resolved fluorescence quenching, *J. Phys. Chem. B*, 2004, 108(12): 3810–3816.
63. Quina, F. H.; Nassar, P. M.; Bonilha, J. B. S.; Bales, B. L. *J. Chem. Phys.*, 1995, 99: 17028.
64. Kaneshina, S.; Kamaya, H.; Ueda, I. Benzyl alcohol penetration into micelles, dielectric-constant of the binding-site, partition-coefficient and high-pressure squeeze-out, *Biochim. Biophys. Acta*, 1984, 777(1): 75–83.
65. Joshi, J. V.; Aswal, V. K.; Bahadur, P.; Goyal, P. S. Role of counterion of the surfactant molecule on the micellar structure in aqueous solution, *Curr. Sci*. 2002 83(1): 47–49.
66. Bales, B. L.; Messina, L.; Vidal, A.; Peric, M.; Nascimento, O. R. Precision relative aggregation number determinations of SDS micelles using a spin probe: a model of micelle surface hydration, *J. Phys. Chem. B*, 1998, 102(50): 10347–10358.
67. Bales, B. L. A definition of the degree of ionization of a micelle based on its aggregation number, *J. Phys. Chem. B*, 2001, 105(29): 6798–6804.
68. Bockstahl, F.; Pachoud, E.; Duplatre, G.; Billard, I. Size of sodium dodecyl sulphate micelles in aqueous NaCl solutions as studied by positron annihilation lifetime spectroscopy, *Chem. Phys.*, 2000, 256(3): 307–313.
69. Jakubowska, A. Effect of electrolytes on the aggregation of sodium dodecyl sulphate: the interactions of different counterions with formed micelles, *Z. Phys. Chem. Int. J. Res. Phys. Chem. & Chem. Phys.*, 2004, 218(11): 1297–1305.

70. Bernheim-Groswasser, A.; Zana, R.; Talmon, Y. Sphere-to-cylinder transition in aqueous micellar solution of a dimeric (gemini) surfactant, *J. Phys. Chem. B*, 2000, 104(17): 4005–4009.

71. Cates, M. E.; Candau, S. J. Statics and dynamics of worm-like surfactant micelles, *J. Phys. Condens. Matter*, 1990, 2(33): 6869–6892.

72. Drye, T. J.; Cates, M. E. Living networks: the role of cross-links in entangled surfactant solutions, *J. Chem. Phys.*, 1992, 96(2): 1367–1375.

73. Briels, W. J.; Mulder, P.; denOtter, W. K. Simulations of elementary processes in entangled wormlike micelles under tension: a kinetic pathway to Y-junctions and shear induced structures, *J. Phys. Condens. Matter*, 2004, 16(38): S3965–S3974.

74. Andreev, V. A.; Victorov, A. I., Molecular thermodynamics for micellar branching in solutions of ionic surfactants, *Langmuir*, 2006, 22(20): 8298–8310.

75. Holland, P. M.; Rubingh, D. N. Mixed surfactant systems: an overview, *ACS Symp. Ser.* 1992, 501: 2–30.

76. Penfold, J.; Staples, E.; Thompson, L.; Tucker, I.; Hines, J.; Thomas, R. K.; Lu, J. R. Solution and adsorption behavior of the mixed surfactant system sodium dodecyl sulfate/*n*-hexaethylene glycol monododecyl ether, *Langmuir*, 1995, 11(7): 2496–2503.

77. Rubingh, D. N. In*Solution Chemistry of Surfactants*, Mittel, K. L. Ed., Plenum Press, New York, 1979.

78. Holland, P. M. Modeling mixed surfactant systems: basic introduction, *ACS Symp. Ser.*, 1992, 501: 31–44.

79. Shiloach, A.; Blankschtein, D. Predicting micellar solution properties of binary surfactant mixtures, *Langmuir*, 1998, 14(7): 1618–1636.

80. Penfold, J.; Staples, E.; Tucker, I.; Thomas, R. K. Surface and solution behavior of the mixed dialkyl chain cationic and nonionic surfactants, *Langmuir*, 2004, 20(4): 1269–1283.

81. Staples, E.; Penfold, J.; Tucker, I. Adsorption of mixed surfactants at the oil–water interface, *J. Phys. Chem. B*, 2004, 104: 606–614.

82. Penfold, J.; Tucker, I.; Thomas, R. K.; Staples, E.; Schuermann, R. Structure of mixed anionic/nonionic surfactant micelles: experimental observations relating to the role of headgroup electrostatic and steric effects and the effects of added electrolyte, *J. Phys. Chem. B*, 2005, 109(21): 10760–10770.

83. Jost, F.; Leiter, H.; Schwuger, M. J. Synergisms in binary surfactant mixtures, *Colloid Poly. Sci*. 1988, 266(6): 554–561.

84. Moore, P. N.; Shiloach, A.; Puvvada, S.; Blankschtein, D. Penetration of mixed micelles into the epidermis: effect of mixing sodium dodecyl sulfate with dodecyl hexa(ethylene oxide), *J. Cosmet. Sci*. 2003, 54(2): 143–159.

85. Acharya, D. P.; Sato, T.; Kaneko, M.; Singh, Y.; Kunieda, H. Effect of added poly(oxyethylene)dodecyl ether on the phase and rheological behavior of wormlike micelles in aqueous SDS solutions, *J. Phys. Chem. B*, 2006, 110(2): 754–760.

86. Bergstrom, M.; Pedersen, J. S.; Schurtenberger, P.; Egelhaaf, S. U. Small-tingle neutron scattering (SANS) study of vesicles and lamellar sheets formed from mixtures of an anionic and a cationic surfactant, *J. Phys. Chem. B* 1999, 103(45): 9888–9897.

INDEX

Multiscale Simulation Methods for Nanomaterials, Edited by Richard B. Ross and Sanat Mohanty